Innovations in Intelligent Microgrid Operation and Control

Innovations in Intelligent Microgrid Operation and Control

Guest Editor

Zhenxiong Wang

MDPI

Basel • Beijing • Wuhan • Barcelona • Belgrade • Novi Sad • Cluj • Manchester

Guest Editor
Zhenxiong Wang
School of Electrical
Engineering
Xi'an Jiaotong University
Xi'an
China

Editorial Office
MDPI AG
Grosspeteranlage 5
4052 Basel, Switzerland

This is a reprint of the Special Issue, published open access by the journal *Electronics* (ISSN 2079-9292), freely accessible at: https://www.mdpi.com/journal/electronics/special_issues/L1R277BLUM.

For citation purposes, cite each article independently as indicated on the article page online and as indicated below:

Lastname, A.A.; Lastname, B.B. Article Title. *Journal Name* **Year**, *Volume Number*, Page Range.

ISBN 978-3-7258-6800-1 (Hbk)
ISBN 978-3-7258-6801-8 (PDF)
https://doi.org/10.3390/books978-3-7258-6801-8

Cover image courtesy of Zhenxiong Wang

Contents

About the Editor

Zhenxiong Wang

Zhenxiong Wang is an Associate Professor and Doctoral Supervisor at the School of Electrical Engineering, Xi'an Jiaotong University, where he is affiliated with the Research Center for High-Quality Power Supply and Intelligent Distribution Network Equipment. His research interests focus on active support and adaptive control of grid-connected converters, power quality analysis and control, microgrid converter control, and intelligent power distribution technologies and equipment, with additional focus on inverter control and microgrid operation. He has presided over and participated in more than 10 projects, including sub-projects of the National Key R&D Program, National Natural Science Foundation of China, provincial and ministerial-level scientific and technological projects, as well as key projects of State Grid Corporation of China and China Southern Power Grid Corporation. With achievements including over 40 published SCI and EI papers, more than 10 applied and authorized national invention patents, 3 social awards, and 1 provincial and ministerial-level award, he has also been selected into Xi'an Jiaotong University's Young Outstanding Talent Support Program, and some of his research technologies have been applied and verified in distribution networks and new energy stations in Fujian, Hebei, Gansu and other regions.

electronics

MDPI

Article

Grid-Connected Active Support and Oscillation Suppression Strategy of Energy Storage System Based on Virtual Synchronous Generator

Zhuan Zhao [1], Jinming Yao [2,*], Shuhuai Shi [3], Di Wang [1], Duo Xu [1] and Jingxian Zhang [1]

[1] Department of Electrical Engineering, Zhengzhou Electric Power College, Zhengzhou 450008, China
[2] College of Automation, Nanjing University of Posts and Telecommunications, Nanjing 210023, China
[3] State Grid Henan Electric Power Research Institute, Zhengzhou 450052, China
* Correspondence: yaojinming@njupt.edu.cn; Tel.: +86-18300700579

Abstract

This paper addresses stability issues, including voltage fluctuation, a frequency offset, and broadband oscillation resulting from the high penetration of renewable energy in a photovoltaic high-permeability distribution network. This paper proposes an active support control strategy which is energy storage grid-connected based on a virtual synchronous generator (VSG). This strategy endows the energy storage system with virtual inertia and a damping capacity by simulating the rotor motion equation and excitation regulation characteristics of the synchronous generator, and effectively enhances the system's ability to suppress power disturbances. The small-signal model of the VSG system is established, and the influence mechanism of the virtual inertia and damping coefficient on the system stability is revealed. A delay compensator in series with a current feedback path is proposed. Combined with the damping optimization of the LCL filter, the instability risk caused by high-frequency resonance and a control delay is significantly suppressed. The novelty lies in the specific configuration of the compensator within the grid–current feedback loop and its coordinated design with VSG parameters, which differs from traditional capacitive–current feedback compensation methods. The experimental results obtained from a semi-physical simulation platform demonstrate that the proposed control strategy can effectively suppress voltage fluctuations, suppress broadband oscillations, and improve the dynamic response performance and fault ride-through capability of the system under typical disturbance scenarios such as sudden illumination changes, load switching, and grid faults. It provides a feasible technical path for the stable operation of the distribution network with a high proportion of new energy access.

Keywords: virtual synchronous generator (VSG); photovoltaic high-permeability distribution network; active support control; broadband oscillation suppression; delay compensation control

1. Introduction

As noted in [1], driven by a carbon peak and carbon neutrality targets, the scale of grid-connected renewable energy, particularly photovoltaics, has experienced rapid expansion. The high proportion of photovoltaic access has led to profound changes in the operating characteristics of the distribution network, and the traditional power system dominated by synchronous generators is facing severe challenges. The intermittence and volatility of photovoltaic power generation, coupled with the relatively weak grid structure of the distribution network, make the voltage fluctuation problem increasingly prominent. When the penetration rate exceeds 30%, the system inertia is significantly reduced and the voltage regulation ability is weakened, which has become a technical bottleneck restricting the consumption of a high proportion of new energy. To address these challenges, "active support technology" in this context refers to the capability of grid-connected converters to proactively provide voltage and frequency regulation, inertia response, and damping to the grid, rather than merely following grid conditions. The problem of voltage fluctuation is manifested in three aspects: a voltage flicker caused by the short-term fluctuation of the photovoltaic output, a voltage over-limit caused by a day and night power reversal, and an insufficient voltage support capacity during fault. The research in Reference [2] shows that the grid following the control mode adopted by traditional photovoltaic inverters is difficult to provide necessary support for in terms of the grid voltage and frequency. The fundamental reason is that the control mode leads to the lack of rotational inertia and damping characteristics of the system. When the light or load changes suddenly, it is easy to cause significant voltage oscillations, which not only affect the power quality, but also pose a threat to the safety and stability of the power grid. In order to compensate for this inertia deficit, virtual synchronous generator (VSG) control technology has been introduced. The technology introduces the rotor motion equation of the synchronous generator into the inverter control algorithm and cooperates with the energy storage unit to make the inverter have the ability to actively support the power grid. The core of VSG technology is to simulate the electromechanical transient characteristics of synchronous generators, so that the energy storage system can provide an effective inertial response and damping support for the power grid, thus realizing the technical evolution from passive adaptation to a grid change to active participation in grid regulation and control. The empirical research in Reference [2] shows that the energy storage system equipped with VSG can reduce the voltage fluctuation by more than 40%, and it also has significant advantages in terms of economic and environmental benefits. In the photovoltaic high-permeability distribution network, the coordination of VSG and energy storage has been proven to be the key technical path. However, its effectiveness is highly dependent on the fineness of the control strategy. In this regard, the academic community has carried out in-depth optimization research: In terms of parameter adaptation, the fuzzy parameter adaptation method proposed in Reference [3] makes the inertia and damping parameters of VSG have a dynamic adjustment ability, which significantly improves the suppression effect of the system on power disturbance. In terms of dealing with complex working conditions, the literature [4] coordinates active power fluctuation suppression and current balance control by reconstructing the negative sequence current and multi-objective optimization for unbalanced grid voltage conditions, and expands the application scenarios of VSG. In terms of system-level coordination, the research in References [5,6] solves the coordination problem between energy storage SOC and grid frequency modulation, as well as the stable operation problem under fault conditions, which consolidates the theoretical basis of the energy storage system participating in the active support of the power grid. The evolution of voltage fluctuation suppression technology is closely related to the design of VSG core parameters. Reference [7] revealed through a small-signal model that the synergy of virtual inertia and the damping coefficient has a nonlinear effect on the stability of the system, and parameter tuning needs to achieve a balance between the dynamic response speed and steady-state accuracy. Reference [8] improved the response speed of the controller through algorithm innovation. However, there are still bottlenecks in the current research: the adaptive mechanism of VSG parameters mostly depends on a single variable, which fails to fully consider the coupling effect of the photovoltaic output randomness and grid impedance, and the dynamic balance between the SOC recovery of energy storage and the

https://doi.org/10.3390/electronics15020020

frequency regulation demand of the power grid also requires more refined optimization algorithms. In summary, the stability problem of the photovoltaic high-permeability distribution network is manifested in the multi-symbiosis of voltage, frequency, and power oscillations. It is difficult for traditional means to meet the urgent needs of the system for 'active support'. The grid-connected ability of the energy storage system through VSG technology is closely related to the core parameters of a VSG and the characteristics of the local power grid [9]. Therefore, this paper proposes an active support control strategy for energy storage which is grid-connected based on a VSG, aiming at improving the stability and power quality of the photovoltaic high-permeability distribution network. The typical three-phase grid-connected converter circuit topology of the technology is shown in Figure 1, which mainly includes an energy storage unit, bidirectional DC/DC converter, DC bus capacitor, three-phase inverter bridge, output filter, distribution network, phase-locked loop (PLL), VSG control loop (including power–frequency and reactive–voltage regulation and virtual impedance, etc.), and PWM generator.

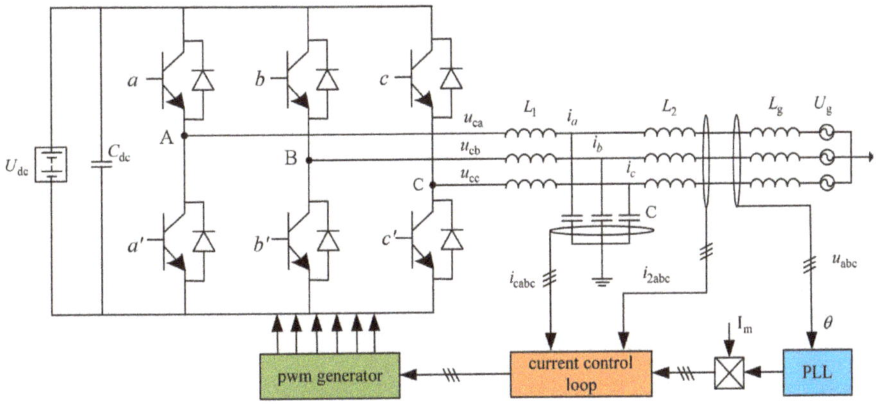

Figure 1. Energy storage VSG control topology diagram.

2. The Control Principle and Small-Signal Model Modeling of Energy Storage VSG

2.1. System Structure and Basic Model

In this study, a lithium battery energy storage system serves as the primary energy storage unit. The system structure is illustrated in Figure 1. Its main structure includes that the output active and reactive power of the grid-connected converter is as follows:

$$\begin{cases} P = \frac{E}{R_g^2+X_g^2}(R_g V_g \cos\delta + X_g V_g \sin\delta - R_g E) \\ Q = \frac{E}{R_f^2+X_g^2}(-R_g V_g \sin\delta + X_g V_g \cos\delta - X_g E) \end{cases} \quad (1)$$

In the formula, E is the output three-phase AC voltage V_{abc} for the grid-connected converter; X_g is the inductance of the line inductance L_g; V_g is the amplitude of the three-phase voltage at the grid-connected point, which is the power angle; the transmission line is inductive, that is, the inductance is much larger than the resistance; and, in the case of the grid-connected energy storage system, there is no phase difference between the output voltage of the grid-connected converter and the grid-side voltage. The primary function of energy storage in VSG is to deliver rapid power support through charging and discharging processes [10], which is mainly described as the dynamic model of both power dynamics and energy dynamics.

https://doi.org/10.3390/electronics15020323

The energy storage responds to the power command P_{ref} through a power electronic converter, and its dynamic characteristics are usually simplified as a first-order inertial link:

$$T_{bat}\frac{dP_{bat}}{dt} + P_{bat} = P_{ref} \tag{2}$$

In the formula, P_{bat} is the actual output power of the energy storage, which is limited by the time constant T_{bat}; P_{ref} is the reference power of the energy storage; and T_{bat} is the response time constant of the energy storage power (typical value is 10 ms–100 ms).

The state of the energy storage dynamic model is directly related to the electromagnetic power P_e of the VSG through the DC bus voltage V_{dc} and the model is as follows:

$$C_{dc}\frac{dV_{dc}}{dt} = P_{bat} - P_{vsg} \tag{3}$$

In the formula, C_{dc} is the DC bus capacitor; P_{vsg} is the active power of the VSG output to the power grid; and the energy storage reflects the energy balance state through the DC-side voltage connected to the DC/AC converter. When the energy storage power P_{bat} is unbalanced with the VSG output power P_{vsg}, the V_{dc} change triggers the VSG to adjust P_e to maintain the energy balance.

2.2. Dynamic Modeling and Time-Varying Characteristics of Lithium Battery Energy Storage System

The dynamic characteristics of the lithium battery energy storage system are not only affected by its power response time constant T_{bat}, but also highly dependent on its time-varying parameters such as the state of charge, internal resistance, open circuit voltage, temperature, and aging degree. In order to more truly reflect the actual dynamic behavior of energy storage in a VSG, this paper adopts the dynamic description of a lithium battery based on an equivalent circuit model.

The equivalent circuit model of the lithium battery (second-order RC model) can be expressed as:

$$V_{bat} = V_{oc}(SOC) - I_{bat} \cdot R_0(SOC) - V_1 - V_2 \tag{4}$$

$$\frac{dV_1}{dt} = \frac{I_{bat}}{C_1} - \frac{V_1}{R_1C_1}, \frac{dV_2}{dt} = \frac{I_{bat}}{C_2} - \frac{V_1}{R_2C_2} \tag{5}$$

In the formula, $V_{oc}(SOC)$ is the open circuit voltage of the battery, which has a nonlinear relationship with SOC; $R_0(SOC)$ is the Ohmic internal resistance, changing with SOC; R_1, C_1 and R_2, C_2 characterize the short-term and long-term polarization effects of the battery, respectively; and I_{bat} is the battery current, limited by the maximum charge and discharge current I_{max}.

The relationship between the energy storage output power P_{bat} and battery terminal voltage and current is as follows:

$$P_{bat} = V_{bat} \cdot I_{bat} \tag{6}$$

The model can more accurately reflect the voltage drop, energy loss, and response delay characteristics of the battery during dynamic power support, and provide a more realistic modeling basis for the inertial support capability of a VSG.

2.3. Operation Constraint and Power Limit Model of Energy Storage System

In actual operation, the power output of the energy storage system is limited by multiple physical constraints, including the upper and lower limits of SOC, the maximum charging and discharging power, the power ramp rate, and the voltage/current safety

https://doi.org/10.3390/electronics15020323

range. These constraints directly affect the virtual inertia response and dynamic adjustment ability of the VSG.

The actual output power R_{ramp} of the energy storage system needs to meet the following constraints:

$$\begin{array}{c} P_{min}(SOC) \leq P_{bat} \leq P_{min}(SOC) \\ \left|\frac{dP_{bat}}{dt}\right| \leq R_{ramp} \end{array} \tag{7}$$

Among them, P_{min} and P_{max} are SOC-related power limits and R_{ramp} is the power ramp rate limit.

Current and voltage constraints:

$$|I_{bat}| \leq I_{max}, V_{min} \leq V_{bat} \leq V_{max} \tag{8}$$

In addition, the allowable operating range of SOC is as follows:

$$SOC_{min} \leq SOC \leq SOC_{min} \tag{9}$$

These constraints will be incorporated into the subsequent small-signal modeling and control design to ensure the safety and feasibility of the system in actual operation.

2.4. *Control Principle of Energy Storage VSG*

In the VSG control system, the key parameters such as the capacitor voltage V_f and output current i_g need to be sampled for the power calculation. The reference voltage of the system can be obtained by active and reactive reference power P_{ref} and Q_{ref}. The reference value of the voltage and current is obtained by double-loop control, and finally, the PWM drive signal is output to realize the output control of the system [11].

A virtual synchronous machine is an advanced control technology that simulates the dynamic behavior of a synchronous motor through a power electronic converter. Its core goal is to provide the power grid with inertia, damping, and power regulation capabilities similar to traditional synchronous generators, especially for power systems with a high proportion of renewable energy access [12].

The VSG achieves dynamic characteristics by simulating the rotor motion equation and excitation control equation of the synchronous motor:

$$\begin{cases} J\frac{d\omega}{dt} = T_m - T_e - D_p\Delta\omega \\ \frac{d\delta}{dt} = \omega - \omega_0 = \Delta\omega \end{cases} \tag{10}$$

In the formula, T_m and T_e are the mechanical torque (provided by energy storage) and electromagnetic torque of the motor, respectively; J is the moment of inertia; D_p is virtual damping; and there is also the actual angular frequency of the system and the rated angular frequency of the system.

The dual-loop control of a VSG aims to balance the grid-friendly dynamic characteristics and fast disturbance rejection capability. The two-layer loop has a clear division of labor and coordinated operation. The specific configuration is as follows.

The schematic diagram of double-loop control is shown in Figure 2: the outer loop includes a P-f adjustment module and a Q-V adjustment module, and the voltage reference signal is generated based on the VSG core dynamic equation; the voltage PI regulator and the current PI regulator are connected in series in the inner loop, and the closed-loop tracking is realized by sampling V_f and i_g. The VSG output end is connected to the power grid through the LC filter (matching with the sampling parameter V_f) to ensure the smooth transmission of the signal.

https://doi.org/10.3390/electronics15020323

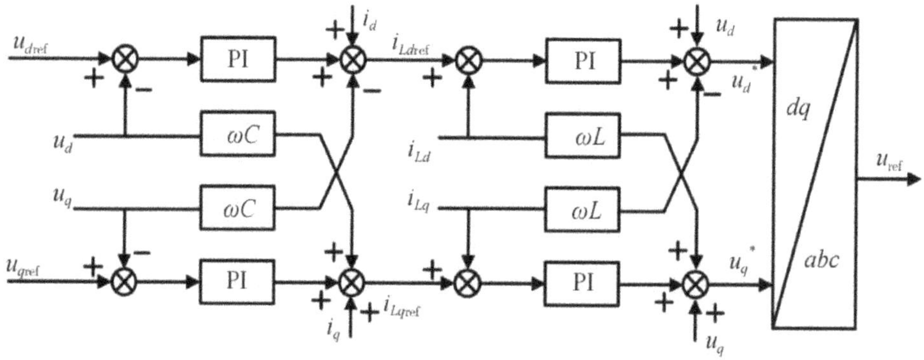

Figure 2. Voltage–current double-loop decoupling control block diagram in *dq* synchronous rotating coordinate system.

In order to ensure the frequency change within the allowable fluctuation range, the relationship between the active power of the generator and the frequency is called the active power–frequency static characteristics of the generator set. Referring to the principle of the synchronous generator governor, the relationship between the virtual synchronous generator *P-f* is as follows:

$$P_m = P_{ref} - K_P(\omega - \omega_0) \tag{11}$$

In the formula, K_p is the active power adjustment coefficient. By combining with Equation (11), the active-frequency control equation of VSG can be obtained, which is as follows:

$$\frac{P_m - P_e}{\omega} = J\frac{d\omega}{dt} + D_p\Delta\omega \tag{12}$$

In the formula, P_e is the electromagnetic power of the system. The damping of the VSG system is improved so that the energy storage system has the ability to suppress power oscillation in power regulation.

The mathematical model of the VSG excitation link is based on the virtual excitation regulator, and the reactive power–voltage regulation is realized by simulating the electromagnetic characteristics of the synchronous motor, which can be expressed as:

$$E = E_0 + n(Q_{ref} - Q_e) + K_v(U_{ref} - U) \tag{13}$$

In the formula, E_0 is the VSG output no-load voltage amplitude and K_v and n are the reactive voltage proportional coefficient and reactive droop coefficient.

In actual control, the real-time state information of the energy storage system is uploaded to the VSG control layer through the battery management system (BMS). This information is used to dynamically adjust the power instructions P_{ref} and Q_{ref} of the VSG to ensure that the energy storage system provides inertial support and damping adjustment within a safe range. Therefore, the energy storage dynamics and energy availability have been integrated into the VSG control logic through the outer loop power command generation mechanism, rather than directly presented in the topology as an independent signal path.

2.5. Derivation of Double-Loop Control Transfer Function

The outer loop realizes the mapping from the power reference value to the voltage reference value, and the transfer function is deeply coupled with the VSG dynamic characteristics.

Combining Formulas (11) and (12) and considering the relationship between the electromagnetic torque and active power, it is derived that:

$$Gp_w(s) = \frac{\Delta\omega(s)}{P_{ref}(s)} = \frac{1}{Js^2 + D_p s + K_p} \tag{14}$$

where K_p is the active power regulation coefficient. The denominator comprehensively reflects the inertia J, damping D_p, and static adjustment ability K_p of the system.

Based on the excitation control Equation (14), the Q-V regulation transfer function is simplified:

$$G_{QV}(s) = \frac{V_{f,ref}(s)}{Q_{ref}(s)} = \frac{K_v}{1 + \frac{n}{X} V_{fN}} \tag{15}$$

In the formula, K_v is the proportional coefficient of the reactive power and voltage, n is the reactive power droop coefficient, and V_{fN} is the rated capacitor voltage, which ensures that the reactive power deviation is effectively converted into the voltage reference value adjustment.

The PI regulator is used to realize the tracking without a static error in the inner loop. The transfer function is constructed based on the LC filter model (L is the output inductance and C is the filter capacitance).

The voltage loop takes V_f as the feedback signal and the transfer function of the PI regulator is as follows:

$$G_{PI-V}(S) = K_P V + \frac{K_i V}{S} \tag{16}$$

The open-loop transfer function of the voltage loop is as follows:

$$G_V(s) = G_{PI-V}(s) \cdot \frac{1}{Ls + \frac{1}{Cs}} \tag{17}$$

The voltage loop compares $V_{f,ref}$ with the sampling V_f, and the output current reference value $i_{g,ref}$.

The current loop uses i_g as the feedback signal and the transfer function of the PI regulator is as follows:

$$G_{PI-I}(S) = K_{PI} + \frac{K_{iI}}{s} \tag{18}$$

The open-loop transfer function of the current loop is as follows:

$$G_I(s) = G_{PI-I}(s) \cdot \frac{1}{Ls} \tag{19}$$

By comparing $i_{g,ref}$ and sampling i_g, the current loop outputs the PWM modulation signal to suppress inductor current fluctuation and load disturbance.

2.6. VSG Small-Signal Model

In order to further illustrate the influence of the moment of inertia and the damping coefficient on the stability of the virtual synchronous machine system, it is necessary to establish a small-signal model of a virtual synchronous control system [13]. According to the (1) energy storage system grid converter output active power, the reactive power can obtain the system output complex power S expression, as shown in the following:

$$S = \frac{ER_g V_g \cos\,\delta + X_g V_g \sin\,\delta E - R_g E^2}{R_g^2 + X_g^2} + j\frac{-R_g V_g \sin\,\delta E + X_g V_g \cos\,\delta E - X_g E^2}{R_f^2 + X_g^2} \tag{20}$$

https://doi.org/10.3390/electronics15020323

An energy storage system grid-connected converter active small-signal model:

$$\Delta P = \Delta\delta\frac{-R_g V_g \sin\,\delta E + X_g V_g \cos\,\delta E}{R_g^2 + X_g^2} + \Delta V_g\frac{ER_g \cos\,\delta + X_g \sin\,\delta E}{R_g^2 + X_g^2} \tag{21}$$

An energy storage system grid-connected converter reactive small-signal model:

$$\Delta Q = \Delta\delta\frac{R_g V_g \cos\,\delta E + X_g V_g \sin\,\delta E}{R_g^2 + X_g^2} + \Delta E\frac{R_g V_g \sin\,\delta - X_g V_g \cos\,\delta}{R_g^2 + X_g^2} \tag{22}$$

When the energy storage grid-connected converter is in the grid-connected state, the virtual angular velocity of the VSG control system is the rated virtual angular velocity, that is, the combined (21) and (22). The small-signal model of VSG control can be derived as follows:

$$\begin{cases} -\Delta P = s(J\omega_0 s + D\omega_0)\Delta\delta \\ s(1 + T_a S)\Delta E = -(k_v s + k_p s + k_i)\Delta Q \end{cases} \tag{23}$$

In the formula, s is the Laplacian operator; T_a is the delay time constant; k_p is the reactive power proportional coefficient; and k_i is the reactive power integral coefficient.

Define matrix Y as:

$$Y = \left(\Delta\delta', \Delta E', \Delta\delta, \Delta E\right) \tag{24}$$

In the formula, Δ is the derivative of $\Delta\delta$, $\Delta E'$ is the derivative of ΔE, and the small-signal model of the combined Formulas (23) and (24) and the VSG control system of the grid-connected converter in the stable grid-connected operation state is as follows:

$$Y' = \begin{bmatrix} -\frac{D}{J} & 0 & -\frac{\partial P/\partial\delta}{J\omega_0} & -\frac{\partial P/\partial E}{J\omega_0} \\ \frac{A}{K_s} & \frac{B}{K_s} & \frac{C}{K_s} & \frac{D}{K_s} \\ 1 & 0 & 0 & 0 \\ 0 & 1 & 0 & 0 \end{bmatrix} Y \tag{25}$$

The expressions of the variables A, B, C, and D in the formula are as follows:

$$\begin{cases} A = -(k_v + k_p)\frac{\partial Q}{\partial\delta} \\ B = -(k_v + k_p)\frac{\partial Q}{\partial E} \\ C = -k_i\frac{\partial Q}{\partial\delta} \\ D = -k_i\frac{\partial Q}{\partial E} \end{cases} \tag{26}$$

According to Equation (23), the root locus diagrams of virtual inertia J and virtual damping D of the system are drawn, respectively.

To precisely analyze the influence of parameters J and D on system stability via the root locus method, the characteristic equation of the system is derived from the linearized small-signal model. The root locus is plotted based on the following open-loop transfer function form:

$$G_{open}(s) = K\frac{G_{other}(s)}{s(Ts + 1)} \tag{27}$$

where the variable gain K represents the key parameters under investigation (i.e., the virtual inertia J or the virtual damping coefficient D). The term $s(Ts + 1)$ in the denominator corresponds to the dominant dynamics of the power loop, primarily reflecting the inertial and damping characteristics. $G_{other}(s)$ encompasses the aggregated dynamics of other fixed parts in the system, such as the current control inner loop (modeled as a first-order delay), measurement filters, and the linearized coefficients from matrices A, B, C, and D in Equation (24). The stability analysis using this model focuses on the outer power/voltage control loop of the VSG, with the inner current loop dynamics being included in $G_{other}(s)$.

https://doi.org/10.3390/electronics15020323

The model describes the dynamic behavior of the entire VSG control system under grid-connected steady-state operating conditions, covering active and reactive power control loops (including the current inner loop). Although the root locus is drawn based on the whole system model, the analysis conclusion clearly points out that the root cause of the system oscillation instability lies in the insufficient equivalent damping of the current control loop, and the configuration of the virtual inertia and damping parameters directly affects the equivalent damping characteristics. Therefore, the root locus essentially reflects the role of the equivalent damping of the current loop in the whole system dynamics, which provides a theoretical basis for the subsequent improvement of the control strategy of enhancing the damping of the current loop through the series compensator.

The root locus analysis of Figure 3a,b shows that the root cause of the system oscillation instability lies in the insufficient equivalent damping characteristics of the current control loop, which is mainly due to the improper configuration of the virtual inertia and the virtual damping parameters: if the virtual inertia is too large or too small, the system stability margin or inertial support capacity will be reduced, while the lack of virtual damping directly weakens the system's ability to dissipate oscillation energy and suppress disturbances [14]. In order to directly solve the core problem of the insufficient equivalent damping of this current loop, a method of connecting a specific compensator in series to the grid current feedback path and improving the current controller is adopted. The core of the improved method is that a controllable additional damping effect is actively injected into the current feedback signal by a compensator, and its physical essence is to introduce a designable negative feedback mechanism. This is equivalent to the damping ratio of the dynamic enhanced current loop, which pulls the unstable poles that may be close to the imaginary axis back to the stable region and effectively suppresses the oscillation. At the same time, by properly designing the compensator parameters, the overdamping effect caused by excessive damping can be avoided while providing sufficient damping to ensure stability, thereby optimizing the dynamic performance of the system. Therefore, in view of the oscillation problem caused by an improper parameter configuration, the current controller is improved, and the scheme of the series compensator to the current feedback path is improved, and the effective compensation is realized by accurately improving the equivalent damping capacity of the current loop. Different from the idea of suppressing oscillation mainly through external impedance remodeling in Reference [15], the series compensator proposed in this paper directly acts on the current feedback loop and stabilizes the system pole by increasing the equivalent damping, avoiding the increase in the control complexity and performance degradation caused by additional impedance. At the same time, compared with the inertial simulation and impedance remodeling on the DC side in [16], the method in this paper takes into account the low-frequency inertial support and high-frequency resonance suppression in the AC grid-connected scenario, which is more suitable for the broadband stability requirements of the photovoltaic high-permeability distribution network.

https://doi.org/10.3390/electronics15020323

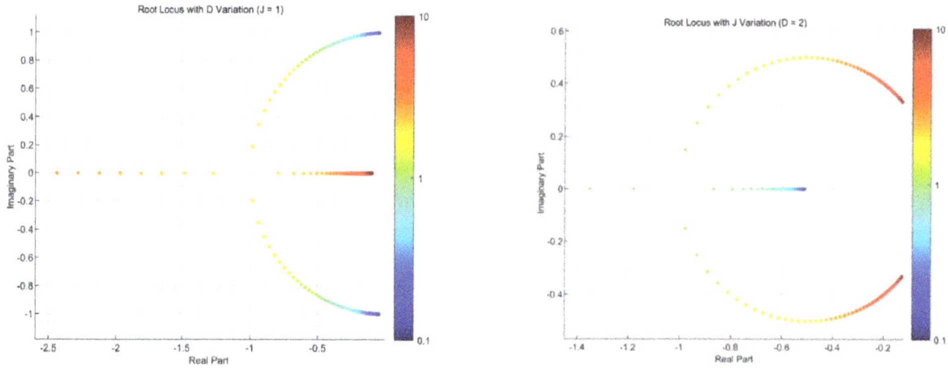

Root Locus with D Variation
(a)

Root Locus with J Variation
(b)

Figure 3. The root locus variation diagram of virtual inertia *J* and virtual damping *D*.

3. The Influence of Battery Limitation on VSG Inertial Response Is Analyzed

In order to quantitatively analyze the influence of energy storage system constraints on the VSG virtual inertial response, this chapter discusses the dynamic response characteristics of a VSG during large grid frequency disturbances under different *SOC* conditions, current constraints, and power ramp constraints through simulation and sensitivity analyses.

The *SOC* state of the energy storage system directly affects its available power support capability. When the *SOC* is close to the lower limit (such as 20%), the energy storage discharge capacity is limited, resulting in insufficient inertial support power provided by the VSG during the frequency drop, and the system frequency recovery time is prolonged. On the contrary, energy storage under a high *SOC* state can provide a sufficient inertial response, but attention should be paid to overcharge protection.

In order to quantify the influence of *SOC*, the virtual inertia effective coefficient is defined:

$$\eta J(SOC) = \begin{cases} 1, & SOC_{opt} \le SOC \le SOC_{max} \\ \frac{SOC - SOC_{min}}{SOC_{opt} - SOC_{min}} & SOC_{min} \le SOC < SOC_{opt} \end{cases} \tag{28}$$

SOC_{opt} is the optimal operating point (usually 50–80%). The actual virtual inertia can be corrected to the following:

$$J_{eff} = J \cdot \eta J(SOC) \tag{29}$$

4. Influence of Current Limitation and Power Ramp on Dynamic Response

The current limitation I_{max} and power ramp rate R_{ramp} directly affect the power response speed of the VSG. In the early stage of a large disturbance, if the energy storage cannot output enough power quickly, the system frequency drop will be more serious.

Considering the current limit, the actual output power of the energy storage is corrected to the following:

$$P_{bat},\ actual = \begin{cases} P_{bat}, cmd & |I_{bat}| \le I_{max} \\ sgn(P_{bat}, cmd) \cdot V_{bat} \cdot I_{max}, & |I_{bat}| > I_{max} \end{cases} \tag{30}$$

https://doi.org/10.3390/electronics15020323

The power ramp constraint is approximated by the following first-order hysteresis:

$$\frac{P_{bat,out}(s)}{P_{bat,actural}(s)} = \frac{1}{\tau_{ramp} + 1} \tag{31}$$

5. Modeling Design of Grid-Side Converter

The LCL filter of the grid-side converter is the key connection between the unit and the grid. This paper also considers the influence of the time-varying characteristics of the energy storage system on the DC-side dynamics to ensure the integrity of the system-level analysis.

$$\frac{di_1(t)}{dt} = \frac{1}{L_1}u_{inv}(t) - \frac{1}{L_1}i_2(t) \tag{32}$$

$$\frac{du_c(t)}{dt} \equiv \frac{1}{C}i_1(t) = \frac{1}{C}i_2(t) \tag{33}$$

$$\frac{di_2(t)}{dt} = \frac{1}{L_g}u_c(t) = \frac{1}{L_g}u_g(t) \tag{34}$$

The transfer function of the output voltage $u_{inv}(s)$ and the grid-connected current $i_2(s)$ is calculated by the Laplace transform as follows:

$$G_{LCL}(s) = \frac{1}{sL_1L_2C(s^2 + \omega_r^2)} \tag{35}$$

$$\omega_r = \sqrt{\frac{L_1 + L_2}{L_1L_2C}} \tag{36}$$

$$f_r = \omega_r / 2\pi \tag{37}$$

In the formula, $s^2 + \omega_r^2$ is the resonant unit of the system, ω^r is the resonant angular frequency, and f_r is the resonant frequency.

In actual operation, the output voltage $u_{inv}(s)$ is affected by the time-varying characteristics of the energy storage battery. According to Section 2.2 (4), there is a coupling relationship between the DC-side voltage V_{dc} and the battery terminal voltage V_{bat}. The time-varying relationship is transmitted to the grid-side converter through the DC bus capacitance equation, thus affecting the dynamic response of the system.

Consider that the filter will have a resonant peak at the resonant frequency f_r, resulting in the high-frequency oscillation of the system.

$$G_{LCL\_d}(s) = \frac{1}{sL_1L_2C(s^2 + 2\varsigma\omega_r + \omega_r^2)} \tag{38}$$

In the formula, ς is the damping coefficient and the damping effect of the system can be obtained by changing this parameter.

In the photovoltaic-rich distribution network, the *SOC* state of the energy storage affects the system damping demand. When the *SOC* is low, the battery internal resistance (*SOC*) increases and the system damping characteristics change. Therefore, the active damping control proposed in this paper considers the energy storage state, and the damping coefficient is designed as:

$$\zeta eff = \zeta_0 \cdot \left[1 + \alpha \cdot (1 - \frac{SOC}{SOC_{mom}})\right] \tag{39}$$

https://doi.org/10.3390/electronics15020323

where α is the adjustment coefficient to ensure that the oscillation can be effectively suppressed under different *SOC* conditions.

In this paper, the commonly used capacitive current feedback active damping method is selected to analyze the system. The diagram annotates the primary inputs and feedback signals of the system: the power references $P_{ref}(s)$ and $Q_{ref}(s)$, the grid current feedback $i_g(s)$, the capacitor voltage feedback $V_f(s)$, and the grid voltage $V_g(s)$. The control chain comprises the VSG outer loop, the voltage loop PI controller, the current loop quasi-proportional-resonant controller $G_i(s)$, the control delay $G_d(s)$, the zero-order hold $G_h(s)$, the PWM and inverter module, and the LCL filter with damping optimization. The arrow directions indicate the signal flow. The current outer loop controller uses a quasi-proportional resonant controller $G_i(s)$ to achieve zero steady-state error tracking of the grid-connected current. The expression is as follows:

$$G_i(s) = k_p + \frac{2\omega_c k_r s}{s^2 + 2\omega_c s + \omega_0^2} \tag{40}$$

In the formula, ω_0 and ω_c represent the basic angular frequency and the resonant angular frequency, respectively, and k_p and k_r are the proportional coefficient and the resonant coefficient of the quasi-proportional-resonant controller, respectively.

$G_d(s)$ is the total control delay of the system, which mainly includes the delay in the control process of the current outer loop and the damping inner loop and $G_h(s)$ is a zero-order holder, which is used to realize the conversion from a digital signal to an analog signal after sampling. The expressions of the two are as follows:

$$G_d(s) = e^{-sT_s} \tag{41}$$

$$G_h(s) = \frac{1 - e^{-sT_s}}{s} \tag{42}$$

In the formula, T_s is the sampling period. Using $j\omega$ instead of s, the frequency response of Formula (10) can be expressed as shown in Formula (43):

$$G_h(j\omega) = \frac{1 - e^{-j\omega T_s}}{j\omega} = \frac{\sin(0.5\omega T_s)}{0.5\omega} e^{-0.5j\omega T_s} \approx T_s e^{-0.5j\omega T_s} \tag{43}$$

Substituting $s = j\omega$ into (43), the approximate expression of the zero-order holder is finally obtained, as shown in Formula (43). It can be seen that the zero-order holder introduces a half-sampling period delay to the system.

6. Design of Delay Compensator

Under the condition of a weak grid, the calculation, sampling, and PWM delay in the control system will seriously deteriorate the phase margin of the system and threaten the stability. The traditional method uses the transcendental function to compensate, but there is a problem of physical implementation difficulties. Therefore, an improved delay compensation scheme is proposed in this paper. The core is to connect the compensator $G_P(z)$ in series with the grid current feedback path and realize the effective compensation of the control delay by reconstructing the feedback path structure. The system structure is shown in Figure 5. Compared to Figure 4, a delay compensator $G_p(z)$ is connected in series within the grid current feedback path. The compensator input is the grid current feedback signal $i_g(s)$, and its output is the compensated signal $i_{g,comp}(s)$, which is then fed into the current loop controller. This structure is designed to counteract the phase lag introduced

https://doi.org/10.3390/electronics15020323

by the control delay $G_d(s)$ and the zero-order hold $G_h(s)$, thereby enhancing system stability under high grid inductance conditions.

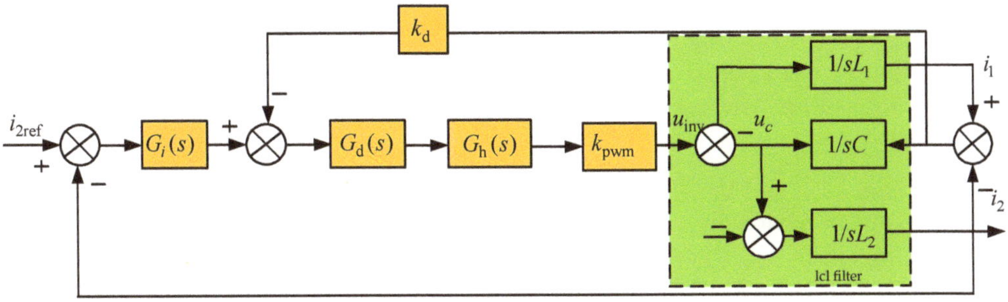

Figure 4. Block diagram of the non-compensated control system.

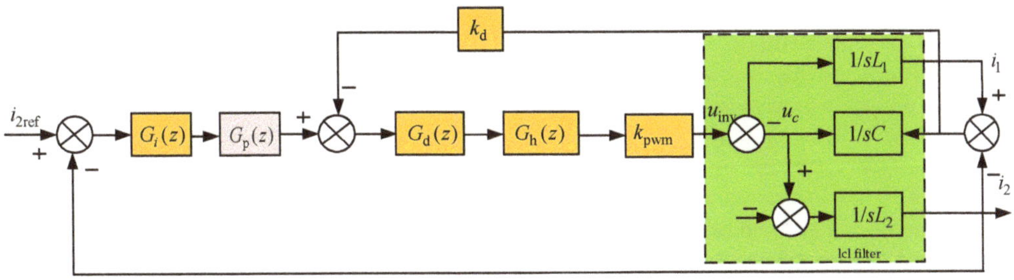

Figure 5. Block diagram of control system with compensation.

The expression of the compensator $G_p(z)$ obtained by a calculation is as follows:

$$G_p(z) = \frac{z^3 + A_1 z^2 + A_2 z + A_3}{z^3 + B_1 z^3 + B_2 z + B_3} \tag{44}$$

$$\text{Of which}: \begin{cases} A_1 = -2\cos\omega_r T_s \\ A_2 = 1 + \frac{k_d k_{pwm} \sin(\omega_r T_s)}{\omega_r L_1} \\ A_3 = -\frac{k_d k_{pwm} \sin(\omega_r T_s)}{\omega_r L_1} \end{cases} \begin{cases} B_1 = \frac{k_d k_{pwm}}{\omega_r L_1} \sin(\gamma\omega_r T_s) - 2\cos(\omega_r T_s) \\ B_2 = 1 + \frac{k_d k_{pwm} \sin[(1-\gamma)\omega_r T_s]}{\omega_r L_1} - \frac{k_d k_{pwm} \sin(\gamma\omega_r T_s)}{\omega_p L_1} \\ B_3 = -\frac{k_d k_{pwm} \sin[(1-\gamma)\omega_r T_s]}{\omega_r L_1} \end{cases} \tag{45}$$

According to the control theory, the closed-loop poles of the system are determined by the roots of the characteristic equation:

$$1 + G(z)H(z)G_p(z) = 0 \tag{46}$$

Before compensation, the characteristic equation is as follows:

$$1 + G(z)H(z) = 0 \tag{47}$$

After compensation, the compensator $G_p(S)$ enters the loop in a multiplicative manner, and the characteristic equation becomes the following:

$$1 + G(z)H(z) \times \frac{z^3 + A_1 z^2 + A_2 z + A_3}{z^3 + B_1 z^2 + B_2 z + B_3} = 0 \tag{48}$$

By carefully designing the positions of its zeros (molecular roots) and poles (denominator roots) in the z-plane, $G_p(z)$ can provide the required precise phase advance at key

 https://doi.org/10.3390/electronics15020323

frequency points (such as near ω_r) to offset the inherent phase lag caused by the total delay T_{delay} and the object characteristic L_1 in the system.

Pole shift: Comparing the above new characteristic equation with the original equation, it is equivalent to a dynamic correction link $G_p(z)$ in series on the original open-loop transfer function $G(z)H(z)$. In the root locus analysis, this is equivalent to the introduction of a new open-loop zero pole, thereby changing the shape of the root locus and guiding the closed-loop pole to move from the position that might otherwise be close to the unstable boundary to a stable region with a sufficient damping ratio.

7. Experimental Verification and Analysis of Energy Storage Grid-Connected System

In order to verify and evaluate the effectiveness of the proposed VSG-based energy storage grid-connected active support control strategy, this experiment uses a semi-physical simulation analysis method. This method combines hardware equipment and a simulation model. By embedding the core controller of the energy storage converter (implementing the VSG algorithm) as a hardware-in-the-loop (HIL) into the real-time simulation platform, it can simulate the dynamic behavior of the energy storage system in the real distribution network environment with high precision in the laboratory environment. Compared with a pure software simulation, this method can effectively reduce the model deviation caused by simplifying the power electronic switching process and controller hardware delay, so that the test results are closer to the actual engineering application, so as to effectively evaluate the performance of this strategy in improving the stability of the photovoltaic high-permeability distribution network (such as voltage support, frequency regulation, and power oscillation suppression).

The operating conditions of the photovoltaic high-permeability distribution network are significantly affected by the fluctuation of the light intensity and the light changes are random and intermittent. Therefore, in order to effectively evaluate the active support ability of the proposed VSG control strategy under typical disturbances, the following key scenarios are set up in this experiment.

A light mutation scene: the step change of the photovoltaic output is simulated in a short time (such as cloud occlusion) and the fast power response and inertia support effect of the VSG are tested.

Load switching or fault scenarios of the distribution network: the voltage support capability, fault ride-through capability, and damping effect of the VSG on system oscillation are tested by simulating a local load mutation or short-circuit fault on the grid side.

In the scenario of a sudden illumination change (the photovoltaic output step decreases from an 80% rated power to 50% rated power), the voltage waveform at the key bus, the system frequency variation curve, and the active and reactive power dynamic response waveform of the VSG output are shown in Figure 6 when comparing the energy storage system with conventional PQ control and the energy storage system with the proposed VSG active support control.

In the load switching scenario, the voltage recovery waveform of the key bus is shown in Figure 6 as shown in Figure 7 with the proposed VSG control method and without the control method. The specific parameters are shown in Table 1 and the parameter pairs are shown in Table 2.

https://doi.org/10.3390/electronics15020323

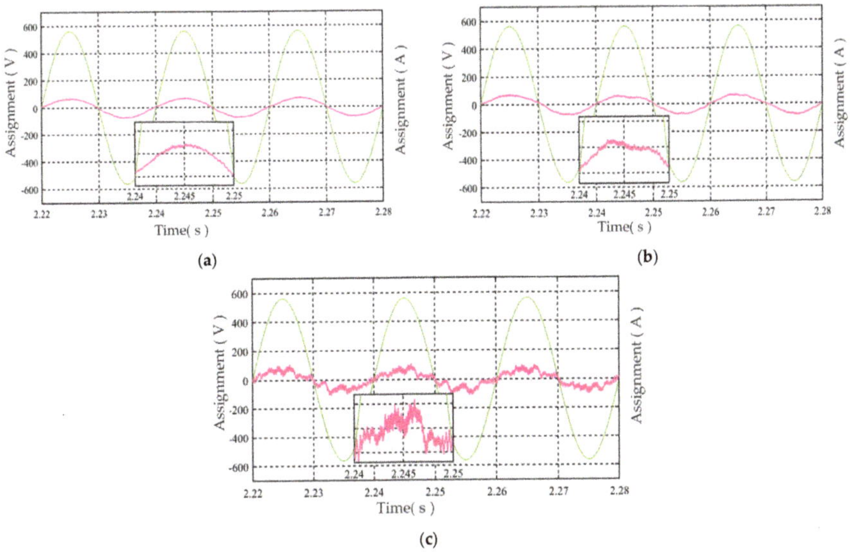

Figure 6. Steady-state waveforms of grid-connected current and voltage under uncompensated control. (**a**) The steady-state waveform of phase A grid-connected current and voltage when the inductance of the previous method is 0 mH. (**b**) The steady-state waveform of phase A grid-connected current and voltage when the inductance of the previous method is 1 mH. (**c**) The steady-state waveform of phase A grid-connected current and voltage when the inductance of the previous method is 2 mH.

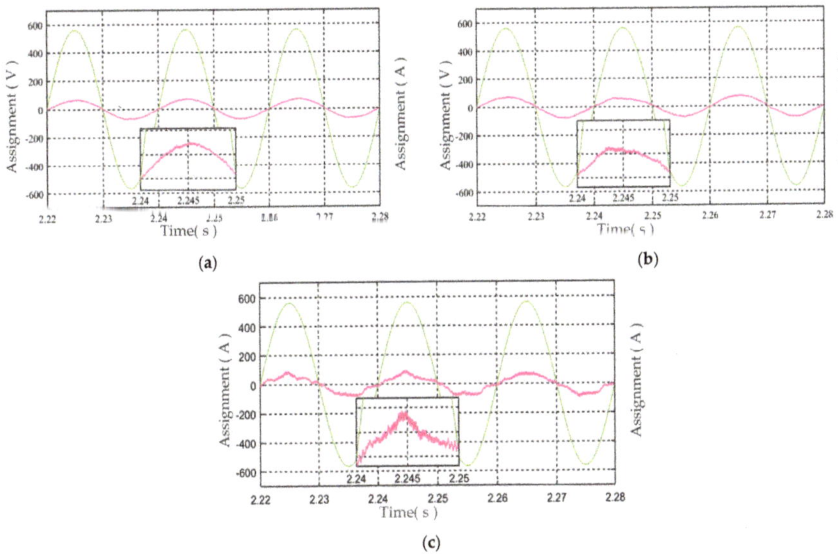

Figure 7. Steady-state waveforms of grid-connected current and voltage under compensation control. (**a**) When the inductance of the proposed method is 0 mH, the steady-state waveforms of phase A grid-connected current and voltage are obtained. (**b**) When the inductance of the proposed method is 1 mH, the steady-state waveforms of phase A grid-connected current and voltage are obtained. (**c**) When the inductance of the proposed method is 2 mH, the steady-state waveforms of phase A grid-connected current and voltage are obtained.

https://doi.org/10.3390/electronics15020323

Table 1. Parameters of the semi-physical simulation platform.

Parameter	Value
Rated Power	10 KVA
DC Bus Voltage	700 V
Grid Voltage (L-N, RMS)	220 V
Grid Frequency	50 Hz
Inverter-side Inductor	1.8 mH
Grid-side Inductor	0.6 mH
Filter Capacitor	20 μF
Virtual Inertia	0.5 kg·m^2 [*]
Virtual Damping	20 N·m·s/rad [*]
Active Droop Coefficient	1×10^{-4}
Reactive Droop	1×10^{-3}
Sampling/SwitchingFrequency	10 kHz
Current Loop PRController	2200

* Note: The values of J and D were selected through root locus analysis (Figure 3) to ensure system stability with adequate damping ratio (>0.7) for the nominal operating point, while providing sufficient inertia time constant.

Table 2. Performance comparison between conventional PQ control and proposed VSG control.

Performance Indicator	Conventional PQ Control	Proposed VSG Control	Improvement
Voltage Dip duringIllumination Change	15.2%	8.5%	44% reduction
Frequency Nadir after LoadSwitch	49.3 HZ	49.7 Hz	0.4 Hz higher
Settling Time for VoltageRecovery	0.45 s	0.18 s	60% faster
Oscillation Damping Ratio (estimated)	0.25	0.68	Significantly more damped

Based on the experimental results shown in Figures 6 and 7, when the inductance is set to L_g = 0 mH, the grid-connected current remains in phase with the grid voltage under appropriate operating parameters—whether or not the delay compensation controller is activated—thereby ensuring system stability. The grid strength is proportional to the inductance value. With L_g = 1 mH and without compensation control, high-frequency resonant components in both the grid-connected current and voltage gradually increase, yet the system can still maintain a stable operating condition. However, as the inductance rises further to L_g = 2 mH, significant high-frequency resonant oscillations appear in the grid current and voltage. If this trend persists, the grid-connected system will eventually become unstable. In Figure 7, after introducing the delay compensation controller, the grid-connected current and voltage remain in phase even as the inductance increases, and the high-frequency resonant components are effectively suppressed. This demonstrates that the proposed control strategy can successfully mitigate broadband oscillations in the energy storage grid-connected system. Even under adverse grid conditions, this control method maintains system stability and exhibits a favorable dynamic performance.

8. Conclusions and Prospect

Aiming at the problems of voltage fluctuation, frequency deviation, and power oscillation in the photovoltaic high-permeability distribution network, this paper studies the active support technology of the energy storage grid connection based on the VSG. By constructing a VSG control strategy, the inertia and damping characteristics of the synchronous generator are simulated to make up for the lack of system inertia caused by a grid connection of the power electronic equipment. A small-signal model is established to analyze the cooperative adjustment mechanism of virtual inertia and the damping coefficient and to reveal the nonlinear influence of the parameter configuration on stability. The current feedback compensator and LCL filter damping optimization scheme are de-

 https://doi.org/10.3390/electronics15020323

signed to effectively suppress the instability risk caused by the high-frequency resonance and control delay. Semi-physical simulation experiments show that the technology can significantly reduce voltage fluctuations, suppress broadband oscillations, and improve the system stability and power quality under sudden illumination changes, load switching, and fault scenarios.

Aiming at the stable operation of a photovoltaic high-permeability distribution network, this paper proposes an active support and oscillation suppression strategy which is energy storage grid-connected based on a VSG. Compared with the existing representative work, the innovation of this paper is mainly reflected in the following.

(1) A delay compensator in series with the current feedback path is proposed. Combined with the damping optimization of the LCL filter, the cooperative suppression of the high-frequency resonance and control delay is realized, and the effective frequency band of oscillation suppression is broadened.

(2) The small-signal model of the VSG system including energy storage dynamics is established, and the nonlinear relationship between virtual inertia, the damping coefficient, and system stability is clarified, which provides a theoretical basis for a parameter adaptive design.

(3) The comprehensive advantages of the proposed strategy in voltage support, frequency regulation, and broadband oscillation suppression under various disturbances such as illumination mutation, load switching, and a grid fault are verified by a semi-physical simulation, which provides a new solution for the active stability control of a high-proportion new-energy distribution network.

Several limitations remain in the current research, such as the dependence of a VSG parameter adaptation mechanism on a single variable, the dynamic balance algorithm of energy storage SOC recovery, and the frequency regulation demand needs to be optimized. In the future, it is necessary to focus on exploring the parameter adaptive strategy of multivariable coupling and integrate the randomness of the photovoltaic output and the impedance characteristics of the power grid to achieve precise parameter tuning. The integration of a VSG and energy storage should be deepened and the energy management algorithms that take into account the dynamic response and economy should be developed; the application verification of the extended technology in practical scenarios such as a complex grid topology and high proportion of a new energy cluster grid connection offers enhanced technical support for the safe and stable operation of the new power system.

Author Contributions: Conceptualization, S.S.; Methodology, S.S.; Software, D.W. and D.X.; Validation, J.Y.; Formal analysis, Z.Z. and D.W.; Resources, S.S.; Data curation, Z.Z., J.Y. and D.X.; Writing—original draft, Z.Z. and J.Z.; Writing—review & editing, Z.Z. and J.Z.; Visualization, S.S. and D.X.; Project administration, D.W.; Funding acquisition, J.Y. and D.W. All authors have read and agreed to the published version of the manuscript.

Funding: This paper is supported by Henan Science and Technology Key Project in 2025 (252102240122) and Henan University Key Scientific Research Project in 2025 (25B470020).

Data Availability Statement: The original contributions presented in this study are included in the article. Further inquiries can be directed to the corresponding author.

Conflicts of Interest: The author Shuhuai Shi was employed by the company State Grid Henan Electric Power Research Institute. The remaining authors declare that the research was conducted in the absence of any commercial or financial relationships that could be construed as a potential conflict of interest.

https://doi.org/10.3390/electronics15020323

References

1. 'New Power System Development Blue Book' Writing Group. *Blue Book on New Power System Development*; China Electric Power Press: Beijing, China, 2023.
2. Dua, M.; Gupta, P. Virtual Inertia Emulation in Standalone Microgrid with Photovoltaic-Battery Energy Storage System. In Proceedings of the 2024 IEEE Region 10 Symposium (TENSYMP), New Delhi, India, 27–29 September 2024; IEEE: New York, NY, USA, 2024; pp. 1–6.
3. Shuai, W.; Yajun, L.; Lin, L.; Lingfei, X. Research on MMC-HVDC Oscillation Suppression Characteristics Based on Virtual Synchronous Generator Control. In Proceedings of the 2024 6th International Conference on Power and Energy Technology (ICPET), Beijing, China, 12–15 July 2024; IEEE: New York, NY, USA, 2024; pp. 509–513.
4. Wu, L.; Yu, X.; Jin, Q.; Zhang, B. Asymmetric Fault Ride-through Strategy Considering Optical Storage Output in A Weak Grid. In Proceedings of the 2024 6th International Conference on Power and Energy Technology (ICPET), Beijing, China, 12–15 July 2024; IEEE: New York, NY, USA, 2024; pp. 1416–1421.
5. Liu, H.; Xie, X.; He, J.; Xu, T.; Yu, Z.; Wang, C. Subsynchronous interaction between direct-drive PMSG based wind farms and weak ac networks. *IEEE Trans. Power Syst.* **2017**, *32*, 4708–4720. [CrossRef]
6. Ouyang, Y.; Sun, D.; Wen, F.; Zou, Y. High Frequency Resonance Mechanism Analysis of LCL Types Grid-connected Inverter without Damping. In Proceedings of the 2022 IEEE 5th International Electrical and Energy Conference (CIEEC), Nanjing, China, 27–29 May 2022; IEEE: New York, NY, USA, 2022; pp. 816–820.
7. Wang, Q.; Zhou, D.; Yin, S.; Lei, Y.; He, T. Improved adaptive inertia and damping coefficient control strategy of VSG based on optimal damping ratio. In Proceedings of the 2022 International Power Electronics Conference (IPEC-Himeji 2022-ECCE Asia), Himeji, Japan, 15–19 May 2022; IEEE: New York, NY, USA, 2022; pp. 102–107.
8. Tan, Y.; Zhang, L.; Ma, T.; Liu, Q. Signal-phase virtual synchronous generator control strategy for energy storage converters. In Proceedings of the 2022 IEEE 5th International Electrical and Energy Conference (CIEEC), Nanjing, China, 27–29 May 2022; IEEE: New York, NY, USA, 2022; pp. 1113–1118.
9. Sato, T.; Asharif, F.; Umemura, A.; Takahashi, R.; Tamura, J. Cooperative Virtual Inertia and Reactive Power Control of PMSG Wind Generator and Battery for Improving Transient Stability of Power System. In Proceedings of the 2020 IEEE International Conference on Power and Energy (PECon), Virtual Event, 7–8 December 2020; IEEE: New York, NY, USA, 2020; pp. 101–106.
10. Shi, R.; Lan, C.; Zhong, Z.; Zhang, Q.; Zhou, Q.; Liu, W. Active power feed-forward compensation based active power oscillation suppression strategy for energy storage VSG grid-connected. *Power Syst. Prot. Control.* **2023**, *51*, 118–126.
11. Tao, J.; Li, H.; Jiang, K.; Ye, X.; Wang, J. Wind storage microgrid energy management system with integrated improved vsg control. In Proceedings of the 2021 IEEE 5th Conference on Energy Internet and Energy System Integration (EI2), Taiyuan, China, 22–24 October 2021; IEEE: New York, NY, USA, 2021; pp. 2048–2053.
12. Chen, G.; Yang, B.; Li, H.; Wang, H.; Wang, B. Research on Virtual Synchronous Generator Control for Parallel Connection of Multiple Grid-Forming Energy Storage Converters. In Proceedings of the 2024 9th Asia Conference on Power and Electrical Engineering (ACPEE), Shanghai, China, 11–13 April 2024; IEEE: New York, NY, USA, 2024; pp. 630–634.
13. Chen, C.; Wu, H.; Xiong, X. Modeling and Small Signal Stability Analysis of Virtual Synchronous Generator Control with Field Winding Transient. In Proceedings of the 2024 IEEE 10th International Power Electronics and Motion Control Conference (IPEMC2024-ECCE Asia), Chengdu, China, 17–20 May 2024; IEEE: New York, NY, USA, 2024; pp. 568–573.
14. Wang, W.; Liu, Z.; Lou, J.; Cheng, C.; Song, Q.; Shi, Y. Modeling of sequence impedance of new energy synchronous generator and its suppression effect on broadband oscillation. *Grid Technol.* **2024**, *48*, 3595–3603.
15. Li, G.; Chen, Y.; Luo, A.; He, Z.; Wang, H.; Zhu, Z.; Wu, W.; Zhou, L. Analysis and Mitigation of Subsynchronous Resonance in Series-Compensated Grid-Connected System Controlled by a Virtual Synchronous Generator. *IEEE Trans. Power Electron.* **2020**, *35*, 11096–11107. [CrossRef]
16. Deng, F.; Zhang, L.; Xiao, Z.; Yao, Z.; Tang, Y. Inertia Emulation in Droop-Based DC Microgrids with Equivalent Converter Impedance Reshaping. *IEEE Trans. Power Electron.* **2024**, *39*, 15206–15216. [CrossRef]

electronics

MDPI

Article

Capacitor State Monitoring Based on Haar Wavelet Transform and Enhanced Kalman Filter

Tao Zhang, Zhiyao Lu, Wenjie Liu *, Yu Ding, Shengfei Wang and Weilin Li

School of Automation, Northwestern Polytechnical University, Xi'an 710129, China
* Correspondence: wenjieliu@nwpu.edu.cn

Abstract: Aviation electrification is an inevitable trend leading the development of future aviation technology, and its development cannot be separated from high-performance onboard power systems. As a key equipment of the system, the DC converter plays a core role in energy conversion, and its operational reliability directly affects the stability of the entire system. As the core component of the converter, capacitors have become a weak link in system reliability due to their high failure rate. Therefore, accurate monitoring of their health status is of great significance. To achieve fast, high-precision online monitoring of capacitors, this paper proposes an intelligent monitoring strategy that integrates Haar wavelet transform and Kalman filter. This method only requires the collection of inductance current and output voltage signals during regular operation, without the need for additional installation sensors. The capacitance current is reconstructed and used to accurately identify the capacitance value (C) and equivalent series resistance (ESR) throughout the entire life cycle in strong noise environments. The simulation and experimental results show that the strategy has good robustness under different operating conditions, with recognition errors of C and ESR controlled within 3% and 2%, respectively, demonstrating the feasibility of the proposed method.

Keywords: DC-DCconverter; aluminum electrolytic capacitor; wavelet transform; current reconstruction; Kalman filter

1. Introduction

More Electric Aircraft, as a representative of the new generation of aviation electrical architecture, gradually replaces traditional mechanical transmission methods such as hydraulic and pneumatic systems with electrical energy integration technology, significantly improving the comprehensive utilization efficiency of onboard energy [1,2]. In this system, DC–DC power converters undertake the core function of energy conversion, and their operational reliability has a decisive impact on the entire onboard power system [3].

In actual flight missions, DC–DC converters often face rapidly changing operating conditions. The key components such as internal power semiconductors, capacitors, and inductors are prone to performance degradation under the coupling effect of electrical thermal multi-physical field stress, which in turn affects the overall reliability of the system. Aluminum electrolytic capacitors (AEC) have become key components of DC–DC converters due to their excellent filtering and voltage stabilization performance [4]. However, as shown in Figure 1, the fault statistics of the converter indicate that the failure rate of AEC is as high as about 30%, which is considered the weakest link in the converter and is therefore commonly used as a reference indicator to evaluate its overall lifespan [5].

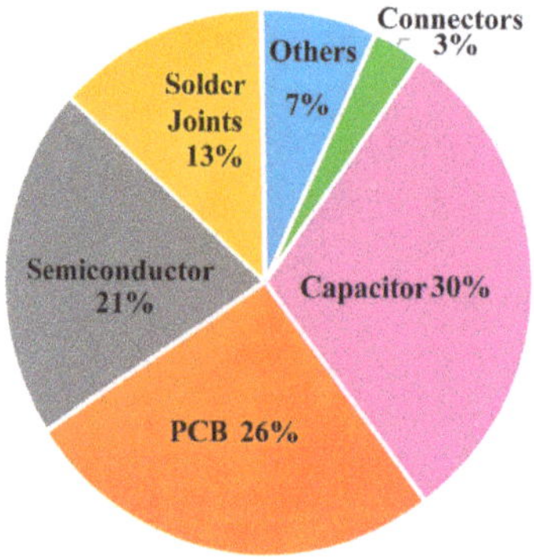

Figure 1. Distribution diagram of main failed components [5].

The capacitance (C) and equivalent series resistance (ESR) are key electrical parameters for evaluating the aging state of capacitors [6,7]. It is generally believed that when the capacitance drops to 80% of the initial value or the ESR rises to twice the initial value, the capacitor fails. State monitoring technology is the core means of identifying the aging status of capacitors and achieving fault warning [8,9]. According to whether the system operation needs to be interrupted during the monitoring process, existing methods can be mainly divided into two categories: offline monitoring and online monitoring [10].

Offline monitoring methods typically require pausing system operation in order to accurately obtain key health status parameters of capacitors. In this method, using an LCR tester to directly measure the C and ESR is a typical means, with a measurement accuracy of up to 0.1%. Therefore, it is often used as a benchmark to evaluate the accuracy of other parameter identification methods. In order to simplify the configuration of measurement equipment, an experimental method based on an RC circuit was proposed in [11]. This method analyzes the phase relationship between the capacitor terminal voltage and the input voltage and combines graphic processing and model simplification to achieve ESR extraction of capacitors in the frequency range of 100 Hz to 100 kHz. To further reduce the influence of human errors in the graphical analysis process, the Newton–Raphson numerical iteration method was introduced to achieve accurate estimation of ESR and capacitance values at a low frequency of 120 Hz by solving the nonlinear equation system describing the voltage relationship [12]. Although the offline monitoring method has the advantage of high accuracy, its operation usually requires the removal of capacitors from the circuit, which is costly to implement and makes it difficult to meet the requirements of continuous operation and real-time evaluation of the system in practical engineering.

Online monitoring technology, as a non-invasive detection method, can obtain real-time operation status and performance parameters of capacitors under continuous circuit operation conditions. A real-time monitoring method for input and output capacitors with ESR estimation was developed for boost converters, and the influence of load, duty cycle, inductance, and temperature on ESR estimation accuracy was systematically analyzed [13]. Another study proposes a non-invasive online monitoring technology that achieves state assessment by analyzing the correlation characteristics between capacitor voltage and current during transient processes [14]. For flyback converters, existing studies have constructed online models suitable for continuous conduction mode and intermittent conduction mode, which can infer the capacitance value C and ESR parameters based on the voltage ripple of the capacitor [15,16]. However, this type of method is complex in analyzing the model, sensitive to changes in operating mode, and has limited adaptability.

In response to the above shortcomings, this paper proposes a fusion monitoring strategy based on Haar wavelet transform and Kalman filter: the switch sequence is accurately extracted from the inductor current through Haar wavelet transform, and the capacitor current can be reconstructed without additional hardware; furthermore, by utilizing the Kalman filtering algorithm, high-precision and full lifecycle online identification of capacitors C and ESR was achieved in strong noise environments. This method only requires the collection of conventional inductor current and output voltage signals, significantly improving the practicality, robustness, and engineering applicability of the monitoring system. Finally, the effectiveness of the proposed strategy under different operating conditions was verified through simulation and experiments.

The rest of the paper is organized as follows. Section 2 analyzed the mechanism of capacitor aging and changes in key parameters. The basic working principle of the boost converter is elaborated on in Section 3. Section 4 focuses on a capacitor current reconstruction method based on the Haar wavelet transform. The estimation of the C and ESR parameters using a Kalman filter algorithm is introduced in Section 5. Simulation verification and analysis are conducted in Section 6, followed by a systematic discussion and evaluation of the experimental results in Section 7. Finally, the conclusion of this work is presented in Section 8.

2. Aging Mechanism and Key Parameters of Capacitors

The aging of aluminum electrolytic capacitors is a complex electrochemical process, and its main failure modes are the decrease in C and the increase in equivalent series resistance ESR. Understanding the physical nature of these parameter changes is crucial for accurately monitoring their health status.

2.1. Aging Mechanism of Capacitors

The internal structure of aluminum electrolytic capacitors is mainly composed of anode aluminum foil, electrolyte, and cathode aluminum foil, and its dielectric layer is an aluminum oxide film generated on the surface of the anode aluminum foil through anodic oxidation. The capacitance value is determined by (1), [17].

$$C = \varepsilon \frac{A_s}{d_s} \tag{1}$$

where ε is the dielectric constant, A_s is the effective area, and d_s is the thickness of the dielectric layer. During the lifespan of a capacitor, its geometric dimensions (A_s, d_s) remain essentially unchanged. The decrease in capacitance value is mainly attributed to the decrease in effective dielectric constant. This is not due to changes in the properties of the stable alumina dielectric layer itself, but rather because the electrolyte used as the cathode gradually evaporates and decomposes under working temperature and electric field stress. The reduction in electrolyte leads to a decrease in its contact area with the dielectric layer and changes the interface characteristics, which is equivalent to a decrease in the effective dielectric constant of the overall system forming the capacitor, resulting in a continuous decrease in the measured capacitance value.

ESR is the equivalent of all active losses of a capacitor in a series equivalent model. These losses mainly include: the resistance of the electrolyte itself (the main contribution); contact resistance between electrode foil and electrolyte; and ohmic resistance of electrode leads and connection points. The volatilization and decomposition of the electrolyte directly lead to a decrease in its ionic conductivity, significantly increasing the body resistance and contact resistance, thereby causing an overall increase in ESR. Therefore, the increase in ESR directly characterizes the increase in internal power loss of capacitors.

2.2. Loss Tangent Analysis

The tangent of the loss angle (tanδ) is a classic and important parameter for evaluating the insulation state of electrical equipment. It is defined as the ratio of active power loss to reactive power in an insulating medium. In the series equivalent circuit model of capacitors, tanδ can be expressed as follows [18]:

$$\tan\delta = \frac{I^2 \cdot \mathrm{ESR}}{I^2/(\omega \cdot \mathrm{C})} = \omega \cdot \mathrm{C} \cdot \mathrm{ESR} = 2\pi f \cdot \mathrm{C} \cdot \mathrm{ESR} \quad (2)$$

As shown in (2), tanδ is directly proportional to the product of C and ESR. As the capacitor ages, the C value decreases while the ESR value increases, and the combined effect will cause a significant change (usually an increase) in tanδ. Therefore, tanδ is a sensitive parameter that can comprehensively reflect the degradation of capacitor insulation performance.

2.3. Complex Impedance Analysis

The aging state of capacitors is fully reflected in the frequency characteristics of their complex impedance $Z(f)$. As described in Section 2.2, in the series model, the modulus and phase angle are [19]

$$|Z| = \sqrt{(\mathrm{ESR})^2 + \left(\frac{1}{\omega \cdot \mathrm{C}}\right)^2}, \ \arg(Z) = -\arctan\left(\frac{1}{\omega \cdot \mathrm{C} \cdot \mathrm{ESR}}\right) \quad (3)$$

In theory, by accurately measuring the impedance modulus and phase angle of a capacitor at a certain frequency, both C and ESR can be calculated simultaneously. This is exactly the basic principle of many offline monitoring instruments. However, online and non-invasive monitoring in DC–DC converters faces challenges: On the one hand, the excitation signal is limited, and the system itself does not provide pure sine excitation for sweep frequency measurement. On the other hand, capacitor current and voltage ripple are non-sinusoidal signals containing switching noise, and it is very difficult to accurately extract impedance phase angle information directly from them.

In response to the above challenges, this article proposes a robust monitoring strategy based on time-domain analysis in strong noise and non-sinusoidal excitation environments. By accurately tracking the changes in C and ESR, this method effectively avoids the problem of extracting impedance phase angle online and can also be used for indirectly detecting the insulation state of capacitors.

3. Working Principle of Boost Converter

Boost converters have been widely used in the energy distribution of aircraft power systems due to their efficient energy conversion and flexible voltage regulation characteristics [20]. Figure 2 shows the topology of a typical boost converter, which mainly consists of input power supply u_{in}, power inductor L, switch tube S, diode Q, output capacitor C and its ESR, and load R.

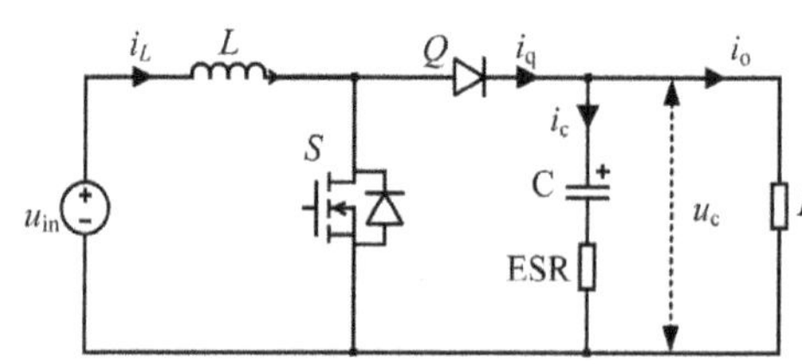

Figure 2. Topology of boost converter.

When the switch tube is conducting, the switch tube S is approximately in a short-circuit state. The input power u_{in} charges the inductor L, and the inductor current i_L increases, resulting in an increase in inductor energy storage; at the same time, the output capacitor C independently supplies power to the load R to maintain the load current i_c, resulting in a gradual decrease in the output voltage u_c.

When the switch tube is turned off, the switch tube S is in an open-circuit state, and the energy stored in the inductor L is released through the freewheeling diode Q, and together with the input power source u_{in}, it supplies power to the load while charging the output capacitor C. During this process, the inductor current i_L decreases and the output voltage u_c increases accordingly.

This paper focuses on the steady-state operating characteristics of converters in continuous conduction mode (CCM). The steady-state operating waveform of C and ESR in series is shown in Figure 3, revealing the dynamic correlation between switch signals, inductor current, and output voltage. It is crucial to note that the monitoring methodology developed in this work is specifically designed for and validated under this CCM regime. The reason for choosing CCM is that the core requirements for high efficiency, high reliability, high power density, and low electromagnetic interference in aviation applications fundamentally contradict the inherent characteristics of discontinuous conduction mode (DCM), such as large current ripple, high component stress, and significant electromagnetic interference. Due to the above advantages, CCM has become the preferred solution for this type of application. Furthermore, the subsequent capacitor current reconstruction strategy fundamentally relies on the continuous and periodic nature of the inductor current in CCM. Its direct application to DCM is precluded, as the DCM operation, characterized by periods of zero inductor current, violates this core signal prerequisite.

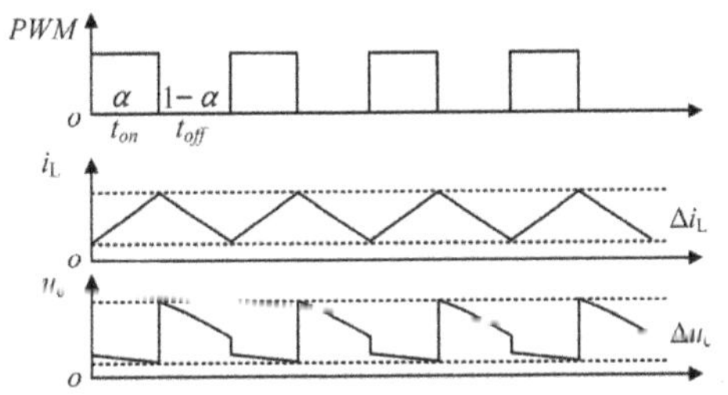

Figure 3. Steady-state operating waveform of CCM.

4. Capacitor Current Reconstruction Based on Haar Wavelet Transform

4.1. Haar Wavelet Transform for Extracting Switch Sequences

The variation in inductor current is directly driven by the switching sequence, and the information of this sequence can be indirectly obtained from the ripple of inductor current. Therefore, this paper uses wavelet transform as a time-frequency analysis tool to extract switch signals from inductor currents.

Wavelet transform decomposes non-stationary time series into different time-frequency domains through a set of scalable and translational basis functions (wavelet function and scale function) [21–23]. This process generates two types of coefficients: approximate coefficient A and detail coefficient D. The approximation coefficient A represents the contour or trend component of the signal in the low-frequency range, reflecting the macroscopic structure of the signal; the detail coefficient D captures the details or abrupt components of the signal in the high-frequency range. The wavelet transform utilizes low-pass and high-pass filter banks corresponding to the scale function φ(n) and wavelet function ψ(n), respectively, to decompose the signal into the low-frequency approximation coefficient $A[i,$

$\tau]$ that characterize its contour and high-frequency detail coefficient $D[i, \tau]$ that contains its details, as shown in (4) and (5). By changing the scale parameter i and translation parameter τ, the time-frequency window of the analysis can be adaptively adjusted to achieve multi-scale signal analysis [24].

$$A[i,\tau] = \frac{1}{\sqrt{2^i}} \sum_n I_L(n) \varphi_{\alpha,\tau}\left(\frac{n-\tau}{2^i}\right) \tag{4}$$

$$D[i,\tau] = \frac{1}{\sqrt{2^i}} \sum_n I_L(n) \psi_{\alpha,\tau}\left(\frac{n-\tau}{2^i}\right) \tag{5}$$

The scale function of the Haar wavelet basis has tight support characteristics, and its domain is limited to a finite unit interval, usually taken as [0, 1), with a function value of 1 within this interval and a constant value of 0 outside the interval, as shown in (6). Power switching transistors cause changes in inductor current when they are turned on and off. The selection of the Haar wavelet for locating switching transients is driven by its minimal support set, limited to only two data points, which provides essential benefits for real-time implementation. It introduces a latency of just one sampling period, a delay that is negligible in the context of state monitoring yet critical for maintaining high-speed operational responsiveness. Computationally, the Haar transform requires only simple addition and subtraction operations, resulting in an extremely light processing load suitable for resource-constrained embedded platforms. Although wavelets with longer support such as those from the Daubechies family can achieve better noise suppression, they inevitably introduce greater delays and require more complex calculations. Given that the primary objective is accurate detection of switching instants rather than optimal signal denoising, the Haar wavelet offers the most favorable trade-off among speed, simplicity, and practically sufficient accuracy [25–27].

$$\varphi(n) = [1,1],\ \psi(n) = [1,-1] \tag{6}$$

Substituting (6) into (4) and (5) yields

$$A(n) = \frac{I_L(n) + I_L(n+1)}{2\sqrt{2}} \tag{7}$$

$$D(n) = \frac{I_L(n) - I_L(n+1)}{2\sqrt{2}} \tag{8}$$

Under a steady-state operating condition, the inductance current of the boost converter is analyzed. According to Equations (7) and (8), the waveform of the inductance current during charging and discharging can be represented as shown in Figure 4 after Haar wavelet transform. During the inductor charging phase (corresponding to the t_2 period), the inductor current shows an upward trend, and at this time, its approximation coefficient A and detail coefficient D satisfy (9).

$$\begin{cases} i_L(t) > i_L(t-1) \\ A(t) = \frac{i_L(t-1)+i_L(t)}{2\sqrt{2}} > 0 \\ D(t) = \frac{i_L(t-1)-i_L(t)}{2\sqrt{2}} < 0 \end{cases} \tag{9}$$

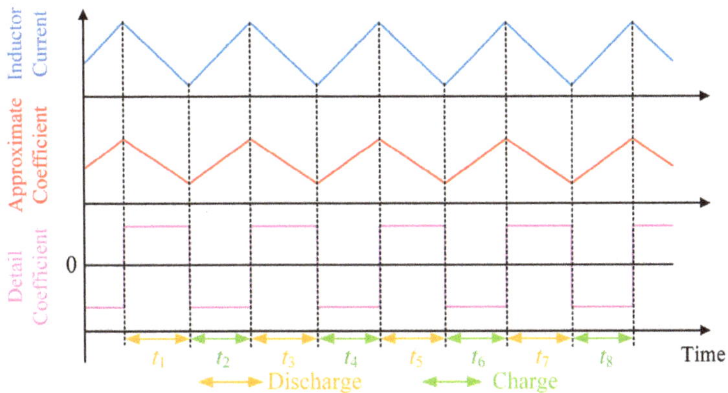

Figure 4. Capacitor current waveform after wavelet transform.

During the inductor discharge stage (corresponding to the t_1 period), the inductor current gradually decreases, and the corresponding approximate and detailed parameters comply with (10).

$$\begin{cases} i_L(t) < i_L(t-1) \\ A(t) = \frac{i_L(t-1)+i_L(t)}{2\sqrt{2}} > 0 \\ D(t) = \frac{i_L(t-1)-i_L(t)}{2\sqrt{2}} > 0 \end{cases} \tag{10}$$

We performed wavelet transform on the inductor current under charging and discharging conditions, and the decomposition results are shown in Table 1. Analysis shows that the approximation coefficient A is always positive, reflecting the low-frequency trend of the signal; the detail coefficient D effectively characterizes the system dynamics: it is negative during charging and positive during discharging. By utilizing this polarity feature, the detail coefficient can be used to extract switch sequences (charging as 1, discharging as 0). It should be noted that the extraction process is based on the calculation of the current difference between adjacent sampling points, and the resulting switch sequence has a delay of one sampling period compared to the actual driving signal. This is an inherent characteristic of the algorithm and will be considered in subsequent analysis.

Table 1. Signal characteristics and reconstructed switch sequence after wavelet transform.

Parameter	Charge	Discharge
Approximate coefficient A	$A > 0$	$A > 0$
Detail coefficient D	$D < 0$	$D > 0$
Switch sequence	1	0

4.2. Capacitor Current Reconstruction

The key to capacitor status monitoring lies in accurately obtaining capacitor current information. However, installing additional current sensors not only increases system complexity and cost, but may also introduce additional measurement errors and electromagnetic interference. Therefore, this paper proposes a capacitor current reconstruction method based on inductor current and its extracted switch sequence, without the need for a dedicated current sensor. The derivation of this method is based on the circuit topology and switch operation mode of boost converter. The specific implementation process is as follows:

Consider the main circuit topology of the boost converter shown in Figure 1. According to the KCL law, there exists the following current relationship at the node where the capacitor is connected to the load:

$$i_c(t) = i_q(t) - i_o(t) \tag{11}$$

The conduction state of the diode is determined by the switching signal $S_{PWM}(t)$ of the switching transistor. In an ideal situation, when $S_{PWM}(t)$ = 0 (switch off), the diode conducts and its current can be expressed as

$$i_q\ (t) = i_L\ (t) \tag{12}$$

When $S_{PWM}(t)$ = 0 (switch on), the diode is turned off, and its current can be expressed as

$$i_q\ (t) = 0 \tag{13}$$

Therefore, the diode current can ideally be expressed as the product of the inductor current and the inverse code of the switch signal:

$$i_q(t) = i_L(t) \cdot \overline{S_{PWM}(t)} \tag{14}$$

Given the inherent delay in digital control systems, the inductor current at the current sampling moment is actually determined by the switch state of the previous cycle. If the current switch signal is directly multiplied by the current, it will cause phase errors, which will affect the convergence and accuracy of subsequent parameter identification algorithms. Therefore, it is necessary to perform timing correction on the diode current formula in the discrete time domain as follows:

$$i_q(t) = i_L(t) \cdot \overline{S_{PWM}(t-1)} \tag{15}$$

When the circuit is in steady-state operation, the average value of the capacitor current is zero within one switching cycle. Therefore, the load current i_o is equal to the average value of the diode current i_q, that is,

$$i_o(t) = avg\left(i_L(t) \cdot \overline{S_{PWM}(t-1)}\right) \tag{16}$$

By substituting (15) and (16) into (11), the capacitance current can be calculated as follows:

$$i_c(t) = i_L(t) \cdot \overline{S_{PWM}(t-1)} - avg\left(i_L(t) \cdot \overline{S_{PWM}(t-1)}\right) \tag{17}$$

In summary, the capacitor current in the boost converter can be determined by both the inductor current and the switching sequence, and the required delayed switching sequence is exactly the same as the switching sequence extracted after wavelet transform.

5. Capacitor State Monitoring Based on Kalman Filter

To achieve accurate monitoring of capacitor C and ESR, it is necessary to establish a discrete relationship model between capacitor terminal voltage and current in the time domain. Firstly, based on the working principle of the boost circuit, the transfer function between the capacitor terminal voltage and current can be derived, as shown in (18).

$$G(s) = \frac{U(s)}{I(s)} = \frac{\mathrm{ESR} \cdot \mathrm{C}s + 1}{\mathrm{C}s} \tag{18}$$

Further using the Tustin transform method shown in (19), discretize (18) to obtain the discrete form shown in (20), where Δt represents the sampling period.

$$s = \frac{2}{\Delta t}\frac{z-1}{z+1} \tag{19}$$

$$G(z) = \frac{\left(\mathrm{ESR} + \frac{\Delta t}{2C}\right)z + \frac{\Delta t}{2C} - \mathrm{ESR}}{z-1} \tag{20}$$

Finally, transform the variable relationships described in (20) into a structural form suitable for parameter identification:

$$U_{c,k} = U_{c,k-1} + \left(\mathrm{ESR} + \frac{\Delta t}{2C}\right)I_{c,k} + \left(\frac{\Delta t}{2C} - \mathrm{ESR}\right)I_{c,k-1} \tag{21}$$

According to the parameter identification form established in (21), it can be incorporated into the framework of Kalman filtering (KF) for solving. This form is highly consistent with the standard state space model of KF, where the terminal voltage of the capacitor can be used as an observed variable, and the capacitance C to be accurately estimated and the equivalent series resistance ESR together form the state vector of the system.

The KF algorithm mainly consists of two equations: the state equation and the measurement equation, as shown in (22) and (23), respectively [28,29]:

$$X(k) = A(k)X(k-1) + T(k)\omega_k \tag{22}$$

$$Y(k) = H(k)X(k) + v_k \tag{23}$$

Among them, $X(k)$ represents the state vector at time k; $A(k)$ represents the state transition matrix, which is used to establish an iterative relationship between the state variables at the previous $k-1$ time and the current k time; $T(k)$ represents the noise driven matrix; ω_k represents process noise excitation; $Y(k)$ represents the observation vector; $H(k)$ represents the observation matrix; and v_k represents observation noise excitation. It should be noted that ω_k and v_k are uncorrelated Gaussian white noise with a mean of 0 and follow a normal distribution, namely $\omega_k \sim N(0, Q_k)$, $v_k \sim N(0, R_k)$, where Q_k is the covariance matrix of process noise and R_k is the covariance matrix of observation noise [30].

To achieve capacitor parameter identification, this study collected inductor current and voltage data during the operation of the boost converter, extracted capacitor current using Haar wavelet transform, and then performed state estimation based on KF method. In this model, matrices A and T are identity matrices, and the terminal voltage U_c is the observation vector Y. The noise covariance R and Q are adjusted according to the system accuracy to obtain. The state vector X and observation matrix H of the KF can be derived from (21), corresponding to (24) and (25), respectively:

$$X(k) = [k_1\ k_2\ k_3]^T \tag{24}$$

$$H(k) = [U_{c,k-1}\ I_{c,k}\ I_{c,k-1}] \tag{25}$$

The relationship between C and ESR can be derived from (21) and (24) as follows:

$$\begin{cases} \mathrm{ESR} = \frac{k_2 - k_3}{2} \\ C = \frac{\Delta t}{(k_2 + k_3)} \end{cases} \tag{26}$$

Construct the online monitoring process shown in Figure 5 based on the derived state and measurement equations. This process is based on the KF algorithm, which recursively estimates the state vector by inputting the capacitor terminal voltage and operating current in real time and directly solves and outputs the online identification results of capacitor C and ESR.

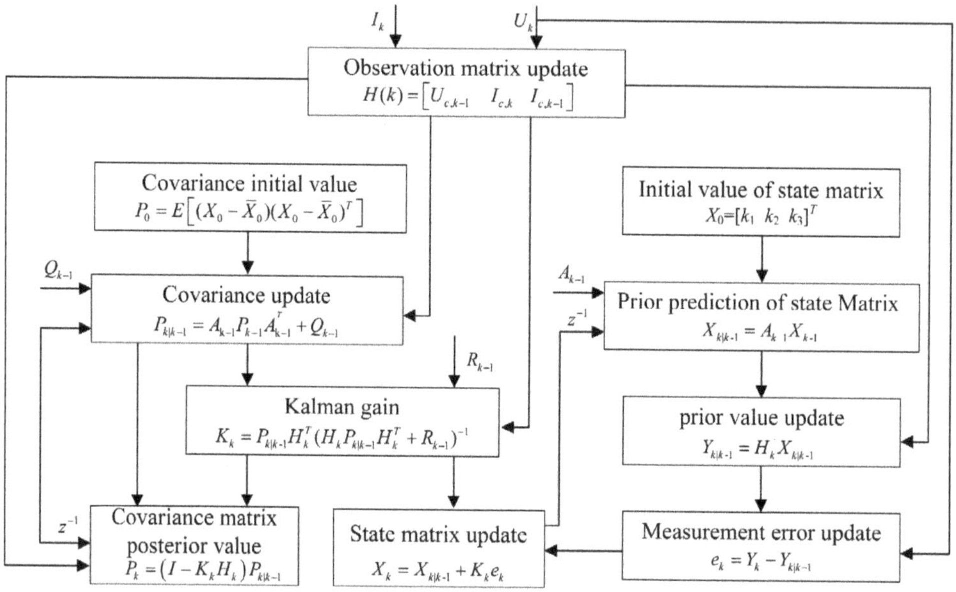

Figure 5. Online identification process of KF algorithm.

The process of the method used in this chapter is shown in Figure 6. Firstly, collect the inductor current and capacitor voltage signals from the boost converter; subsequently, the inductance current was subjected to Haar wavelet transform using (8) to extract its detail coefficient D. According to Table 1, the switch sequence S_{PWM} delayed by one sampling period was derived from the detail coefficient D. Combine the switch sequence with the inductor current and reconstruct the capacitor current through (17). Furthermore, the measured capacitor voltage is combined with the reconstructed capacitor current to construct the state space equation of the system. Finally, based on the KF algorithm, the state space equation is solved to achieve accurate estimation of capacitance C and ESR.

The state space model proposed in this article is derived from the ideal topology of the boost converter, aiming to clearly explain the core theoretical basis of the method. This model has not explicitly considered non-ideal factors such as inductance parasitic resistance, switch voltage drop, PCB routing resistance, etc. However, it should be emphasized that the KF algorithm itself has inherent robustness to a certain degree of model uncertainty. In the state estimation process, the effects of these unmodeled dynamic characteristics and parasitic parameters can be effectively incorporated and absorbed into the process noise covariance matrix Q_k. By adjusting Q_k and R_k appropriately, the filter can compensate for model mismatch to some extent, thus still achieving accurate state estimation.

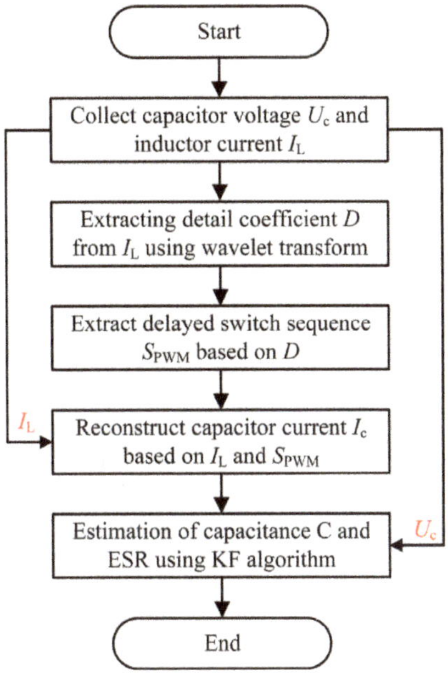

Figure 6. Flow chart of capacitor state monitoring based on wavelet transform and KF algorithm.

6. Simulation Verification

Based on the monitoring algorithm proposed in the previous text, a simulation model of the boost converter was built in the Plecs 4.8 environment. In this model, the output capacitance is characterized in the form of an ideal capacitor series with equivalent resistance, and the system controller adopts a PI control strategy. The data processing adopts Origin 2024b. During the simulation process, the inductor current signal and capacitor voltage signal are synchronously collected for the validation of subsequent capacitor state monitoring algorithms; at the same time, the waveform of the capacitor current is recorded to evaluate the effectiveness of the capacitor current reconstruction method based on Haar wavelet transform.

The simulation model and parameters (listed in Table 2) are established based on a typical set of components from a single manufacturing batch, aiming to validate the fundamental principle of the proposed method. The experimental object is the steady-state operating condition of the boost converter under a 0.4 ms operating cycle, and its key waveform is shown in Figure 7. Among them, Figure 7a is the collected capacitor voltage waveform, which will serve as the basis for subsequent capacitor electrical parameter estimation. The waveform of the collected inductor current is shown in Figure 7b. By performing Haar wavelet transform on the current signal, the approximate coefficients are obtained as shown in Figure 7c. It can be seen that the approximation coefficient preserves the main trend and dynamic characteristics of the original inductor current well. On the other hand, the detail coefficients obtained from wavelet transform are shown in Figure 7d. This detail coefficient contains high-frequency switch information of the system, which can be used to extract a sequence of switch states delayed by one sampling period. Based on the extracted switch sequence and inductor current information, the reconstructed capacitor current waveform is shown in Figure 7e. By comparing and analyzing it with the actual capacitance current obtained through data collection in Figure 7f, it can be seen that both have good agreement in amplitude and phase, and the waveform characteristics are basically the same. The simulation model and parameters (listed in Table 2) are established based on a typical set of components from a single manufacturing batch, aiming to validate the fundamental principle of the proposed method.

Table 2. System simulation parameters.

Parameter	Value
Input voltage U_{in}	140 V
Output voltage U_C	270 V
Inductance L	650 µH
Capacitor C	680 µF
Load R	105 Ω
Switching frequency f_s	20 kHz

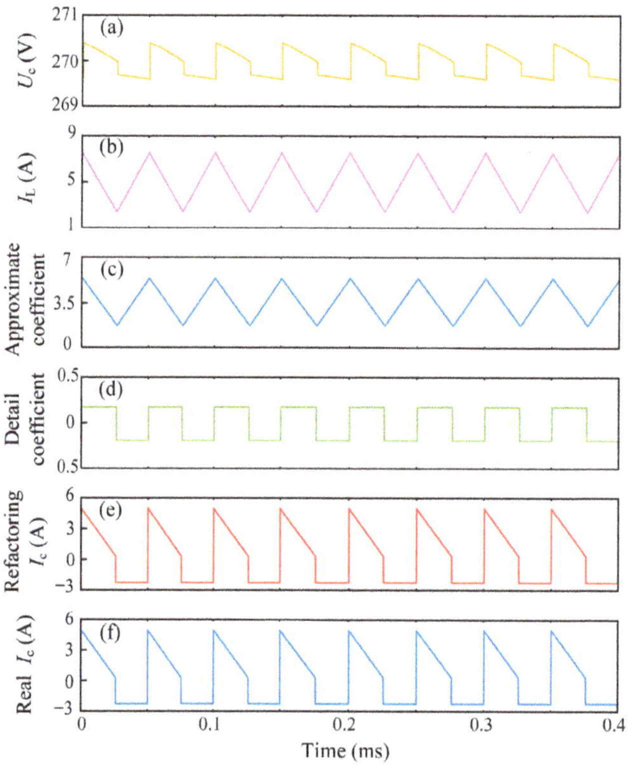

Figure 7. Simulation process of capacitor current reconstruction. (**a**) Capacitor voltage within 0.4 ms. (**b**) Inductor current within 0.4 ms. (**c**) Approximation coefficient after wavelet transform. (**d**) Detail coefficient after wavelet transform. (**e**) Reconstructed capacitor current. (**f**) Real capacitor current.

To verify the accuracy of the KF algorithm in estimating capacitance C and ESR, the capacitance simulation model was set with C at 680 µF and ESR at 0.1 Ω. By collecting the voltage across the capacitor and combining it with the reconstructed capacitor current, the system evaluates the accuracy and robustness of the capacitor state monitoring algorithm under different sampling frequencies, capacitor lifecycles, different input voltages, and noise interference environments.

1. Different sampling frequencies

To evaluate the impact of sampling frequency on the accuracy of electrolytic capacitor parameter identification, this paper conducted comparative experiments using four sampling frequencies: 200 kHz, 400 kHz, 600 kHz, and 800 kHz, based on a switching frequency of 20 kHz. By analyzing the identification results of C and ESR at different sampling frequencies, the effects of sampling rate on the stability and accuracy of parameter estimation were evaluated. The identification results are shown in Table 3. The identification errors of C and ESR are shown in Figure 8.

Table 3. Identification results of different sampling frequency parameters.

Sampling Frequency	C True	C Predicted	Relative Error	ESR True	ESR Predicted	Relative Error
100 kHz	680 μF	749.70 μF	10.25%	100 mΩ	90.13 mΩ	9.87%
200 kHz	680 μF	699.65 μF	2.89%	100 mΩ	98.59 mΩ	1.41%
400 kHz	680 μF	683.88 μF	0.57%	100 mΩ	99.70 mΩ	0.30%
600 kHz	680 μF	677.35 μF	0.39%	100 mΩ	99.86 mΩ	0.14%
800 kHz	680 μF	678.84 μF	0.17%	100 mΩ	100.09 mΩ	0.09%

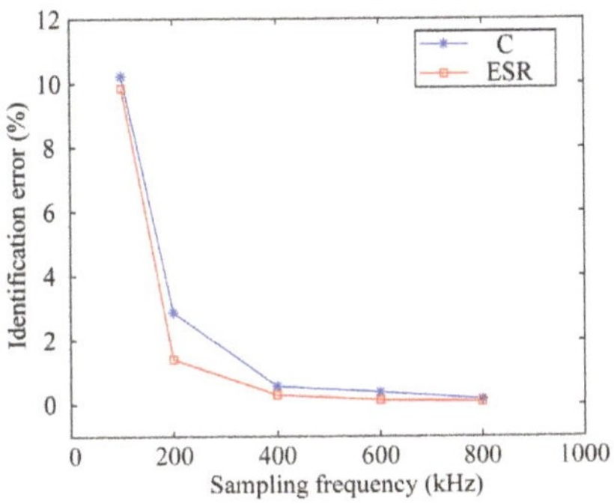

Figure 8. Identification errors of C and ESR at different sampling frequencies.

The identification results show that the estimation error of C and ESR can be controlled within 3% in the frequency range of 200 kHz to 800 kHz. When the sampling frequency is increased to 400 kHz, the identification accuracy of C and ESR is significantly improved; when the sampling frequency is further increased to 800 kHz, the accuracy improvement effect gradually approaches saturation. This phenomenon indicates that moderately increasing the sampling frequency helps to suppress the interference of high-frequency switching noise on parameter identification, but excessively high sampling rates have limited contribution to accuracy improvement. Therefore, a sampling rate of 400 kHz (20 times the switching frequency) is established as a benchmark that effectively balances high identification accuracy with computational load for the proposed method. Given that high sampling frequency significantly increases the data storage and processing burden of the system, and with a focus on finding a balance between accuracy and resource consumption, future work will set the sampling frequency for capacitance parameter identification based on the KF algorithm to 400 kHz. It is noteworthy that for applications where the switching frequency is inherently lower, this 20× ratio principle allows the method to be implemented with a correspondingly lower, more feasible sampling rate on resource-constrained hardware.

2. Capacitor lifecycle

Considering that the C and ESR of electrolytic capacitors undergo significant changes with aging during long-term operation, it is necessary to effectively monitor their entire lifecycle status. This paper takes the nominal capacitance value of 680 μF and ESR of 0.1 Ω as the initial healthy state, and the failure threshold is set as the capacitance value dropping to 80% of the initial value (544 μF) and ESR rising to twice the initial value (0.2 Ω). To verify the effectiveness and robustness of the proposed parameter identification algorithm, testing and analysis were conducted at five typical aging stages: 0% (initial state), 25%, 50%, 75%, and 100% (failure state). The identification results are shown in Table 4.

Table 4. Identification results of capacitor life cycle parameters.

Aging State	C True	C Predicted	Relative Error	ESR True	ESR Predicted	Relative Error
0%	680 μF	683.88 μF	0.57%	100 mΩ	99.70 mΩ	0.30%
25%	646 μF	649.94 μF	0.61%	125 mΩ	124.46 mΩ	0.43%
50%	612 μF	615.24 μF	0.53%	150 mΩ	149.66 mΩ	0.23%
75%	578 μF	580.77 μF	0.48%	175 mΩ	174.37 mΩ	0.36%
100%	544 μF	547.81 μF	0.70%	200 mΩ	201.02 mΩ	0.51%

From the data in Table 4, it can be seen that at different aging stages, the predicted values of capacitor C and ESR are highly consistent with the true values. The maximum relative error of capacitor C is 0.70% (100% aging state), and the minimum is 0.48% (75% aging state); the maximum relative error of ESR is 0.51% (100% aged state) and the minimum is 0.23% (50% aged state). The relative error of C and ESR identification under all operating conditions remains within 1%, indicating that the parameter identification method proposed in this paper has good estimation accuracy and robustness under different aging states and can effectively achieve state monitoring of the entire life cycle of electrolytic capacitors.

3. Different input voltages

Under different power conditions, the monitoring of capacitor status will have an impact on the accuracy of parameter identification due to changes in the circuit operating point. To evaluate the impact of this factor, this paper constructs different power operating conditions by adjusting the input voltages to 80 V, 140 V, and 200 V while ensuring that the output voltage of the boost converter is stable at 270 V. The capacitor parameter identification method is simulated and verified under these conditions. Figures 9 and 10 show the capacitor voltage, inductor current, and switch sequence waveforms corresponding to input voltages of 80 V and 200 V, respectively.

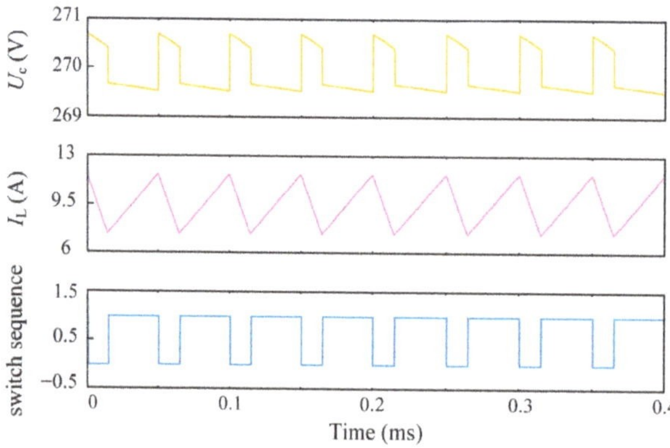

Figure 9. Input voltage 80 V working waveform.

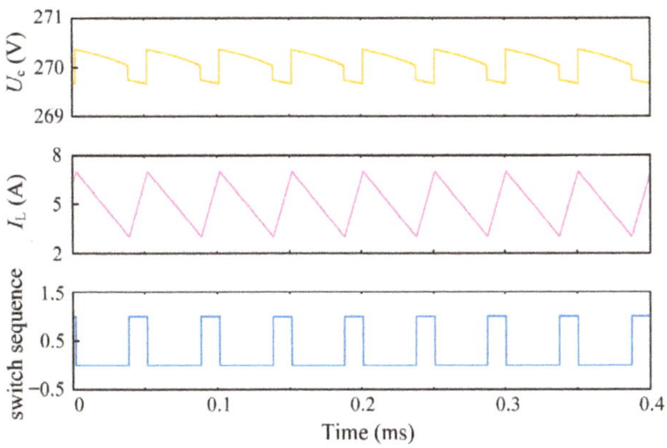

Figure 10. Input voltage 200 V working waveform.

The identification results obtained by setting C to 680 μF and ESR to 0.1 Ω are shown in Table 5. From the identification results, it can be seen that under different input voltage conditions, the identification results of capacitor C and ESR are highly consistent with their actual values, and the maximum relative errors do not exceed 1%, indicating that the parameter identification method proposed in this paper has good accuracy and robustness. However, as the input voltage increases, the identification error shows a gradually increasing trend. Specifically, the identification error of capacitor C increased from 0.31% at 80 V input to 0.73% at 200 V input; the identification error of ESR also increased from 0.19% to 0.51%. The main reason for this phenomenon is that when the input voltage increases, the duty cycle of the boost converter decreases accordingly, resulting in a decrease in the ripple amplitude of the capacitor current and output voltage. Due to the fact that ripple is a dynamic excitation signal in the parameter identification process, its amplitude reduction will weaken the ability to extract key state information, while the measurement noise level remains basically unchanged, resulting in a decrease in the system's signal-to-noise ratio. This change affects the convergence performance of the KF, causing a slight decrease in parameter identification accuracy with increasing voltage.

Table 5. Simulation identification results under different input voltages.

Aging State	C True	C Predicted	Relative Error	ESR True	ESR Predicted	Relative Error
80 V	680 μF	682.11 μF	0.31%	100 mΩ	99.81 mΩ	0.19%
140 V	680 μF	683.88 μF	0.57%	100 mΩ	99.70 mΩ	0.30%
200 V	680 μF	684.96 μF	0.73%	100 mΩ	99.49 mΩ	0.51%

4. Noise interference

Under the Gaussian white noise model, system noise is usually assumed to have uniformly distributed frequency components. Although this assumption is convenient for theoretical analysis under ideal conditions, there is a certain gap between it and the statistical characteristics of noise in actual aviation operating environments. To more accurately evaluate the performance of the proposed method under near real aviation conditions, this study introduces colored noise with strong correlation as the testing condition. The colored noise is generated by a first-order autoregressive model:

$$V_{k+1} = \lambda_{k+1/k} V_k + e_k \tag{27}$$

where $\lambda_{k+1/k} = 1.0001$, $e_k \sim N(0,0.006)$. To eliminate the influence of random errors on the identification results, independent simulations were repeated 100 times for each noise configuration. The current and voltage observation signals obtained each time were input into the KF algorithm for parameter identification, and the identification error distribution was finally calculated as shown in Figure 11.

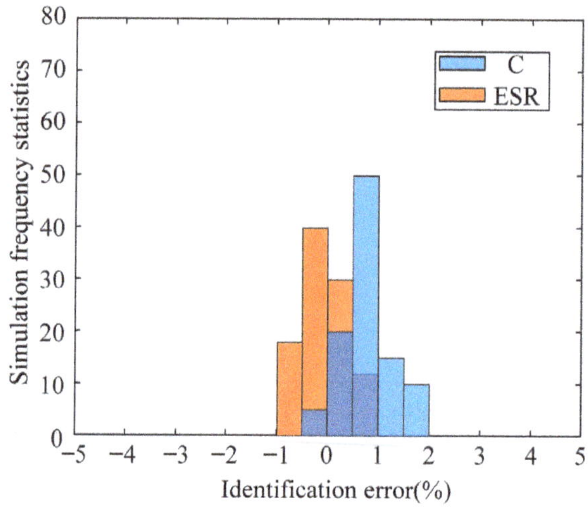

Figure 11. Distribution of identification error.

According to Figure 11, in 100 independent simulation experiments, the identification error of capacitor C is mainly concentrated in the range of −0.5% to 2%, and the identification error of ESR is mainly concentrated in the range of −1% to 1%. The vast majority of error samples are concentrated around zero, with the highest proportion of errors within ± 2%, reflecting the good noise suppression ability of the KF. This error distribution further validates the effectiveness and robustness of the parameter identification method proposed in this paper in noisy environments.

7. Experimental Verification

To verify the performance of the capacitor state monitoring algorithm proposed in this paper, detailed experimental testing and analysis were conducted in this section. The experimental platform structure is shown in Figure 12, where the DC power supply is connected to a bidirectional DC–DC converter operating in boost mode, and the output of the converter is connected to an electronic load to simulate actual working conditions. To obtain key waveform data, use an oscilloscope to synchronously collect the output voltage signal, inductor current signal, and switch drive signal of the converter. The control system builds corresponding control algorithms in the upper computer, deploys them to the rapid prototyping controller through code generation, and finally outputs PWM signals to the DSP of the bidirectional DC–DC converter, achieving closed-loop control and algorithm execution. This section of the experiment validated the same batch of components, and the results were statistical validation of multiple components to ultimately prove their universality. The experimental platform parameters are shown in Table 6, which are basically consistent with the simulation.

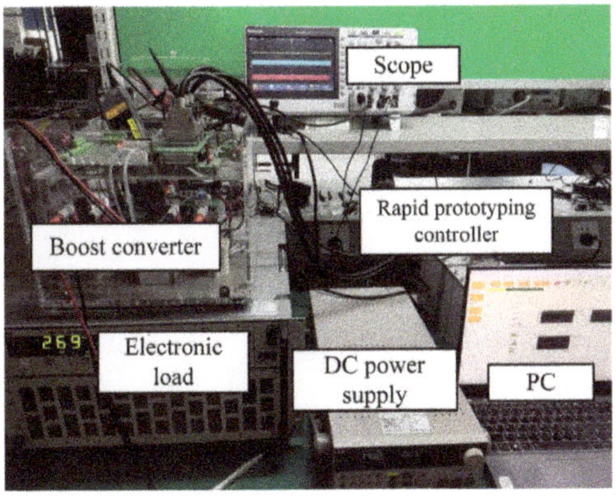

Figure 12. Experimental setup.

Table 6. Experimental platform parameters.

Parameter	Value
Input voltage U_{in}	140 V
Output voltage U_C	270 V
Inductance L	650 μH
Nominal value of capacitor C_1	680 μF
Load R	105 Ω
Switching frequency f_s	20 kHz

7.1. Analysis of Experimental Results

To eliminate the deviation between the actual and nominal values of the capacitor, the TH2830 LCR tester was used to accurately measure the actual parameters of the capacitor. The measured capacitance value C was 667.01 μF, and the equivalent series resistance ESR was 83.57 mΩ. The specific results are shown in Figure 13. Under steady-state conditions, with 140 V as the input voltage, the experimental waveforms of output voltage, inductor current, and switch drive signal were collected through an oscilloscope, as shown in Figure 14.

Figure 13. Measurement of actual capacitance value.

After preprocessing the data collected by the oscilloscope, the Plecs model was imported. The capacitor voltage and inductor current obtained at a sampling frequency of 400 kHz are shown in Figure 15a,b, respectively. Further Haar wavelet transform was applied to the inductor current, and the approximate coefficients obtained are shown in Figure 15c, whose waveform is basically consistent with the simulation results in Figure 7c; the detail coefficient is shown in Figure 15d. Although there is some fluctuation within the effective interval under the influence of actual sampling and noise interference, the overall trend of change still remains consistent with the simulation. Based on the above analysis, the extracted switch sequence is shown in Figure 15e, which is basically consistent with the actual switch sequence shown in Figure 15f, verifying the correctness of the switch sequence extraction algorithm proposed in this paper. Further combining the derived (17), the waveform of the capacitor current is reconstructed, as shown in Figure 15g.

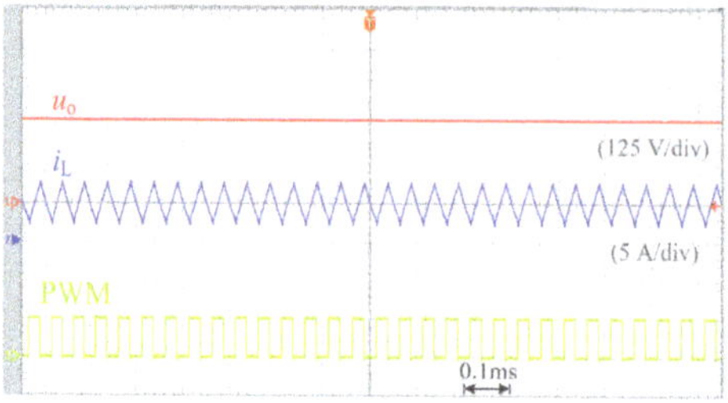

Figure 14. Experimental waveform under 140 V input voltage.

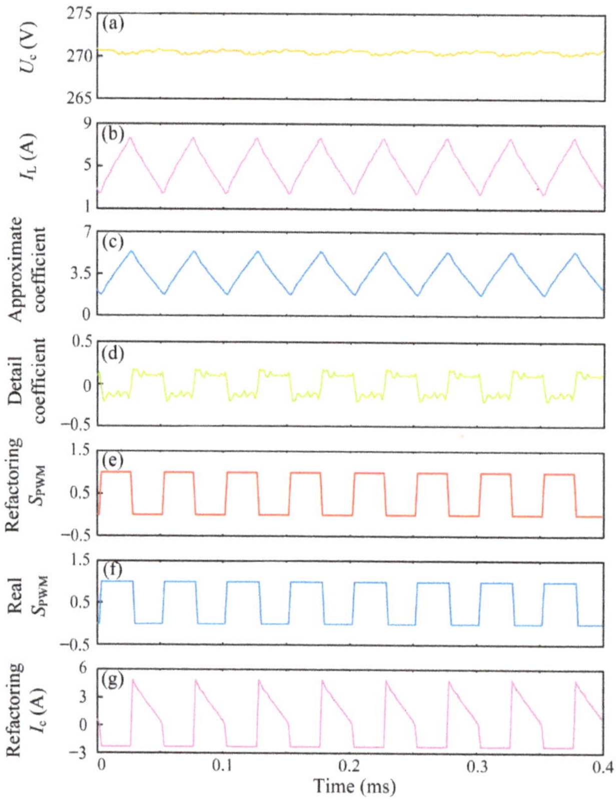

Figure 15. The experimental process of capacitor current reconstruction. (**a**) The capacitor voltage is within 0.4 ms. (**b**) Inductor current within 0.4 ms. (**c**) Approximate coefficient after wavelet transform. (**d**) Detail coefficient after wavelet transform. (**e**) Refactored switch sequence. (**f**) Real switch sequence. (**g**) Refactoring capacitor current.

The experimental waveforms collected by an oscilloscope under input voltages of 80 V and 200 V are shown in Figure 16. The specific process of reconstructing the capacitor current is consistent with the method described in Figure 15 and will not be repeated here. By using the reconstructed capacitor current and capacitor voltage as inputs and importing them into the KF algorithm model, the parameter estimation of capacitor C and ESR can be achieved. The corresponding results are shown in Table 7.

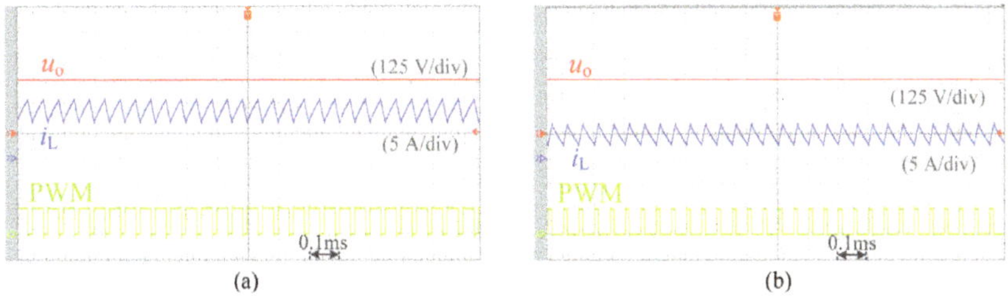

Figure 16. Experimental waveforms at different input voltages. (**a**) 80 V. (**b**) 200 V.

Table 7. Experimental identification results under different input voltages.

Aging State	C True	C Predicted	Relative Error	ESR True	ESR Predicted	Relative Error
80 V	667.01 μF	656.74 μF	1.54%	83.57 mΩ	82.75 mΩ	0.98%
140 V	667.01 μF	652.61 μF	2.16%	83.57 mΩ	82.62 mΩ	1.14%
200 V	667.01 μF	648.21 μF	2.82%	83.57 mΩ	82.11 mΩ	1.75%

To better evaluate the performance of the proposed method from simulation to experiment, simulation verification was conducted again based on the actual capacitance parameters measured on the experimental platform. Specifically, in the simulation model, the capacitance value was set to 667.01 μF, the ESR was set to 83.57 mΩ, and the remaining parameters were consistent with Table 2. Under the condition of 140 V input voltage, the recognition results of the method are C = 669.92μF and ESR = 83.26 mΩ. The corresponding relative errors are 0.43% and 0.37%, respectively. Compared with the simulation results, the parameter identification error in the experimental environment has increased, mainly due to the difference between the ideal simulation model and the actual experimental system. The unmodeled parameters such as circuit and inductance parasitic resistance in the experimental circuit introduce additional losses, while sensor measurement errors and environmental electromagnetic interference also lead to signal quality degradation. In addition, the parameter tolerances and non-ideal characteristics of actual components jointly affect the recognition accuracy. However, the experimental identification errors of capacitor C and equivalent series resistance ESR were still controlled within 3% and 2%, respectively, fully verifying the effectiveness and robustness of this method under non-ideal practical conditions. In addition, the overall identification accuracy of ESR is higher than that of capacitor C, and with the increase in input voltage, the identification accuracy of both shows a decreasing trend, which is consistent with the change pattern reflected in the simulation results.

7.2. Comparative Analysis and Discussion

To fully verify the superiority of this method, this section compares it with the classical frequency-domain impedance analysis method. As described in Section 2.3, impedance analysis can directly calculate C and ESR by measuring the impedance amplitude and phase angle of a capacitor at a specific frequency. This comparative experiment was conducted on the same experimental platform (parameters shown in Table 6) with an input voltage of 140 V. For impedance analysis, the impedance at the first harmonic of the switching frequency (20 kHz) was extracted from the time-domain signal collected by the oscilloscope, which includes the capacitor voltage and reconstructed capacitor current. Specifically, the amplitude and phase angle of the frequency point are estimated by performing DFT analysis on the signal, and then C and ESR are calculated [31]. The comparison results are shown in Table 8.

Table 8. Performance comparison between the proposed method and impedance analysis method.

Method	C Predicted	Relative Error	ESR Predicted	Relative Error
LCR tester (benchmark)	667.01 μF	-	83.57 mΩ	-
Proposed method	652.61 μF	2.16%	82.62 mΩ	1.14%
Impedance analysis method	602.74 μF	9.64%	94.35 mΩ	12.90%

According to Table 8, under the same online experimental environment, the recognition accuracy of our method is much higher than that of the impedance analysis method based on DFT. The main reason for the significant error (>9%) in the impedance analysis method is that the capacitor voltage and current signals in DC–DC converters are not single-frequency sine waves but switch ripples containing rich harmonic components. In the context of non-sinusoidal and strong noise, extracting the phase angle information of a single frequency point through DFT is highly susceptible to interference from spectral leakage and background noise, resulting in significant deviations in amplitude and phase angle estimation, which seriously reduces the accuracy of parameter identification. The time-domain analysis method proposed in this article accurately extracts switch dynamics through Haar wavelet transform and uses KF to estimate the optimal time-domain model parameters, cleverly avoiding the problem of directly conducting high-precision phase angle measurement. Therefore, this method achieves online monitoring with near offline accuracy while maintaining its non-invasive advantages, and its comprehensive performance is significantly better than traditional online impedance analysis methods.

8. Conclusions

This paper proposes an online monitoring method that integrates Haar wavelet transform and KF for the state monitoring requirements of boost converter aluminum electrolytic capacitors in multi-electric aircraft. The research focuses on solving two core problems: non-invasive reconstruction of capacitor current and high-precision identification of key parameters throughout their lifecycle.

Firstly, a switch sequence extraction and capacitor current reconstruction strategy based on Haar wavelet transform is proposed. This method only requires the collection of inductance current signals during regular operation, and by analyzing the polarity changes in its detail coefficients, accurately restores the switch sequence, thereby achieving non-invasive reconstruction of capacitor current without the need for additional hardware. Furthermore, a discrete state space model of the capacitor is established, combined with the KF algorithm, to recursively estimate C and ESR under strong noise background, achieving high robustness online identification of key parameters.

Simulation and experimental results show that the fusion monitoring method exhibits excellent performance under various operating conditions, including different sampling frequencies, aging degrees, input voltages, and noise environments. In simulation, the identification error between capacitor C and ESR can be controlled within 1%; under experimental conditions, the recognition errors of C and ESR were less than 3% and 2%, respectively, verifying the comprehensive advantages of this method in accuracy, robustness, and engineering applicability. It should be pointed out that the experimental validation of this study is currently based on a single capacitor sample. Although the proposed method achieved satisfactory accuracy on this component, future work will include statistical validation on multiple components and units to further confirm the universality and robustness of the method across a wider range of component differences. Furthermore, extending the proposed methodology to handle DCM operation represents another critical direction for future research, aiming to broaden the applicability of this monitoring strategy across all potential converter operating conditions. This study provides a practical and efficient solution for online status monitoring and predictive maintenance of key capacitive components in aviation grade DC–DC converters.

Author Contributions: Conceptualization, T.Z.; methodology, T.Z. and W.L. (Wenjie Liu); software, W.L. (Wenjie Liu) and Z.L.; validation, T.Z. and W.L. (Wenjie Liu); formal analysis, T.Z. and Y.D.; investigation, Z.L.; resources, W.L. (Wenjie Liu) and W.L. (Weilin Li); data curation, T.Z. and S.W.; writing—original draft preparation, T.Z.; writing—review and editing, W.L. (Weilin Li); visualization, Y.D. and S.W.; supervision, W.L. (Wenjie Liu); project administration, W.L. (Weilin Li); funding acquisition, T.Z. All authors have read and agreed to the published version of the manuscript.

Funding: This research was supported by the National Natural Science Foundation of China (No. 52402506).

Data Availability Statement: Data are contained within the article.

Conflicts of Interest: The authors declare no conflicts of interest.

References

1. Zhao, X.; Xu, C.; Ma, K.; Wu, X.; Chen, X.; Li, X.; Wu, X. Development of High-Power DC Solid-State Power Controllers Using SiC FETs for Aircraft Electrical Systems. *Electronics* **2025**, *14*, 4157. [CrossRef]
2. Liang, Y.; Bodnár, D.; Ram Chandra Mouli, G.; Ragni, D.; Bauer, P. Charging Demand Prediction: Small All-Electric Aircraft and Electric Vertical Takeoff and Landing Aircraft. *IEEE Trans. Transp. Electrif.* **2025**, *11*, 2732–2747. [CrossRef]
3. Martinez, A.; Torres, F.; Marin, J.; Rojas, C.A.; Gak, J.; Rommel, M.; May, A.; Wilson-Veas, A.H.; Miguez, M.; Rossi, C.; et al. Analysis and Design of an SiC CMOS Three-Channel DC-DC Synchronous Buck Converter for High-Temperature Applications. *Appl. Sci.* **2024**, *14*, 9789. [CrossRef]
4. Yang, Z.; Xi, L.; Zhang, Y.; Chen, X. An Online Parameter Identification Method for Non-Solid Aluminum Electrolytic Capacitors. *IEEE Trans. Circuits Syst. II* **2022**, *69*, 3475–3479. [CrossRef]
5. Wang, H.; Liserre, M.; Blaabjerg, F. Toward Reliable Power Electronics: Challenges, Design Tools, and Opportunities. *IEEE Ind. Electron. Mag.* **2013**, *7*, 17–26. [CrossRef]
6. Li, C.; Yu, Y.; Yang, Z.; Liu, Q.; Peng, X. ESR Estimation for Aluminum Electrolytic Capacitor of Power Electronic Converter Based on Compressed Sensing and Wavelet Transform. *IEEE Trans. Ind. Electron.* **2022**, *69*, 1948–1957. [CrossRef]
7. Yao, B.; Zhao, S.; Zhang, Y.; Wang, H. A Health Indicator of Aluminum Electrolytic Capacitors Based on Strain Sensing. *IEEE Trans. Power Electron.* **2023**, *38*, 7982–7987. [CrossRef]
8. Zhao, Z.; Davari, P.; Lu, W.; Wang, H.; Blaabjerg, F. An Overview of Condition Monitoring Techniques for Capacitors in DC-Link Applications. *IEEE Trans. Power Electron.* **2021**, *36*, 3692–3716. [CrossRef]
9. Zhang, L.; Liu, Y.; Zhang, J.; Li, Y. A Submodule Insertion Window-Based Capacitor Condition Monitoring Method for Modular Multilevel Converters. *IEEE J. Emerg. Sel. Top. Power Electron.* **2025**, *13*, 871–879. [CrossRef]
10. Ramees, M.K.P.M.; Ahmad, M.W. Advances in Capacitor Health Monitoring Techniques for Power Converters: A Review. *IEEE Access* **2023**, *11*, 133540–133576. [CrossRef]
11. Amaral, A.M.R.; Cardoso, A.J.M. An Experimental Technique for Estimating the ESR and Reactance Intrinsic Values of Aluminum Electrolytic Capacitors. In Proceedings of the 2006 IEEE Instrumentation and Measurement Technology Conference Proceedings, Sorrento, Italy, 24–27 April 2006; pp. 1820–1825.
12. Amaral, A.M.R.; Buatti, G.M.; Ribeiro, H.; Cardoso, A.J.M. Using DFT to Obtain the Equivalent Circuit of Aluminum Electrolytic Capacitors. In Proceedings of the 2007 7th International Conference on Power Electronics and Drive Systems, Bangkok, Thailand, 27–30 November 2007; pp. 434–438.
13. Miao, W.; Liu, X.; Lam, K.; Pong, P. Condition monitoring of electrolytic capacitors in boost converters by magnetic sensors. *IEEE Sens. J.* **2019**, *19*, 10393–10402. [CrossRef]
14. Ahmad, M.W.; Kumar, P.N.; Arya, A.; Anand, S. Non-invasive technique for DC-link capacitance estimation in single-phase inverters. *IEEE Trans. Power Electron.* **2018**, *33*, 3693–3696. [CrossRef]

15. Yao, K.; Cao, C.; Yang, S. Noninvasive online condition monitoring of output capacitor's ESR and C for a flyback converter. *IEEE Trans. Instrum. Meas.* **2017**, *66*, 3190–3199. [CrossRef]
16. Yao, K. A noninvasive online monitoring method of output capacitor's C and ESR for DCM flyback converter. *IEEE Trans. Power Electron.* **2019**, *34*, 5748–5763. [CrossRef]
17. Dissanayake, K.; Kularatna-Abeywardana, D. A Review of Supercapacitors: Materials, Technology, Challenges, and Renewable Energy Applications. *J. Energy Storage.* **2024**, *96*, 112563. [CrossRef]
18. Zhang, X.; Chen, Q. ESR Calutation Model of Metalized Film Capacitor in WPT System. In Proceedings of the 2022 International Conference on Computer Engineering and Artificial Intelligence, Shijiazhuang, China, 22–24 July 2022; pp. 177–180.
19. Ogawa, Y.; Yamasoto, T.; Hasegawa, K. Accurate ESR Monitoring of a DC-Link Capacitor with Switching Ripple Voltage by a Leafony. In Proceedings of the 2024 13th International Conference on Renewable Energy Research and Application, Nagasaki, Japan, 9–13 November 2024; pp. 1521–1525.
20. Goyal, V.K.; Shukla, A. Two-Stage Hybrid Isolated DC–DC Boost Converter for High Power and Wide Input Voltage Range Applications. *IEEE Trans. Ind. Electron.* **2022**, *69*, 6751–6763. [CrossRef]
21. Kudrys, J.; Prochniewicz, D.; Zhang, F.; Jakubiak, M.; Maciuk, K. Identification of BDS Satellite Clock Periodic Signals Based on Lomb-Scargle Power Spectrum and Continuous Wavelet Transform. *Energies* **2021**, *14*, 7155. [CrossRef]
22. Guo, T.; Zhang, T.; Lim, E.; López-Benítez, M.; Ma, F.; Yu, L. A Review of Wavelet Analysis and Its Applications: Challenges and Opportunities. *IEEE Access* **2022**, *10*, 58869–58903. [CrossRef]
23. Jang, W.; Lee, W. Detecting Wireless Steganography with Wavelet Analysis. *IEEE Wirel. Commun. Lett.* **2021**, *10*, 383–386. [CrossRef]
24. Park, S.; Yoo, S.; Jeong, N.; Cha, E. Haar Wavelet-Based Representation Learning for Unpaired Image-to-Image Translation. *IEEE Access* **2025**, *13*, 61821–61832. [CrossRef]
25. Ganguly, P.; Chattopadhyay, S.; Datta, T. Haar-Wavelet-Based Statistical Scanning of Turn Fault in Vehicular Starter Motor. *IEEE Sens. Lett.* **2022**, *6*, 6000904. [CrossRef]
26. Hu, J.; Lin, Y.; Hu, M.; Wang, H. A Test Response Compression Method for Monolithic 3-D ICs Based on 3-D Haar Wavelet Transforms. *IEEE Trans. Instrum. Meas.* **2021**, *70*, 3506412. [CrossRef]
27. Khade, S.; Gite, S.; Thepade, S.D.; Pradhan, B.; Alamri, A. Detection of Iris Presentation Attacks Using Hybridization of Discrete Cosine Transform and Haar Transform with Machine Learning Classifiers and Ensembles. *IEEE Access* **2021**, *9*, 169231–169249. [CrossRef]
28. Fang, Y.; Panah, A.; Masoudi, J.; Barzegar, B.; Fatehi, S. Adaptive Unscented Kalman Filter for Robot Navigation Problem. *IEEE Access* **2022**, *10*, 101869–101879. [CrossRef]
29. Tao, Y.; Yau, S.S.-T. Outlier-Robust Iterative Extended Kalman Filtering. *IEEE Signal Process. Lett.* **2023**, *30*, 743–747. [CrossRef]
30. Guo, Y.; Li, W.; Yang, G.; Jiao, Z.; Yan, J. Combining Dilution of Precision and Kalman Filtering for UWB Positioning in a Narrow Space. *Remote Sens.* **2022**, *14*, 5409. [CrossRef]
31. Amaral, A.M.R.; Cardoso, A.J.M. A Simple Offline Technique for Evaluating the Condition of Aluminum-Electrolytic-Capacitors. *IEEE Trans. Ind. Electron.* **2009**, *56*, 3230–3237. [CrossRef]

electronics

MDPI

Article

Trust Evaluation Framework for Adaptive Load Optimization in Motor Drive System

Ali Arsalan [1,*], Behnaz Papari [1], Grace Karimi Muriithi [1], Asif Ahmed Khan [2], Gokhan Ozkan [2] and Christopher Shannon Edrington [2]

[1] Department of Automotive Engineering, Clemson University, Greenville, SC 29607, USA; bpapari@clemson.edu (B.P.); gmuriit@clemson.edu (G.K.M.)
[2] Holcombe Department of Electrical and Computer Engineering, Clemson University, Clemson, SC 29634, USA; asifahm@clemson.edu (A.A.K.); gokhano@clemson.edu (G.O.); cedring@clemson.edu (C.S.E.)
* Correspondence: aarsala@clemson.edu

Abstract

Electric drive systems (EDSs) are vital for automotive and industrial applications but remain highly vulnerable to cyber and physical anomalies (CPAs), such as inverter open-circuit faults, sensor failures, and malicious cyberattacks. Ensuring reliable EDS operation requires the controller to receive accurate and uncompromised feedback and reference signals continuously. However, many existing data-driven detection and mitigation strategies rely on large training datasets, impose significant computational overhead, and often lose effectiveness under various abnormal operating conditions. To overcome these limitations, this paper introduces a trust evaluation framework that continuously assesses the reliability of all incoming signals to the EDS controller by combining behavioral analysis with historical reliability records. The proposed scheme offers a lightweight and model-independent approach, enabling reliable, adaptive decision-making by leveraging both current and historical signal behavior. To this end, this paper further integrates the resulting trust values into a torque-split optimization algorithm, enabling adaptive load optimization by dynamically reducing the torque contribution of motors operating under abnormal or low-trust conditions, thereby demonstrating clear applicability for automotive drive systems. The framework is validated in a real-time OPAL-RT environment across multiple CPA scenarios, demonstrating accurate anomaly detection and adaptive torque redistribution. Owing to its simplicity and versatility, the proposed method can be readily extended to other safety-critical drive applications.

Keywords: electric drive system; cyber or physical anomalies; open-circuit fault (OCF); trust evaluation; diagnosis; adaptive load optimization

1. Introduction

In recent years, the application of power electronic systems (PESs) has become central to a wide range of industries, including the automotive sector, industrial automation, utility grid power management, and marine power systems. Advanced control schemes associated with these PESs have enhanced their efficiency and enabled higher power density. However, these advancements in control, communication, and power levels have also made them increasingly susceptible to a range of cyber anomalies and physical faults [1,2]. Therefore, various model-based [3–9] and data-driven schemes [10–17] are proposed in the existing literature for the detection of these cyber and physical anomalies. Model-based schemes are favored over data-driven approaches due to their faster diagnostic speed and lower computational requirements. However, their performance is often limited under uncertain operating conditions due to system nonlinearities and high reliance on the accuracy of the system's mathematical model. In contrast, data-driven approaches do not depend on an explicit system model and are capable of handling complex system dynamics through advanced machine learning techniques. Nevertheless, data-driven schemes face challenges related to the scarcity of real-world training data associated with CPAs and the high computational cost involved.

Compared to existing methods for the detection of CPAs, trust evaluation schemes offer a more effective approach to assessing the behavioral reliability of key entities, particularly feedback and reference signals, which are integral to any closed-loop control framework. Moreover, trust evaluation approaches integrate the advantages of both data-driven and model-based schemes. They operate independently of system models, impose a low computational burden, and eliminate the need for large offline training datasets. Moreover, trust evaluation leverages both current and historical operational data, allowing it to assess cumulative behavior over time, leading to more reliable and adaptive decision-making. However, the effectiveness of these frameworks heavily depends on the quality of the reliability assessment metrics and the accurate identification of relevant system parameters.

In this context, ref. [18] introduced a metrics-based trustworthiness scheme for routing protocols in low-power and lossy networks. This approach strengthens security mechanisms against selfish behaviors and internal attackers while achieving lower energy consumption and a higher packet delivery ratio. Moreover, a lightweight and reliable trust mechanism based on multi-source feedback information fusion is proposed in [19] to enhance the adoption of IoT edge devices. This approach effectively mitigates the impact of bad-mouthing attacks caused by malicious feedback providers. Additionally, a multi-dimensional attack-resistant trust model designed for acoustic sensor networks is proposed in [20], utilizing packet error loss, packet loss rate, and link capacity. Given the unreliable communication channels in underwater environments, trust models originally designed for terrestrial wireless sensor networks cannot be effectively adapted for acoustic applications. Similarly, ref. [21] introduces a distributed trust mechanism for wireless sensor network applications, integrating communication trust, energy trust, data trust, recommended reliability, recommended similarity, and indirect trust. Moreover, ref. [22] proposes a context-aware trust evaluation approach that employs a multi-source reputation and trust-based mechanism to assess user trustworthiness in fog-based IoT. In [23], a trust-based mechanism for detecting distributed denial-of-service attacks in vehicular ad hoc networks is presented. The trust evaluation incorporates factors such as residual energy, frequency value statistics, trust policy, trust hypothesis statistics, and data factors. Moreover, a trust assessment scheme [24] is proposed for online social networks, integrating context-specific information, locally generated social trust relationships, and network topological structure data. This approach enhances social trust prediction by up to 5.5% compared to existing methods. Additionally, ref. [25] introduces a mobility trust evaluation model based on communication success rate and sensor node energy consumption to estimate trust values in heterogeneous vehicle networks. In this framework, a mobility strategy is developed for electric vehicles to minimize the total number of transmission hops in trust evaluation. Ref. [26] presents a simplified blockchain-based consensus algorithm for trust evaluation in massive machine-type communication, addressing the lack of an inherent security framework in the Internet of Energy Things and Industrial Internet of Things. Moreover, as federated learning is preferred for its distributed nature, low latency, and high privacy benefits, ref. [27] proposes a trust evaluation framework for federated learning in digital twins for mobile networks based on individual and coordinated trust values for participating users. This framework addresses the limitations of the simplex evaluation factor in conventional coarse-grained trust calculation methods.

To the best of our knowledge, most existing trust evaluation schemes are designed primarily for communication networks or sensor data in distributed systems, where interactions between nodes are governed by trust values. In these systems, the failure of a node does not necessarily compromise overall functionality. However, this is not applicable when centralized control is integrated with various input signals processed over a fixed sampling period for continuous decision-making. In this case, halting the interaction of these incoming signals would disrupt the entire operation. As such, existing schemes lack a methodology specifically tailored to centralized control systems, particularly for applications like EDS. To address this gap, the proposed work introduces a trust evaluation framework for an EV-based EDS, where a decline in trust values triggers a response that reduces the torque demand for the associated EDS under abnormal conditions. Moreover, much like communication networks, these operational technology systems are also vulnerable to various types of CPAs. Therefore, adapting trust-based frameworks from communication networks to cyber-physical systems offers a valuable enhancement to the paradigm of trust evaluation, expanding its application in automotive systems. Moreover, Table 1 illustrates a concise summary that synthesizes the contributions of the proposed scheme and contrasts them with the key limitations of prior studies. The contributions of this work can be summarized as follows:

1. A generalized trust computing method is proposed, applicable to any system with centralized control and incoming feedback and input signals. Since this framework involves several tunable factors, it can be adjusted to meet the specific requirements of different systems.
2. The robustness of the proposed trust method is validated across a diverse range of CPAs. To achieve this, this scheme is integrated with a real-time EDS model simulated on OPAL-RT, effectively simulating various CPA scenarios, including power switch open-circuit faults, current sensor faults, and cyberattacks.
3. This work presents a trust-aware torque-split optimization algorithm that incorporates the trust values of each EDS to further optimize torque distribution across each traction motor, following the principles of adaptive load optimization (ALO).

Table 1. Characteristics and limitations of past studies and the proposed approach.

<table>
<tr><th>Method</th><th>Investigated System</th><th>Anomaly Modes</th><th>Contributions</th><th>Key Limitations</th></tr>
<tr><td>[28]</td><td rowspan="6">Electric drive system</td><td>False data injection (FDI) attacks</td><td>Cyberattack detection and impact analysis on 3-phase current</td><td rowspan="6">• Data-driven schemes rely heavily on the availability and quality of training datasets.
• Most diagnostic methods depend only on instantaneous current values, without considering historical system behavior.
• Existing schemes can typically detect only a limited range of cyber and physical anomalies.</td></tr>
<tr><td>[29]</td><td>FDI and relay attacks</td><td>Cyberattack detection and impact analysis on 3-phase current</td></tr>
<tr><td>[30]</td><td>FDI attacks</td><td>Cyberattack detection and impact analysis on 3-phase current</td></tr>
<tr><td>[31]</td><td>OCFs in single power switch</td><td>Impact analysis on stator current, detection and distinguish OCFs</td></tr>
<tr><td>[32]</td><td>Single and Double switch OCFs in power converter</td><td>Impact analysis on stator current, detection and distinguish OCFs</td></tr>
<tr><td>[17]</td><td>Single and double switch OCFs in power converter and FDI attacks</td><td>Impact analysis, detection and differentiation</td></tr>
</table>

Table 1. *Cont.*

Method	Investigated System	Anomaly Modes	Contributions	Key Limitations
[18]	Low-power and lossy networks	Selfish behaviors and internal attackers	Strengthens security mechanism; lower energy consumption and higher packet delivery ratio	• Most trust evaluation schemes are limited to distributed control and not applied to centralized systems. • Trust evaluation has not yet been explored in the context of electric drive systems. • Current schemes lack a response mechanism to mitigate the impact once anomalies are detected.
[19]	IoT edge devices	Bad mouthing attacks	Lightweight and reliable trust mechanism	
[20]	Acoustic sensor networks	Cyberattacks	Multi-dimensional attack-resistant trust	
[21]	Wireless sensor network	None	Distributed trust mechanism	
[23]	Vehicular ad hoc networks	Denial-of-service (DoS) attacks	Trust-based mechanism for detecting distributed DoS attacks	
[25]	Heterog-eneous vehicle networks	None	Trust evaluation for mobility strategy to minimize the total number of transmission hops	
This work	Electric drive system	Cyberattacks, OCFs, and sensor faults	• A trust-based strategy is developed to detect various cyber and physical anomalies. • The trust score is integrated into the torque-split optimization to minimize anomalies' impact on motor drive characteristics.	

In this article, Section 2 presents the formulation of the trust evaluation framework for EDS and trust-aware torque-split optimization. Section 3 outlines the trust dynamics for various CPAs to validate the proposed approach's applicability and evaluates the performance of trust-aware torque-split optimization under different EDS trust values. Finally, Section 4 provides the conclusion.

2. Trust-Aware Control Framework for Electric Drive System

Modern EVs can generally be classified into single-motor or multi-motor drive systems. Traditional EDSs typically lack a continuous authentication mechanism for feedback and reference input signals sent to the motor controller. This absence of the authentication process makes the EDSs vulnerable to disruptions caused by non-routine operating conditions, such as cyberattacks or physical faults, which can further compromise gate drive signals, thereby jeopardizing the system's security and operational reliability. To address this, a trust evaluation framework is integrated at the controller's input as a security checkpoint, as shown in Figure 1. This framework continuously monitors the integrity of input signals from sensors and the CAN bus, serving as a monitoring mechanism for the EDS. The details of this trust evaluation framework are explained below.

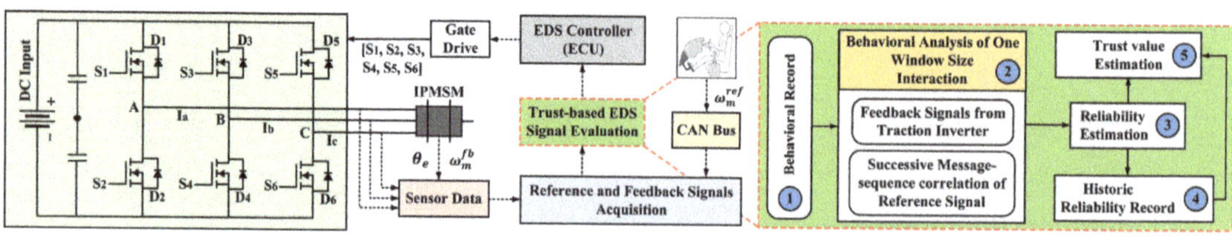

Figure 1. Schematic diagram of an electric motor drive based on trust-aware control.

2.1. Trust Behavior Model

To assess the trust level of incoming signals involved in the EDS controller's decision-making process, the controller must comprehensively record the behavior of these signals. The frameworks supporting the trust evaluation (TE_i) of the incoming signals within the time window (t_i) can be formalized as shown in Equation (1). Here, RB_i represents the recorded behavior of the signals, where i represents the segment number corresponding to data within the sliding window. Since these signals are not validated at every sampling instant, the trust evaluation is performed after collecting data over a specified window size. This approach implicitly assumes that the entities contributing to each window remain trustworthy and continue transmitting data to the EDS controller. Once the window is completed, the trustworthiness of the previously accepted entities is reassessed through re-evaluation. This procedure distinguishes the proposed mechanism from a zero-trust approach, where no entity is ever implicitly trusted and every incoming signal must be validated before entering the controller. In contrast, zero-trust is not practical for EDS applications because reliable behavioral validation cannot be achieved from a single sampling instant; instead, a dataset spanning a defined window size is required for meaningful trust assessment. Additionally, even if a trust mechanism were applied at each sampling instance of EDS, it could introduce computational delays that would compromise the performance of the subsequent controller by increasing the total harmonic distortion. Therefore, this proposed trust mechanism is referred to as a trust mechanism rather than a zero-trust mechanism. This recorded data from each window serves as the input for exploratory data analysis (EDA) techniques. Additionally, RL_i denotes the reliability score of the incoming signals, while RL_{his} captures their historical reliability record, derived by applying exponential normalization to the values obtained from EDA. Together, these parameters are used to quantify the proposed trust evaluation framework.

$$TE_i = \langle RB_i, EDA_i, RL_i, RL_{his} \rangle \tag{1}$$

The recorded behavior is expressed in Equation (2), where the size of each behavioral segment ($bh_{i,f}$) is determined by the length of the extracted features or the window size, denoted as (lg), with (f) representing the number of different features. The features of the EDS (E_f), utilized to compute $bh_{i,f}$, are outlined in Equation (3). Here, $I^m_{d,q}$ and ω^m_m represent the measured current and speed values, respectively, while $I^{ref}_{d,q}$ and ω^{ref}_m denote their corresponding reference values. These values, corresponding to the stator current and speed of the permanent magnet synchronous machine (PMSM), can be obtained via model predictive current control, as presented in [33]. Furthermore, the formulation of $bh_{i,f}$ in Equation (4) incorporates the degree of abnormality ($abnD^{EDA}_{(i,k)}$) observed between measured and reference values during the interaction of incoming signals with the EDS controller. This is achieved by employing EDA techniques, including mean, interquartile range, and Pearson correlation, for each set of values shown in Equation (3). Consequently, the proposed trust evaluation scheme leverages EDA methods to accurately quantify the operational discrepancies in the EDS.

$$RB_i = \langle bh_{i,f} \rangle \in lg \tag{2}$$

$$E_f = \langle I^m_d, I^m_q, I^{ref}_d, I^{ref}_q, \omega^m_m, \omega^{ref}_m \rangle \tag{3}$$

$$bh_{i,f} = abnD^{EDA}_{(i,k)} = EDA_{(i,k)}(E^m_f, E^{ref}_f) \tag{4}$$

Here, E^m_f and E^{ref}_f are the measured and reference values of EDS features, respectively. The estimated values of operational abnormalities, derived through EDA techniques, are utilized to assess the operational reliability of the EDS, with their records systematically maintained as part of the EDS historical reliability database. Moreover, residual values obtained by taking the mean square error between the measured and reference values of $I_{d,q}$ are analyzed through EDA to ensure stationarity. The mean and interquartile range of these residuals are then used to compute $bh_{i,f}$, which is subsequently applied in Equation (5) to evaluate the reliability of the traction inverter's feedback signals (RL_i^{fb}). Conversely, for the speed values, the Pearson correlation coefficient is employed to calculate $bh_{i,f}$, which is subsequently used in Equation (5) to evaluate the reliability values of the EDS input signals (RL_i^{in}). As the reliability is evaluated for all incoming signals of the EDS controller, both input and feedback signals are taken into account. Furthermore, the reliability formulation for the input signal is separated from that of the feedback signals, because this can also help to localize the reliability degradation between feedback and inputs for future work. Different data normalization methods are utilized for RL_i^{fb} and RL_i^{in} because of the nature of the input data. Furthermore, since RL_i^{in} utilizes an EDA approach with values already scaled between 0 and 1, the negative exponent-based scaling is not applied for computing RL_i^{in}. The parameter N_i in Equation (5) denotes the total number of $bh_{i,f}$ values considered in the numerator, which can vary based on the number of EDA techniques utilized. The value of N_i while computing RL_i^{in} will be 1. The reliability values are bounded within the range [0, 1] due to the applied normalization scheme. This method for estimating operational reliability is highly versatile and can be scaled to apply to other systems. Moreover, the overall signal-based reliability of EDS is formalized in Equation (6).

$$RL_i^{fb}(\Delta t) = \frac{\sum e^{-bh_{i,f}}}{N_i}, \quad RL_i^{in}(\Delta t) = \frac{bh_{i,f}}{N_i} \tag{5}$$

$$RL_i^{total} = RL_i^{in}(\Delta t) + RL_i^{fb}(\Delta t) \tag{6}$$

Hence, the recorded behavior of the EDS is used to represent its operational trustworthiness. To evaluate the overall EDS reliability value while considering both hardware-based feedback signals and user-provided input signals, the calculation is formalized in Equation (7). In this formulation, where multiple values are aggregated, normalization plays a crucial role in ensuring balanced evaluation, as shown in Equation (7). This equation also incorporates historical reliability values, which serve as a weighting factor to determine the current reliability. The weighting factor is expressed as $e^{-RL_{hist}}$, where RL_j^{-ive} denotes the historical negative interactions. Additionally, the term $(\sigma.c)$ represents the time-forgetting factor, which gradually diminishes the influence of historical negative interactions on current reliability values over time. This indicates that the historical reliability values based on the far past will not have much value. Moreover, since the historical reliability values account only for negative interactions, a higher RL_{hist} value leads to a smaller outcome from the negative exponent-based weighting factor in (5), thereby reducing the overall system reliability, and vice versa.

$$RL_i(\Delta t) = e^{-RL_{hist}}.RL_i^{total} \tag{7}$$

$$RL_{hist} = \frac{\sqrt{\sum RL_j^{-ive}}}{1 + \sqrt{\sum RL_j^{-ive}}} - \sigma.c, \; j \in \{fb, in\} \tag{8}$$

The variable c represents the number of negative iterations in the past 10 interactions. Additionally, σ corresponds to the variance (var) of the behavioral stability of the EDS with segment length lg, denoted as $S_i(\Delta t)$, which is calculated using Equation (9). The stability value is determined based on the consecutive interaction of E_f for each sampling period. The variance of the captured data is higher in cases of abnormal operational behavior compared to normal conditions. These dynamics are then employed to estimate σ, which influences the changes in reliability values over time under varying operational conditions.

$$S_i(\Delta t) = \frac{E_f^{z+1} - E_f^z}{lg}; \quad \sigma = \sqrt{var(S_i^{lg})} \tag{9}$$

2.2. Trust Evaluation Model

This section explains the proposed trust evaluation criteria for EDSs. The primary objective of this approach is to assess the operational trustworthiness of an EDS, based on its operational reliability values. The resulting trust values will be incorporated into the torque-split optimization algorithm, as illustrated in Figure 2. In this way, the distribution of the required total torque can be further optimized, taking into account the trustworthiness of each EDS and reducing the torque demand from an EDS already operating under abnormal conditions. The trust value, $T_i(e^{-T_f})$, is calculated using Equation (10), where a higher trust factor (T_f) results in a higher trust value. The mathematical formulation in Equation (10) is designed to regulate the rate of change in trust values. It is assumed that trust can neither start at zero nor one; instead, it always begins at 0.5. Furthermore, Equation (10) ensures that trust does not increase abruptly but rather grows gradually over time toward a maximum value of 1, without ever exceeding it. This behavior is also illustrated in the graphical results presented in the Results section. These dynamics primarily depend on the trust factor (T_f), which in turn is governed by the variations in the previously obtained reliability values.

$$T_i(e^{-T_f}) = \begin{cases} 0, & if: e^{-T_f} = 0 \\ max\{0, min(0.5 + (1 - e^{-T_f}), 1)\}, & else. \end{cases} \tag{10}$$

$$T_f = \alpha . \sqrt[\phi]{RL_i(\Delta t).g(t)}, \ \phi = 0.5 \tag{11}$$

Moreover, the value of α, which enforces an acceptable reliability threshold for computing trust, is estimated via Equation (12). Through α, one can distinguish between normal conditions, mild abnormalities, and severe abnormalities based on the current reliability values. Additionally, the function $g(t)$ in Equation (13) modifies the reliability values to regulate the rate at which trust changes over time, ensuring that trust increases gradually. The terms $(\sigma \cdot c)$ and $g(t)$ describe how trust changes over time, showing a quick drop at first and then a slow recovery, as can be seen for various case study scenarios in the Results section. Furthermore, the weighting factors (δ and ζ) can be adjusted to regulate the rate of trust variation across behavioral segments, depending on the specific application.

$$\alpha = \begin{cases} 1, & if: 0.5 < (RL_i^{fb}) \& (RL_i^{in}) < 1 \\ 0, & otherwise. \end{cases} \tag{12}$$

$$g(t) = min\{(1 - e^{-\delta . t^2} . \zeta), 1\} \tag{13}$$

Hence, the main purpose of these formulations is to express reliability values in terms of trust while controlling their dynamics under both normal and abnormal conditions, ensuring that trust decreases sharply when needed but recovers gradually.

2.3. Trust-Aware Torque-Split Optimization

The primary function of the EDS is to deliver the required torque necessary to generate the yaw moment commanded by the vehicle's lateral stability control system. In Figure 2,

Motors 1 and 2 represent the motor drives on the front-right and rear-right wheels, respectively, while Motors 3 and 4 correspond to the front-left and rear-left motor drives. The total torque demand (Trq_{dem}) and speed demand are initially distributed between the right and left motors. These distributed torque and speed values are then used to determine the torque-split ratio (Trq_{sr}) between the front and rear wheels, separately for the right and left sides. In this context, Equation (14) illustrates how the torque is calculated for the right-side wheels.

$$Trq_1 = Trq_{sr}(Trq^r_{(dem)}); \quad Trq_2 = 1 - Trq_1 \tag{14}$$

$Trq^r_{(dem)}$ is the demanded torque for the right-sided wheels of the vehicle. The torque for the two right wheels as a function of associated trust values [$T_{i,1}(\Delta t)$ and $T_{i,2}(\Delta t)$] is defined by Equations (15) and (16), where $\Delta\hat{Trq}_1$ and $\Delta\hat{Trq}_2$ represent the additional torque required to achieve yaw moment (γ), as expressed in Equation (17). Here, D_f and D_r represent the distances from the vehicle's center of mass to the front and rear axles, respectively, L_d denotes the distance between the front and rear axles, and r_w is the dynamic tire radius.

$$Trq_1 = T_{i,1}(\Delta t)(Trq_1 - \Delta\hat{Trq}_1) \tag{15}$$

$$Trq_2 = T_{i,2}(\Delta t)(Trq_2 - \Delta\hat{Trq}_2) \tag{16}$$

$$\Delta\hat{Trq}_1 = \frac{D_f r_w}{(D_f + D_r)L_d}\gamma; \Delta\hat{Trq}_2 = \frac{D_r r_w}{(D_f + D_r)L_d}\gamma \tag{17}$$

Based on the physical constraints, each xth motor ($x = 1, 2$) has associated maximum ($Trq_{x,max}$) and minimum ($Trq_{x,min}$) torque limits, as shown in Equation (18). Similarly, the torque for the left-side motors ($x = 3, 4$) can be estimated in the same manner.

$$Trq_x = \begin{cases} Trq_x, & if :< Trq_x < Trq_{x,max}\} \\ Trq_{x,min}, & if : Trq_x < Trq_{x,min} \\ Trq_{x,max}, & else \end{cases} \tag{18}$$

Furthermore, interior permanent magnet synchronous motors (IPMSMs) are commonly chosen as traction motors in electric vehicles (EVs) due to their high power density and smooth torque production. In this context, Figure 1 illustrates the IPMSM-based configuration of a single EDS. Moreover, the dynamic mathematical model in [33] is utilized to model the model-predictive current control-based IPMSM controller for EDS.

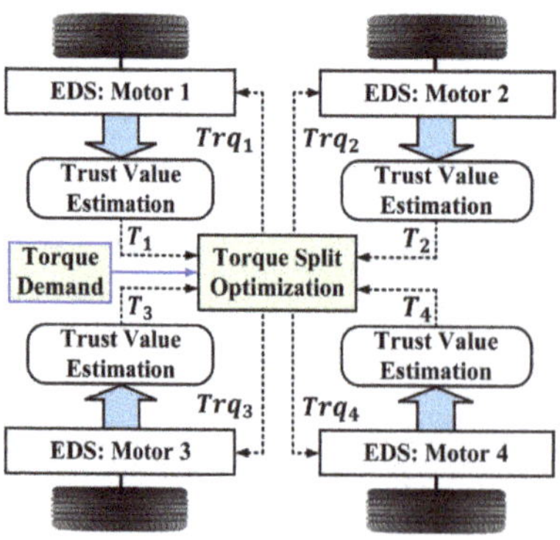

Figure 2. Application of trust-based evaluation on torque-split optimization.

3. Real-Time Experimental Validation

The effectiveness of the proposed trust evaluation approach is validated using a real-time simulation model of the EDS shown in Figure 1. This experimental setup aims to illustrate the viability of the presented framework for real-time applications. The EDS with ATM-based MPC is implemented in a MATLAB/Simulink environment integrated with a real-time digital simulator, i.e., Opal-RT OP4510. The model is executed in RT-LAB with real-time hardware synchronization. Moreover, the detailed implementation of EDS in OPAL-RT is shown in Figure 3, where the power converter and IPMSM are simulated in the FPGA solver with a sampling time of up to 210 ns , and the controller is implemented in the CPU of OPAL-RT with a sampling rate of 25 μs.

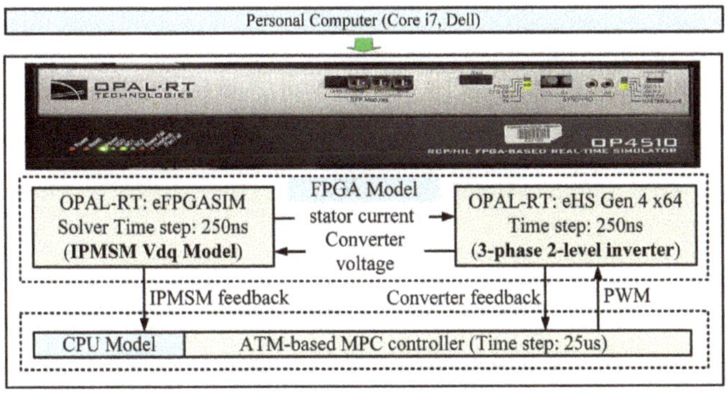

Figure 3. Motor drive system with MPC-based control implementation on OPAL-RT testbed.

3.1. Trust Value Estimation

This study evaluates the effectiveness of the proposed trust estimation scheme for an EDS under non-routine operating conditions, including open-circuit faults (OCFs) in traction inverter power switches, current sensor faults, and cyberattacks, as detailed in Table 2. These cyber and physical anomalies, discussed in [17,34], are selected due to their prominence in existing research. The real-time effects of these anomalies on the electrical, mechanical, and thermal parameters of the EDS are not reiterated here, as they have already been thoroughly analyzed in [17]. The following subsections validate the proposed trust evaluation framework in relation to these cyber and physical anomalies.

Table 2. CPAs considered for the proposed trust evaluation strategy.

Cases	Category	Description
I	SS-OCF [17]	OCF instigated in S4.
II	SDS-OCF [17]	OCF instigated in S3, S4.
III	CDS-OCF [17]	OCF instigated in S3, S6.
IV	PDS-OCF [17]	OCF instigated in S3, S5.
V	Stuck fault [17]	Stuck fault in phase-B current sensor.
VI	Sensor OCF [17]	OCF fault in phase-A current sensor.
VII	Cyberattack [34]	Targeting reference speed of EDS.
VIII	Cyberattack [17]	Targeting Phase B current. [$mul = 2$]
IX	Cyberattack [17]	Targeting Phases B, C currents. [$mul = 2$]
X	Cyberattack [17]	Targeting Phases A, B, C currents. [$mul = 5$]
XI	Cyberattack [17]	Targeting Phases A, B currents. [$mul = -0.75$]
XII	Cyberattack [17]	Targeting Phase A current. [$mul = -1.2$]

3.1.1. Dynamics of Trust Value During Power Switch Faults

The OCF-based physical faults in the traction inverter of an EDS are classified into four categories: single-switch (SS) OCF, series-double-switch (SDS) OCF, cross-double-switch (CDS) OCF, and parallel-double-switch (PDS) OCF [17]. These faults are simulated by permanently setting the gating signal of a specific power switch in the inverter to zero, replicating the conditions of an OCF. In this section, one fault from each category—single and double-switch—is introduced to observe the dynamics associated with reliability and trust value computations using the proposed scheme.

In this context, the reliability is estimated using Equation (6), the weighted reliability via Equation (7), and the trust values using Equation (10) for each OCF case study, as shown in Figure 4. All values are normalized between 0 and 1. The reliability calculated using Equation (6) remains above 0.8 unless an OCF occurs in the EDS. At the moment an OCF occurs, the reliability drops sharply due to the event-driven assessment approach embedded in Equation (6). However, this reliability measure does not consider the cumulative impact of past negative interactions caused by CPAs. As a result, the reliability value returns to its original level once the fault is removed. To address this, the weighted reliability is computed using Equation (7), which incorporates a time-forgetting factor. This factor prevents the reliability from immediately returning to 1 after the fault is cleared. Instead, the reliability gradually recovers, reflecting the lingering impact of past negative events. As a result, the time-forgetting factor ensures a more realistic and gradual restoration of the reliability values over time. Furthermore, the trust values are estimated for each OCF case, with the initial trust value set at 0.5, representing a baseline level of intermediate trust in the EDS at the start. As the EDS operates under normal conditions, the trust value gradually increases over time, reflecting the natural progression of trust in any system. In the event of an OCF, the trust value decreases but subsequently recovers, gradually rising back, based on the dynamics driven by the corresponding reliability values. Moreover, the recovery time associated with each fault category after the fault is cleared varies in each case, as shown in Figure 4, because it is influenced by the variance values highlighted in Equation (9). Moreover, it is assumed that no negative historical interactions exist in this case. As a result, the rates of change in reliability and trust are similar.

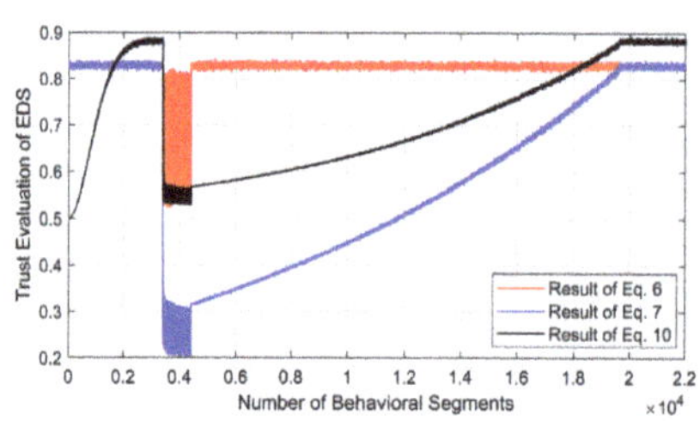

(a)

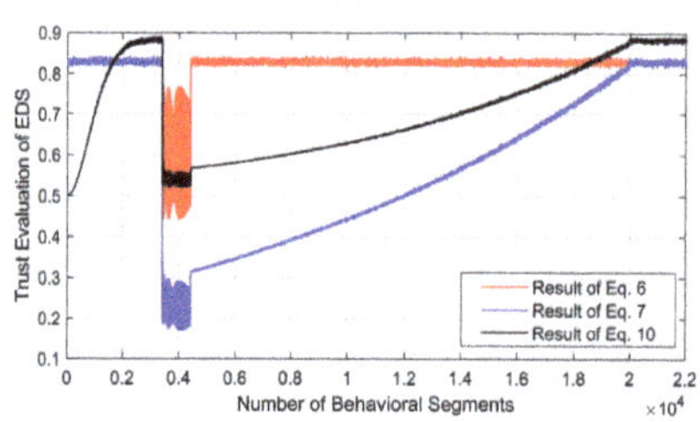

(b)

Figure 4. *Cont.*

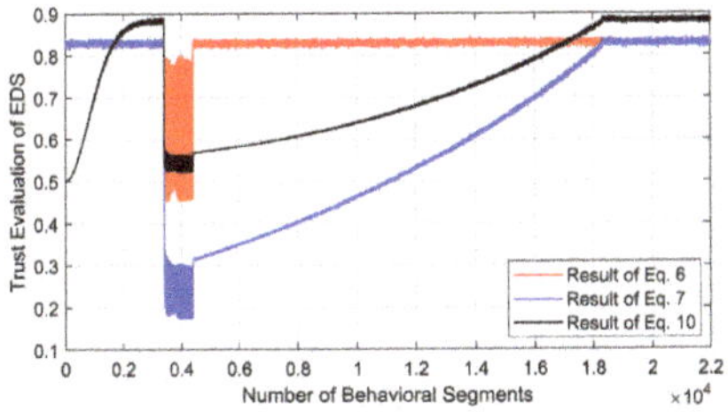

(**c**)

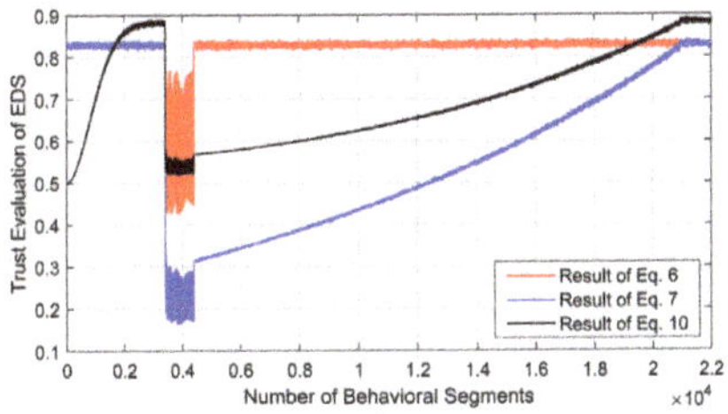

(**d**)

Figure 4. Trust evaluation dynamics during OCF-based physical faults: (**a**) SS-OCF [Case I], (**b**) SDS-OCF [Case II], (**c**) CDS-OCF [Case III], (**d**) PDS-OCF [Case IV].

3.1.2. Dynamics of Trust Value During Sensor Faults

Two types of physical faults in the EDS current sensor are considered: stuck fault and OCF. In the case of a stuck fault, where the phase-B current remains fixed at a constant value, the corresponding impact on reliability and trust quantification is illustrated in Figure 5a. Similarly, when the current sensor associated with phase-A fails and provides zero current feedback to the EDS controller (OCF), the resulting variations in reliability and trust are depicted in Figure 5b. It is evident that the recovery time is longer in the case of an OCF fault due to the increased fluctuation in the stability values, and this behavior is formulated using Equation (9).

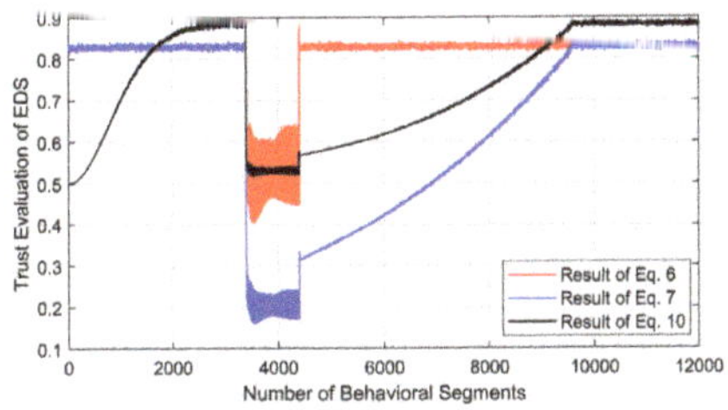

(**a**)

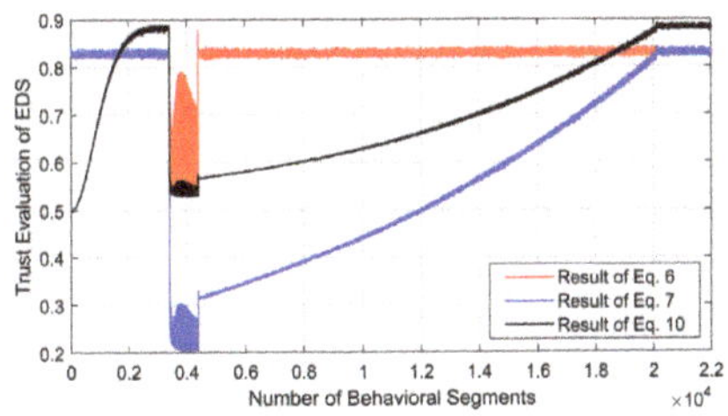

(**b**)

Figure 5. Trust evaluation dynamics during sensor-based physical faults: (**a**) stuck fault [Case V], (**b**) OCF [Case VI].

3.1.3. Dynamics of Trust Value During Cyberattacks

This section analyzes the dynamics of trust values in the EDS under various cyberattacks affecting the reference input speed signal and feedback signals. Specifically, for a cyberattack targeting the reference speed, the Pearson correlation coefficient (PCC) between the message sequence patterns of the input and feedback speed in the CAN bus, as reported in [34], is shown in Figure 6a. The corresponding variations in the trust value of the EDS, computed using Equation (10) and derived from the PCC values, are depicted in Figure 6b. Notably, unlike other cases, attacks that manipulate the reference speed do not increase the total harmonic distortion of the feedback three-phase current. As a result, the impact of a gradual increase in trust is negligible, leading to a trust value that closely follows the observed PCC variations, as shown in Figure 6b. Moreover, cases VIII to XII examine cyberattacks that manipulate the 3-phase feedback current values from the current sensor, thereby affecting the EDS controller. These attacks are formulated as described in [17] and are further classified into positive and negative multiplier-based attacks. Additionally, the reliability and trust values associated with these positive and negative multiplier-based false data injection attacks on three-phase feedback current are illustrated in Figures 7 and 8. Among these cases, case VIII exhibits the shortest trust recovery time, as it involves tampering with only a single phase, whereas cases X and XII have the longest trust recovery times, evaluated based on the number of behavioral segments.

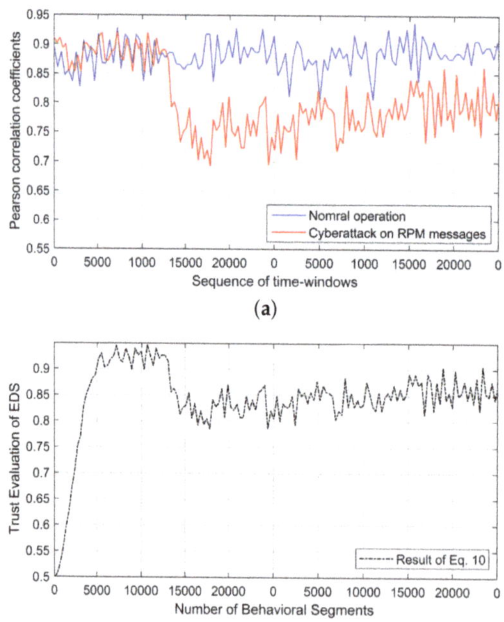

(a)

(b)

Figure 6. Case VII: Trust evaluation dynamics during a cyberattack on reference speed. (**a**) Pearson correlation coefficient [34], (**b**) trust value dynamics.

Hence, estimating the trust values using the proposed model-based scheme validates its effectiveness, demonstrating its applicability to various types of CPAs. Consequently, it can be utilized to monitor the operational status of the EDS, enabling its integration into the torque-split optimization algorithm for real-time updates of the required torque values. Additionally, the proposed scheme can serve as an anomaly detection mechanism. However, implementing this functionality may require a dynamic threshold estimation scheme, which will be the focus of future work. Moreover, the comparison of contributions between proposed and prior studies is highlighted in Table 3.

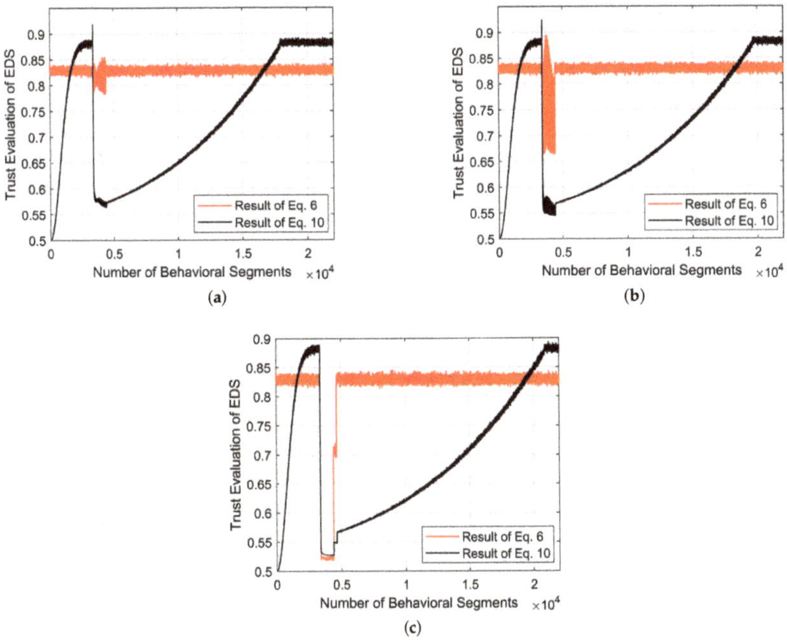

Figure 7. Trust evaluation dynamics under positive multiplier-based cyberattacks: (**a**) Case VIII, (**b**) Case IX, (**c**) Case X.

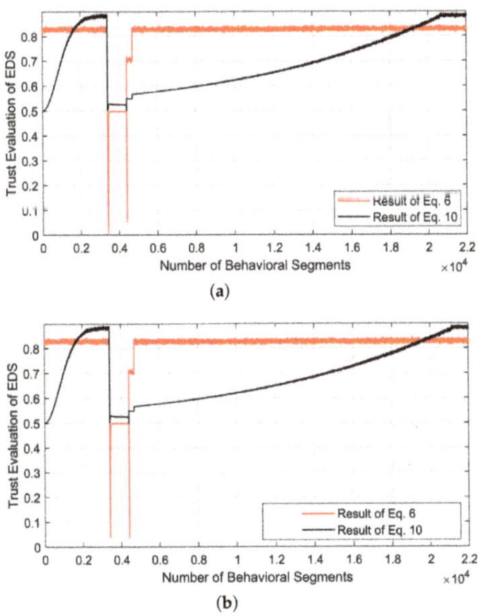

Figure 8. Trust evaluation dynamics under negative multiplier-based cyberattacks: (**a**) Case XI, (**b**) Case XII.

Table 3. Comparison of the proposed contributions with prior anomaly-detection studies.

Research Methodology	CA Detection	OCF Detection	SF Detection	Detection Time (ms)	Accuracy (%)
[35]	T	NT	NT	–	92.10
[30]	T	NT	NT	50.0	99.50
[36]	NT	T	NT	40.0	88.53
[11]	NT	T	T	20.0	98.83
[28]	T	NT	NT	12.5	99.90
[37]	T	NT	NT	10.0	90.00
[29]	T	NT	NT	2.5	98.44
This work	T	T	T	2.5	100

CA: Cyberattack. OCF: Open-Circuit Fault. SF: Sensor Fault. NT: Not Tested. T: Tested.

3.2. *Impact of Trust Values on Torque-Split Optimization*

To evaluate the impact of individual EDS trust values on torque-split optimization, a vehicle model, as described in [38], is utilized. The drive cycle used for this vehicle model is shown in Figure 9. Additionally, the optimized torque demands for each motor in a four-motor drive system are presented in Figure 10. Under normal operation, when the trust values exceed 0.8, each EDS is considered fully trustworthy based on its operational performance. As a result, during this phase, the torque-split optimization algorithm operates normally. The torque values associated with each motor (M1, M2, M3, and M4) during this normal operation are shown as original torque values in Figure 10, corresponding to the drive cycle illustrated in Figure 9. To investigate the impact of the previously calculated trust values, it is assumed that EDS-1 and EDS-4 experience CPAs, which are detected by the proposed trust evaluation framework. As a result, these trust values are input into the torque-split optimization algorithm to reduce the torque demand from the affected EDSs, as they are under stress due to the CPA event. Specifically, the trust values for EDS-1 and EDS-4 are reduced to 0.6 and 0.7, respectively. Consequently, the torque values for M1 and M4 are also reduced, as shown by the updated torque in Figure 10. The average percentage difference between the original and updated torque for M1 and M4 is 40% and 30%, respectively. Moreover, Figure 11 illustrates the residual values obtained using the reference (I_{dq}^{ref}) and measured (I_{dq}^{m}) current values. The results demonstrate that reducing the demanded torque during a CPA event effectively minimizes system-level fluctuations, thereby mitigating the impact of CPA. In this context, anomalies associated with physical faults, sensor faults, and cyberattacks are considered in Figure 11 to validate this mitigation approach.

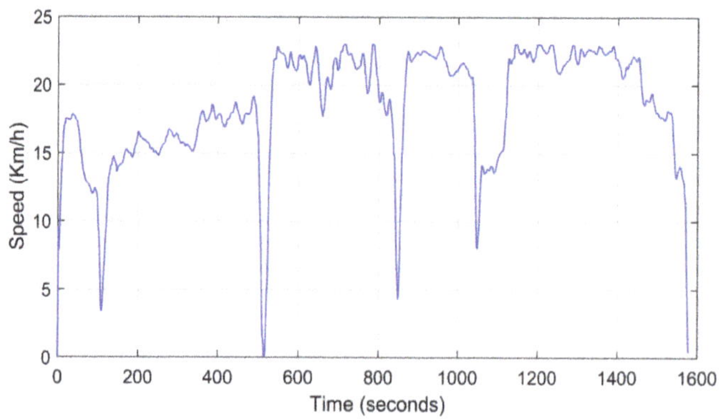

Figure 9. Drive cycle utilized for the vehicle model.

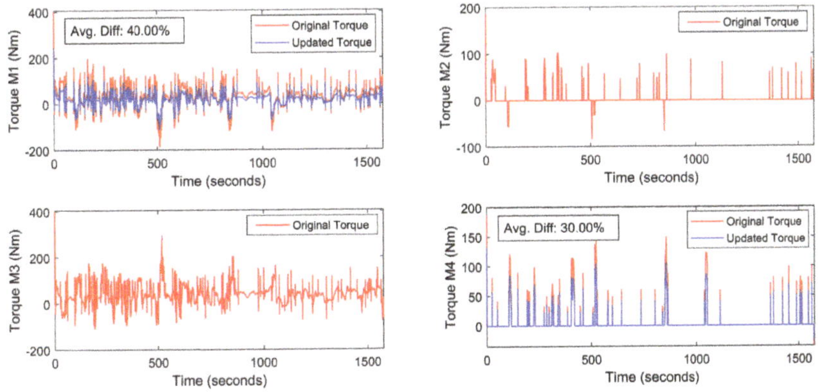

Figure 10. Torque values corresponding to each electric motor in an EV.

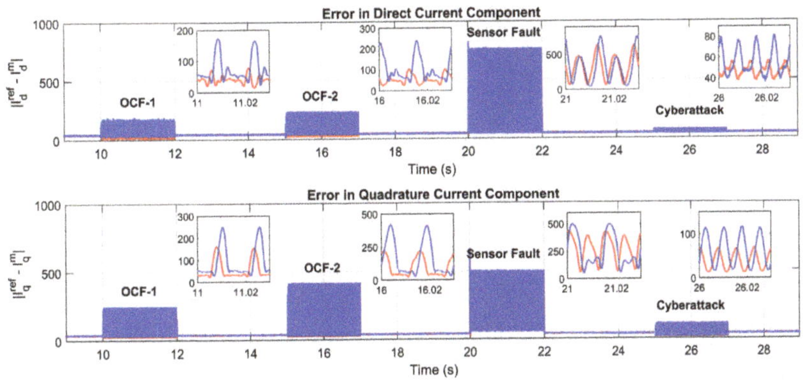

Figure 11. Impact of ALO on system fluctuations during various CPAs: with ALO [red], without ALO [blue].

Hence, this approach to reducing torque demand can mitigate the impact of CPAs on system performance. By lowering the torque demand, the EDS reduces the current drawn from the battery. As a result, the impact of the associated CPA on system performance is lessened. This reduction in current not only minimizes the effect on the battery's state of charge (SoC) but also decreases the likelihood of further failures that may stem from the initial CPA event.

4. Conclusions

The proposed trust evaluation framework for EDS monitors the reliability of the controller's input signals using EDA schemes, which are subsequently utilized to formulate trust values. This scheme facilitates the detection of various CPAs and enhances the resilience of EDS by continuously assessing and monitoring reliability. Furthermore, integrating trust values with torque-split optimization serves as a mitigation strategy, reducing the impact of anomalies on system-level fluctuations through ALO.

Various case studies involving open-circuit faults, sensor malfunctions, and cyber threats validate the effectiveness of the proposed framework in identifying and mitigating these anomalies while demonstrating its applicability and versatility across different scenarios. This approach can be further refined to distinguish between different types of anomalies, thereby enabling precise localization. Moreover, the scale of disruption can be characterized based on its impact on the trust values. Moreover, since an electric vehicle can be equipped with multiple EDSs, performing trust evaluation for each system enables researchers to develop health monitoring applications and implement responsive control strategies in the event of reliability degradation in any individual EDS.

Additionally, the integration of the proposed scheme with torque split optimization can be further enhanced for different realistic driving conditions while further incorporating vehicle dynamics, which will be the focus of our future work.

Author Contributions: Methodology, A.A.; Validation, A.A. and G.O.; Investigation, A.A., B.P., G.K.M., A.A.K. and C.S.E. All authors have read and agreed to the published version of the manuscript.

Funding: This study was supported by the Simulation-Based Reliability and Safety Program for modeling and simulation of military ground vehicle systems, under the technical services contract No. W56HZV-17-C-0095 with the U.S. Army DEVCOM Ground Vehicle Systems Center (GVSC). DISTRIBUTION STATEMENT A. Approved for public release; distribution is unlimited. OPSEC9429.

Data Availability Statement: The original contributions presented in this study are included in the article. Further inquiries can be directed to the corresponding author(s).

Acknowledgments: The authors would like to express their gratitude for the valuable resources and facilities provided by the Secure Energy and Automation Laboratory (SEAL) and Real Time Controls and Optimization Laboratory (RTCOOL). The authors would like to acknowledge the Warren H. Owen Distinguished Professorship Endowment for its support of the research effort.

Conflicts of Interest: The authors declare no conflicts of interest.

References

1. Global Automotive Cybersecurity Report 2023. Available online: https://upstream.auto/reports/global-automotive-cybersecurity-report/ (accessed on 17 August 2024).
2. Zhou, X.; Sun, J.; Li, H.; Lu, M.; Zeng, F. PMSM open-phase fault-tolerant control strategy based on four-leg inverter. *IEEE Trans. Power Electron.* **2019**, *35*, 2799–2808. [CrossRef]
3. Zhou, D.; Qiu, H.; Yang, S.; Tang, Y. Submodule voltage similarity-based open-circuit fault diagnosis for modular multilevel converters. *IEEE Trans. Power Electron.* **2018**, *34*, 8008–8016. [CrossRef]
4. Zhou, D.; Yang, S.; Tang, Y. A voltage-based open-circuit fault detection and isolation approach for modular multilevel converters with model-predictive control. *IEEE Trans. Power Electron.* **2018**, *33*, 9866–9874. [CrossRef]
5. Zhao, H.; Cheng, L. Open-switch fault-diagnostic method for back-to-back converters of a doubly fed wind power generation system. *IEEE Trans. Power Electron.* **2017**, *33*, 3452–3461. [CrossRef]
6. Gou, B.; Ge, X.L.; Liu, Y.C.; Feng, X.Y. Load-current-based current sensor fault diagnosis and tolerant control scheme for traction inverters. *Electron. Lett.* **2016**, *52*, 1717–1719. [CrossRef]
7. Giraldo, J.; Urbina, D.; Cardenas, A.; Valente, J.; Faisal, M.; Ruths, J.; Tippenhauer, N.O.; Sandberg, H.; Candell, R. A survey of physics-based attack detection in cyber-physical systems. *ACM Comput. Surv.* **2018**, *51*, 1–36. [CrossRef]
8. Mishra, S.; Shoukry, Y.; Karamchandani, N.; Diggavi, S.N.; Tabuada, P. Secure state estimation against sensor attacks in the presence of noise. *IEEE Trans. Control Netw. Syst.* **2016**, *4*, 49–59. [CrossRef]
9. Mo, Y.; Sinopoli, B. Secure control against replay attacks. In Proceedings of the 2009 47th Annual Allerton Conference on Communication, Control, AND Computing (Allerton), Monticello, IL, USA, 30 September–2 October 2009 ; IEEE: New York, NY, USA, 2009; pp. 911–918.
10. Gao, Z.; Cecati, C.; Ding, S.X. A survey of fault diagnosis and fault-tolerant techniques—Part II: Fault diagnosis with knowledge-based and hybrid/active approaches. *IEEE Trans. Ind. Electron.* **2015**, *62*, 3768–3774. [CrossRef]
11. Gou, B.; Xu, Y.; Xia, Y.; Deng, Q.; Ge, X. An online data-driven method for simultaneous diagnosis of IGBT and current sensor fault of three-phase PWM inverter in induction motor drives. *IEEE Trans. Power Electron.* **2020**, *35*, 13281–13294. [CrossRef]
12. Cai, B.; Zhao, Y.; Liu, H.; Xie, M. A data-driven fault diagnosis methodology in three-phase inverters for PMSM drive systems. *IEEE Trans. Power Electron.* **2016**, *32*, 5590–5600. [CrossRef]
13. Wang, T.; Xu, H.; Han, J.; Elbouchikhi, E.; Benbouzid, M.E.H. Cascaded H-bridge multilevel inverter system fault diagnosis using a PCA and multiclass relevance vector machine approach. *IEEE Trans. Power Electron.* **2015**, *30*, 7006–7018. [CrossRef]

14. Ye, S.; Jiang, J.; Li, J.; Liu, Y.; Zhou, Z.; Liu, C. Fault diagnosis and tolerance control of five-level nested NPP converter using wavelet packet and LSTM. *IEEE Trans. Power Electron.* **2019**, *35*, 1907–1921. [CrossRef]
15. Jan, S.U.; Lee, Y.D.; Shin, J.; Koo, I. Sensor fault classification based on support vector machine and statistical time-domain features. *IEEE Access* **2017**, *5*, 8682–8690. [CrossRef]
16. Arsalan, A.; Timilsina, L.; Papari, B.; Muriithi, G.; Ozkan, G.; Kumar, P.; Edrington, C.S. Cyber Attack Detection and Classification for Integrated On-board Electric Vehicle Chargers subject to Stochastic Charging Coordination. *Transp. Res. Procedia* **2023**, *70*, 44–51. [CrossRef]
17. Arsalan, A.; Papari, B.; Timilsina, L.; Muriithi, G.; Moghassemi, A.; Rahman, S.M.I.; Buraimoh, E.; Ozkan, G.; Edrington, C.S. Enhanced Real-Time ATM-Based MPC for Electric Vehicles With Cyber–Physical Security Aspect. *IEEE Trans. Transp. Electrif.* **2025**, *11*, 4698–4716. [CrossRef]
18. Djedjig, N.; Tandjaoui, D.; Medjek, F.; Romdhani, I. Trust-aware and cooperative routing protocol for IoT security. *J. Inf. Secur. Appl.* **2020**, *52*, 102467. [CrossRef]
19. Yuan, J.; Li, X. A reliable and lightweight trust computing mechanism for IoT edge devices based on multi-source feedback information fusion. *IEEE Access* **2018**, *6*, 23626–23638. [CrossRef]
20. Han, G.; Jiang, J.; Shu, L.; Guizani, M. An attack-resistant trust model based on multidimensional trust metrics in underwater acoustic sensor network. *IEEE Trans. Mob. Comput.* **2015**, *14*, 2447–2459. [CrossRef]
21. Jiang, J.; Han, G.; Wang, F.; Shu, L.; Guizani, M. An efficient distributed trust model for wireless sensor networks. *IEEE Trans. Parallel Distrib. Syst.* **2014**, *26*, 1228–1237. [CrossRef]
22. Hussain, Y.; Zhiqiu, H.; Akbar, M.A.; Alsanad, A.; Alsanad, A.A.A.; Nawaz, A.; Khan, I.A.; Khan, Z.U. Context-aware trust and reputation model for fog-based IoT. *IEEE Access* **2020**, *8*, 31622–31632. [CrossRef]
23. Poongodi, M.; Hamdi, M.; Sharma, A.; Ma, M.; Singh, P.K. DDoS detection mechanism using trust-based evaluation system in VANET. *IEEE Access* **2019**, *7*, 183532–183544. [CrossRef]
24. Jiang, N.; Wen, J.; Li, J.; Liu, X.; Jin, D. Gatrust: A multi-aspect graph attention network model for trust assessment in osns. *IEEE Trans. Knowl. Data Eng.* **2022**, *35*, 5865–5878. [CrossRef]
25. Wang, T.; Luo, H.; Zeng, X.; Yu, Z.; Liu, A.; Sangaiah, A.K. Mobility based trust evaluation for heterogeneous electric vehicles network in smart cities. *IEEE Trans. Intell. Transp. Syst.* **2020**, *22*, 1797–1806. [CrossRef]
26. Garcia, N.; Hammad, E.; Farraj, A. Soft-Trust Based Architecture for NextG IIoT/IoET Security, Authentication and Authorization. In Proceedings of the 2023 IEEE Texas Power and Energy Conference (TPEC), College Station, TX, USA, 13–14 February 2023; IEEE: New York, NY, USA, 2023; pp. 1–6.
27. Guo, J.; Liu, Z.; Tian, S.; Huang, F.; Li, J.; Li, X.; Igorevich, K.K.; Ma, J. TFL-DT: A trust evaluation scheme for federated learning in digital twin for mobile networks. *IEEE J. Sel. Areas Commun.* **2023**, *41*, 3548–3560. [CrossRef]
28. Yang, B.; Ye, J.; Guo, L. Fast detection for cyber threats in electric vehicle traction motor drives. *IEEE Trans. Transp. Electrif.* **2021**, *8*, 767–777. [CrossRef]
29. Guo, L.; Ye, J.; Yang, B. Cyberattack detection for electric vehicles using physics-guided machine learning. *IEEE Trans. Transp. Electrif.* **2020**, *7*, 2010–2022. [CrossRef]
30. Yang, B.; Wu, S.; Hu, K.; Ye, J.; Song, W.; Ma, P.; Shi, J.; Liu, P. Enhanced Cyber-Attack Detection in Intelligent Motor Drives: A Transfer Learning Approach With Convolutional Neural Networks. *IEEE J. Emerg. Sel. Top. Ind. Electron.* **2024**, *5*, 710–719. [CrossRef]
31. Guo, L.; Wang, K.; Wang, T. Open-Circuit Fault Diagnosis of Three-Phase Permanent Magnet Machine Utilizing Normalized Flux-Producing Current. *IEEE Trans. Ind. Electron.* **2024**, *71*, 3351–3360. [CrossRef]
32. Hang, J.; Shu, X.; Ding, S.; Huang, Y. Robust Open-Circuit Fault Diagnosis for PMSM Drives Using Wavelet Convolutional Neural Network With Small Samples of Normalized Current Vector Trajectory Graph. *IEEE Trans. Ind. Electron.* **2023**, *70*, 7653–7663. [CrossRef]
33. Zhang, X.; Zhang, L.; Zhang, Y. Model predictive current control for PMSM drives with parameter robustness improvement. *IEEE Trans. Power Electron.* **2018**, *34*, 1645–1657. [CrossRef]
34. Jedh, M.; Ben Othmane, L.; Ahmed, N.; Bhargava, B. Detection of Message Injection Attacks Onto the CAN Bus Using Similarities of Successive Messages-Sequence Graphs. *IEEE Trans. Inf. Forensics Secur.* **2021**, *16*, 4133–4146. [CrossRef]
35. Jawdeh, S.A.; Choi, S.; Liu, C.H. Model-Based Deep Learning for Cyber-Attack Detection in Electric Drive Systems. In Proceedings of the 2022 IEEE Applied Power Electronics Conference and Exposition (APEC), Houston, TX, USA, 20–24 March 2022; pp. 567–573. [CrossRef]
36. Xia, Y.; Xu, Y. A Transferrable Data-Driven Method for IGBT Open-Circuit Fault Diagnosis in Three-Phase Inverters. *IEEE Trans. Power Electron.* **2021**, *36*, 13478–13488. [CrossRef]

37. Yang, B.; Ye, J. Data-driven detection of physical faults and cyber attacks in dual-motor ev powertrains. In Proceedings of the 2022 IEEE Transportation Electrification Conference & Expo (ITEC), Anaheim, CA, USA, 15–17 June 2022; IEEE: New York, NY, USA, 2022; pp. 991–996.
38. Muriithi, G.; Papari, B.; Moghassemi, A.; Sundar, A.; Arsalan, A.; Buraimoh, E.; Timilsina, L.; Ozkan, G.; Edrington, C. Vulnerability Assessment and Detection of Stealthy Sequential Cyberattacks in Hybrid Tracked Vehicles. *IEEE Trans. Transp. Electrif.* **2024**, *11*, 6472–6489. [CrossRef]

electronics

MDPI

Review

Cybersecurity in MAS-Based Adaptive Protection for Microgrids—A Review

Armando J. Taveras Cruz [1,2,*], Miguel Aybar-Mejía [1,2], Carlos G. Colon-González [3], Deyslen Mariano-Hernández [1], Jesús C. Hernandez [4], Fabio Andrade-Rengifo [5] and Luis Hernández-Callejo [6,*]

1 Engineering Area, Instituto Tecnológico de Santo Domingo, Santo Domingo 10602, Dominican Republic; miguel.aybar@intec.edu.do (M.A.-M.); deyslen.mariano@intec.edu.do (D.M.-H.)
2 Programa de Doctorado de la Universidad de Jaén, Energías Renovables, Departamento de Ingeniería Eléctrica Edificio A3, Escuela Politécnica Superior de Jaén, 23071 Jaén, Spain
3 Graduate School, Polytechnic University of Puerto Rico, San Juan 00918, Puerto Rico; colon_30077@students.pupr.edu
4 Department of Electrical Engineering, University of Jaén, 23071 Jaén, Spain; jcasa@ujaen.es
5 Department of Electrical and Computer Engineering, University of Puerto Rico-Mayagüez, Mayagüez, PR 00680, USA; fabio.andrade@upr.edu
6 Departamento Ingeniería Agrícola y Forestal, Universidad de Valladolid, 42004 Soria, Spain
* Correspondence: armando.taveras@intec.edu.do (A.J.T.C.); luis.hernandez.callejo@uva.es (L.H.-C.)

Abstract: With the ever-growing reliance on digital communication networks in microgrids equipped with digital control systems and highly distributed energy resources, the threat of cyberattacks is more present than ever. Therefore, a robust cybersecurity response framework could be in place to secure smart grids, including microgrids, against cyberattacks. Adaptive protection systems, which are crucial for microgrid reliability and resilience, are also vulnerable. On the other hand, multi-agent systems are often employed in microgrid adaptive protection, providing a decentralized and cooperative framework where intelligent agents can monitor system conditions, exchange information, and detect anomalies. Many researchers in the literature have focused on addressing microgrid protection with multi-agent systems against physical faults in scenarios with various degrees of distributed energy resource penetration. Other research efforts have leveraged multi-agent systems, as well as technologies such as artificial intelligence, machine learning, advanced encryption, and authentication, to enhance the capabilities of microgrids for maintaining resilient operation under cyberattacks. However, both physical and cybersecurity anomalies have rarely been tackled in the same scheme. This paper aims to provide a systematic review of the use of cybersecurity strategies for multi-agent-based adaptive protection schemes. From the results of this study, it was found that most research efforts do not address microgrid protection with an integrated approach, considering both physical and cybersecurity threats, as well as the application of established industry communication and cybersecurity standards. All of this, while maintaining scalability and performance, is crucial.

Keywords: microgrid; multi-agent; adaptive protection; cybersecurity

1. Introduction

Microgrids (MGs) are localized groups of power sources and loads that can operate in connection with the primary power grid or independently in an 'islanded' mode. They integrate various distributed energy resources (DERs) such as solar panels, wind turbines, and battery storage systems [1,2]. A key feature of alternating current (AC) microgrids is their ability to operate autonomously, which enhances resilience against power outages and natural disasters. This autonomy in MGs is achieved through a hierarchical control structure that includes primary, secondary, and tertiary controls [3,4]. Microgrids enhance the operational flexibility and resilience of power systems by enabling the integration of renewable energy sources and allowing for self-sustained operation during grid failures [1,2]. Microgrids support the global shift towards sustainable and clean energy by integrating renewable energy sources, thereby reducing the carbon footprint associated with power generation [1]. They provide energy independence and increased robustness against long-lasting outages, which is particularly beneficial for critical facilities like military installations [1].

Despite their advantages, conventional protection techniques for microgrids cannot cope with the dynamic nature of microgrid operations and the increasing number of DERs and features of smart grids, exposing the limitations of these more static protection designs [5]. Various researchers have proposed adaptive protection relay strategies that dynamically adjust protection settings in response to changes in the microgrid's operational conditions. This is crucial for maintaining reliability and stability, particularly given the variability introduced by distributed generation sources such as solar and wind [6,7]. An approach presented by various authors is adaptive protection based on multi-agent systems (MASs). In this method, either software or hardware agents continuously monitor the microgrid's status and adjust relay settings accordingly. This ensures that protection schemes remain effective under varying operational conditions, such as changes in load or generation [6,8]. An MAS applies advanced communication protocols, such as IEC 61850 GOOSE [9], based on Ethernet and standard communication protocols used for high-speed and high-priority communication, to facilitate fast and reliable data exchange among agents [10–12]. This ensures coordinated protection actions and enhances the system's responsiveness to faults and network changes, while reducing dependency on a central control unit and improving the system's resilience and flexibility [8,13].

The cybersecurity landscape for microgrids has become increasingly complex. The 2025 ERO Reliability Risk Priorities Report by North American Electric Reliability Corporation (NERC) emphasizes that rapid infrastructure growth and digitization—including information technology (IT) and operational technology (OT) convergence, cloud-based technology reliance, artificial intelligence (AI) emergence, and dispersed management systems such as DER aggregators and Internet of Things (IoT) devices—significantly expand the cyberattack surface for MGs [14]. The report projects a 30% growth in electrical demand by 2050.

Inverter-based resources (IBRs) and DERs represent as much as 90% of new generation capacity added annually [14]. However, they rely heavily on information and communication technologies (ICTs), transforming these systems into complex cyber-physical energy systems (CPESs) that are inherently vulnerable to cyber threats [15]. Critical components, including smart meters, digital metering devices, advanced metering infrastructure (AMI), phasor measurement units (PMUs), and industrial electronic devices (IEDs), represent vulnerabilities when inadequately protected [15]. These distributed components constantly communicate with control centers through multiple channels, increasing the probability of cyberattacks.

Consequently, despite their operational advantages, microgrids and MAS-based adaptive protection schemes face significant cybersecurity challenges due to their communication dependencies. These threats can target both control and communication entities, making microgrids susceptible to attacks such as false data injection (FDI) and denial-of-service (DoS) [12,16]. The absence of centralized DER monitoring, combined with the widespread adoption of inverter-based DERs and software-intensive ICT infrastructure, further complicates the cybersecurity landscape [17]. Addressing these multifaceted challenges is crucial for maintaining the integrity of modern power systems [15,18]. The NERC reports that 69% of online threats evaluated by its Electricity Information Sharing and Analysis Center (E-ISAC) aim to sabotage the power grid. Therefore, given the increasing number of cyber threats, there is an urgent need to develop strategies for the resilient management of microgrids. This includes proactive and comprehensive responses to cyberattacks, ensuring reliable operation [15].

This paper aims to provide a systematic review of the challenges, current advances, and future trends in cybersecurity strategies for MG MAS-based adaptive protection schemes, as well as the technologies used to develop them. The main contributions are as follows:

(i) A synthesis and critical evaluation of existing cybersecurity approaches for MAS-based adaptive protection in microgrids.
(ii) An identification of the research gaps and a proposal for future research directions.

The remainder of this paper is organized as follows. Section 2 discusses the applied methodology. Section 3 provides an overview of AC microgrids and MAS-based adaptive protection. In Section 4, the cybersecurity threats to MAS-based adaptive protection in AC microgrids are described, followed by a more specific discussion on existing solutions and current research for securing MASs in adaptive protection in Section 5. Finally, Sections 6 and 7 conclude this review by presenting the conclusions and outlining future research directions, respectively.

2. Methodology

A detailed search was conducted on four principal scholarly databases: Scopus, IEEE, Web of Science, and Google Scholar. MAS-based adaptive protection schemes and cybersecurity in the literature from the last five years were selected, while some foundational works were also included from earlier periods. A structured combination of keywords was employed, including "microgrid," "smart grid," "adaptive protection," "multi-agent systems," "cybersecurity," and "cyber-attacks." The data collected was subjected to a critical analysis, resulting in the conclusions presented at the end of this review. Initially, the authors analyzed the contributions of the relevant works, and subsequently, the most recent primary research papers on each subtopic were identified and subjected to the same analysis strategy. Figure 1 presents a conceptual diagram of the methodology followed.

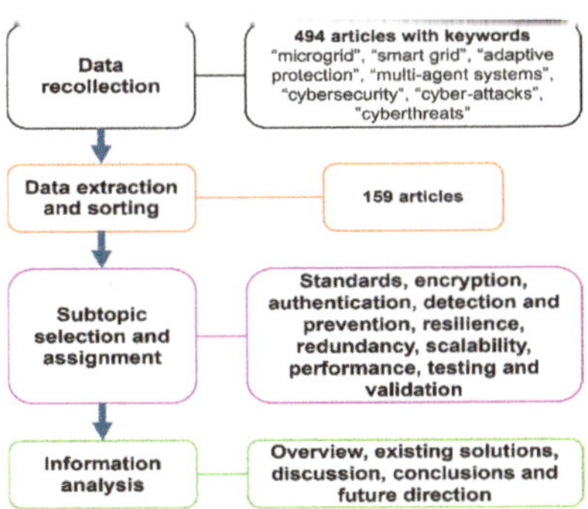

Figure 1. Methodology.

3. Overview of AC Microgrids and MAS-Based Adaptive Protection

3.1. Structure and Components of AC Microgrids

AC microgrids are localized power systems that integrate various DERs, energy storage systems (ESSs), loads, and control units. They can operate in both grid-connected and islanded modes, providing flexibility and reliability in power supply.

Several key components comprise the fundamental structure of a microgrid. The first are distributed generation (DG) units. These can be from renewables such as solar panels, wind turbines, and other renewable energy sources connected to the AC grid via inverters [19,20], but can also come from non-renewable sources, such as diesel generators and other conventional power sources, which can also be part of the DG units, directly connected to the AC bus [19]. Another component can be energy storage systems (ESSs). These can include lithium-ion batteries, lead-acid batteries, sodium-based batteries, vanadium redox flow batteries, and hydrogen storage systems [21]. ESSs are typically connected to the local AC bus through direct current (DC)-to-AC converters with inductance capacitance (LC) filters to ensure stable voltage and power quality [22]. Loads are considered part of the component lineup within microgrids. These can be residential, commercial, or industrial loads that the microgrid serves. Loads can vary in their power requirements and operational characteristics [23,24]. Subsequently, we have control units, which manage the immediate power balance and voltage/frequency regulation within the microgrid. This includes inverter control for renewable sources and synchronization mechanisms for grid connections [19,25]. These units ensure long-term stability and optimal operation by dispatching energy sources according to predefined setpoints and participation factors [25].

A microgrid operates in two primary modes: grid-connected and islanded modes. In grid-connected mode, the microgrid operates in synchronization with the primary grid, allowing for power exchange and enhanced overall grid stability [19]. In islanded mode, the microgrid operates independently from the primary grid, relying on its internal DG units and ESS to maintain power supply [19,26]. A synchronization process is required to ensure smooth transitions between these modes to prevent large power flows and potential damage to loads [19].

AC MGs are complex systems that integrate various components and require sophisticated control strategies to ensure reliable and efficient operation. The combination of renewable and non-renewable energy sources, along with advanced energy storage and control systems, makes them a versatile solution for modern power needs. Figure 2 shows an MG system with the point of common coupling (PCC) with the grid, protection relays, circuit breakers (CBs), and DER, as well as the agents of an MAS.

Networked microgrids face a multitude of general challenges, primarily stemming from their dynamic nature and the integration of diverse distributed energy resources (DERs). A central issue is the dynamic and complex operating conditions introduced by the intermittent and fluctuating output of renewable energy sources, leading to difficulties in system stability and protection [5]. This is further complicated by sophisticated and dynamic microgrid topologies that frequently change, coupled with bidirectional power flow. This renders conventional unidirectional protection schemes inadequate and can cause issues like unwanted tripping or protection blinding [27]. Finally, the growing intelligence and connectivity of microgrid components directly expose them to an increasing landscape of cyber-physical security threats, making robust cybersecurity measures and data security paramount for seamless and secure operation [27,28].

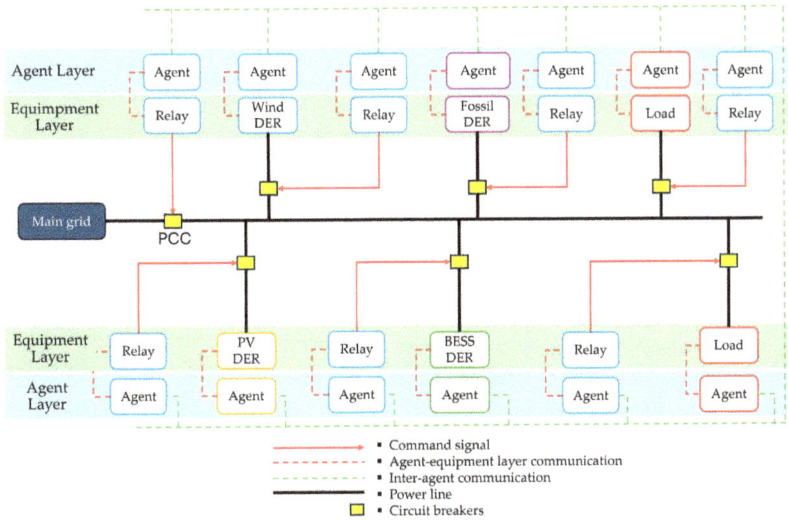

Figure 2. MG structure with local components and agents.

3.2. MAS in MG Protection and Cybersecurity

Microgrids face unique challenges due to their ability to operate in both grid-connected and islanded modes. This dual-mode operation requires significant reconfiguration of electrical protections to maintain selectivity and reliability, as fault conditions can change drastically between modes [15]. Additionally, inverter-based distributed generators in microgrids often provide low short-circuit currents, making fault detection challenging [29]. Consequently, adaptive protection has the objective of dynamically adjusting protection settings in response to changing conditions within a microgrid, isolating faults quickly and efficiently, thereby minimizing the impact on the rest of the network. This approach ensures that protection schemes remain effective under varying operational scenarios, such as transitioning between grid-connected and islanded modes [15,30].

Adaptive protection plays a vital role in enhancing the cyber resilience of microgrids. By dynamically adjusting to operational changes and potential cyber threats, adaptive protection helps maintain reliable operation even in the face of unforeseen incidents [31]. Multi-agent systems (MASs) are an effective method for implementing adaptive protection schemes, particularly in a decentralized structure, offering advantages such as higher speed, reliability, and scalability compared to centralized approaches [32,33]. In multiple research papers, the focus of applying MAS has been to enhance the accuracy of fault detection and location [7,8,33–36]. They coordinate the operation of relays and circuit breakers to ensure effective fault isolation, which is crucial for adaptive protection schemes [37–39]. The MAS-based protection system utilizes intelligent electronic devices (IEDs) as agents. Agents in the MAS collect real-time data on current and voltage from the distribution system [34]. These agents possess capabilities such as communication, decision-making, and data processing, which are essential for adaptive protection in microgrids [39].

An MAS fundamentally relies on the communication infrastructure for its operation and effectiveness [5,13]. The communication infrastructure enables agents, regardless of their organization into layers or zones, to communicate and cooperate with their environment [10,40]. Communication protocols, such as IEC 61850 (which includes GOOSE technology), are frequently employed to facilitate high-speed and secure data transmission between intelligent electronic devices (IEDs) that act as agents [10,41]. Whether structured hierarchically or on a peer-to-peer basis, reliable and fast communication is deemed essential for effective MAS operation in real-time power system applications [42,43]. Therefore, dependable and fast communication is deemed a critical enabling factor for effective MAS operation in real-time power system applications [44,45]. A consolidated comparison of the reviewed approaches is presented in Table 1 below.

Table 1. Summary of recent MAS-based approaches to adaptive protection.

Ref.	VLD	CMP	CSCAD
[46]	JADE	ACL	•
[47]	MATLAB/Simulink	•	•
[48]	DIgSILENT MATLAB	•	Eliminates communication dependency between agents for fault isolation
[49]	PSCAD/EMTD	•	•
[41]	ETAP OPAL-RT	IEC 60870 [50] /Modbus	•
[39]	MATLAB/Simulink JADE	TCP/IP	Protection against physical faults and cyberattack detection
[40]	MATLAB/Simulink	IEC 61850	Reducing telecommunication risk and minimizing time delay for single-event faults by using offline calculations
[42]	MATLAB JADE	Binary data exchange between agents	Enhances robustness against cyberattacks and one-point failures by operating in a fully distributed manner
[36]	PSCAD/EMTDC JADE	FIPA	Reduces the risk of communication failures and delays by having a backup offline protection strategy
[10]	ETAP	IEC 61850	•
[51]	ETAP MATLAB	IEC 61850	Self-healing scheme that adapts based on virtual local area network (VLAN) segmentation and communication mapping
[38]	Not detailed	•	•
[44]	JADE MATLAB	Blockchain for information exchange	•
[52]	Simulink AnyLogic	•	•
[53]	ETAP	IEC 61850	•
[7]	MATLAB	•	•
[54]	RSCAD JADE	•	•
[55]	MATLAB	•	•
[56]	MATLAB	IEC 61850	•

Legend: CMP = Communication protocols, CSCAD = Cybersecurity-related challenges addressed, VLD = Validation method, ACL = Agent communication language, • = Unclear or not addressed.

An MAS also plays a role in enhancing cybersecurity by monitoring cyberattacks and implementing countermeasures, thereby maintaining the integrity and reliability of the protection system [39,57,58]. Khatana et al. [59] developed a distributed malicious agent detection scheme to enable honest DERs to detect and isolate the communication links of maliciously behaving DERs in their neighborhood during distributed power allocation. This approach represents a novel method for intruder detection and isolation; however, its robustness needs to be validated under a broader range of attack scenarios. Karanfil et al. [60] introduced a security monitoring platform based on the IEC 62351-7:2017 standard [61] for network and system management (NSM). This represents the first implementation of a microgrid-specific security monitoring solution using the IEC NSM framework for real-time cyberattack detection—additionally, Roy et al. [62] designed a machine learning-based MAS framework incorporating a Master Agent (MA) for cyberattack detection and Slave Agents (SAs) for localized mitigation. Detection is conducted centrally via a one-class classifier (OCC) trained on secure data. At the same time, mitigation is managed in a decentralized fashion using Support Vector Regression (SVR) models informed by local measurements. The system targets previously unseen attacks in automatic generation control (AGC) and high-voltage direct current (HVDC) systems. Albarakati et al. [39] proposed an MAS-based adaptive protection mechanism for cyberattack detection and fault response in distribution networks. The framework enables fault location, line isolation, system reconfiguration, and differentiation between cyber-induced and natural faults, thus reducing relay malfunctions and improving reliability.

In addition to detection, an MAS can autonomously mitigate attacks by leveraging distributed response mechanisms. Al-dulaimi et al. [63] introduced a data-driven cooperative stochastic control system for islanded AC microgrids, featuring a lightweight prevention model to mitigate false data injection (FDI) and a communication-based controller. The proposed system enhances performance in frequency restoration and power sharing in sparse networks, while also reducing reliance on centralized control schemes that are vulnerable to compromise. A cybersecure distributed secondary control strategy for inverter-based islanded microgrids was introduced by Bidram et al. [16]. The approach utilizes the Weighted Mean Subsequence Reduced (WMSR) algorithm to eliminate corrupted information from compromised agents, leveraging time-varying communication graphs to enhance security without relying on static assumptions.

Both detection and mitigation can also be combined. Hu et al. [64] presented a decentralized consensus decision-making (DCDM) approach leveraging blockchain technologies for cybersecurity protection in multi-microgrid (MMG) systems. This design eliminates the need for a central authority, thereby addressing the single point of failure (SPoF) challenge inherent in centralized architectures. Zhou et al. [65] proposed a hierarchical MAS-based detection and mitigation (MHDM) scheme with three layers of rule-based anomaly detection. This structure enables the classification of operational states using cyber-physical and fault-specific data, enhancing the detection of diverse cyberattack vectors. Choi et al. [66] developed an MAS-based cyberattack detection and mitigation framework for distribution automation systems (DASs). The system employs message authentication for data integrity and validates control commands using power system domain knowledge, effectively distinguishing cyberattacks from physical faults by leveraging both cyber and physical characteristics.

An MAS can also support cybersecurity risk management by enabling trust-based decision frameworks. Boakye-Boateng et al. [67] proposed a trust management system where substation risk posture is computed based on trust scores from intelligent electronic devices (IEDs) and supervisory control and data acquisition (SCADA) human–machine interfaces (HMIs). This system enables the dynamic assessment of security states based on device behavior over time.

To strengthen the cyber-resilience of microgrid systems, MAS-based frameworks can be focused on adapting control schemes under adversarial conditions. Wang et al. [68] proposed a cooperative control strategy for bidirectional interlinking converters (BICs) in networked AC/DC microgrids. The method uses adaptive control to coordinate agents' responses during false signal injection attacks. It is resilient to node compromises that may be undetectable by traditional intrusion detection systems. Abianeh et al. [69] developed a multi-agent deep reinforcement learning (RL) approach for cybersecurity in DC microgrids. This framework automatically identifies weaknesses in conventional detection methods and generates stealthy FDI attacks, enabling a more robust vulnerability analysis and offering a novel complementary detection mechanism for the secondary control layer.

In summary, MAS frameworks offer significant benefits for microgrid cybersecurity, including enhanced reliability, scalability, autonomy, flexibility, and resilience [16,65,70]. Table 2 provides a comparative overview of the reviewed MAS-based MG cybersecurity approaches.

Table 2. Overview of MAS-based MG cybersecurity approaches.

Ref.	Contribution	CStd	FC	ATT	EU	Key Tools		CSCAD
[59]	Distributed detection for malicious DER	•	DT	FDI	•	OPAL-RT, UDP	–	Detecting and isolating malicious DER; deviant behavior
[63]	Integrates distributed stochastic control and intelligent secondary frequency control	•	MT	FDI, DoS	•	MATLAB	–	Cyberattacks on communication links and local controls
[67]	MAS-based trust management system for substation risk	•	DT	DoS	•	Docker, Python, JADE, Modbus TCP/IP	–	Protocol-based attacks on IEDs and SCADA HMIs
[60]	Hybrid rule-based and ML anomaly detection	IEC 62351	DT	MitM	✓	OPAL-RT, IEC 61850 IEC-104 [9,50]	–	Cyberattack detection via a security monitoring platform with ML
[69]	Multi-agent deep RL for vulnerability ID and detection of FDI attacks	•	ID	FDI	•	MATLAB/Simulink, dSPACE	–	Maintaining stealth under compromised links; overcoming detection failure for stealthy FDIs
[62]	ML-based MAS for unseen data integrity and availability attacks	•	DT	FDI	•	Unspecified	–	Detecting and neutralizing unseen cyberattacks; detection using only secured training data
[39]	MAS for fault location, isolation, reconfiguration, and cyberattack detection	•	DT	FDI	•	MATLAB/Simulink, TCP/IP, JADE	–	Protection against physical faults and cyberattack detection
[64]	Decentralized consensus decision-making (DCDM) with blockchain for cybersecurity	•	DM	•	BLK	Blockchain	–	Timely mitigation of cyberattack risks; decentralized consensus
[68]	Distributed resilient control for BICs in AC/DC microgrids	•	CO	FDI	•	MATLAB/Simulink OPC-UA	–	Managing coordinated node attacks; resilient control not reliant on prompt detection/isolation
[65]	Three-level hierarchical MAS for detecting and mitigating cyberattacks/physical faults	•	DM	LAN MitMDoS	•	MATLAB/Simulink, TCP/IP, JADE	–	Detecting and mitigating cyberattacks; distinguishing from physical faults
[16]	Secure intrusion mitigation for microgrid distributed control systems	•	MT	FDI	•	MATLAB/Simulink	–	Secure intrusion mitigation; ensuring connectivity under attack; restoring frequency/voltage
[66]	MAS for cybersecurity-enhanced DAS with anomaly detection/mitigation	IEC 62351	DM	MitM DoS	✓	MATLAB, IEC 61850	–	Detecting and dealing with cyberattacks in distribution system; maintaining resiliency during outage

Legend: CStd = Cybersecurity standard, ATT = Attack, EU = Encryption and authentication, FC = Focus, CSCAD = Cybersecurity-Related Challenges Addressed, DT = Detection, CO = Coordination, MT = Mitigation, ID = Identification, BLK = Blockchain, MitM = Man in the middle, FDI = False data injection, DoS = Deniel of service, • = Unclear, ✓ = Applied.

4. Cybersecurity Threats in AC Microgrids

4.1. Cyber Threats in Power Systems

With increased dependence on communication in microgrid management, cyberattacks pose a more prevalent threat. Most studies provide simulations, models, or taxonomies rather than empirical breakdowns of the various types of cyberattacks. As a result, precise counts or percentage shares of malware, DDoS, and FDIA, among others, across real-world power systems and microgrids are not systematically reported in the reviewed literature. Nonetheless, in the literature, it has been found that power systems, such as microgrids, can be subjected to multiple identified attack strategies that exploit their vulnerabilities.

A significant vulnerability of cyber-physical systems like microgrids and smart grids stems from the heavy dependency on communication infrastructure for real-time monitoring, protection, and operational interoperability, which makes these systems prone to attacks that manipulate data or disrupt physical systems [28]. Key vulnerable components include intelligent electronic devices (IEDs) such as sensors, gateways, smart meters, and smart relays, which attackers can infiltrate with malware or access physically/remotely to send false commands, statuses, or inaccurate measurements [28,71]. The inherent interdependence of cyber and physical systems means that cyberattacks can trigger physical damage, such as component overloading, fires, and blackouts, and physical disruptions can, in turn, create new cyber vulnerabilities. These interconnected vulnerabilities pave the way for sophisticated coordinated cyber-physical attacks, such as coordinated FDI against communication links or Electrical Internet of Things (EIoT) botnet attacks, which can cause widespread frequency instability, cascading failures, and significant operational and economic impacts across the grid [72].

Building upon the classification framework of [73], this section adopts a taxonomy-driven structure to categorize these threats, aligning them with their attack vectors, targeted layers, and potential impacts on microgrid operations [73], providing a foundation for the following subsections. Figure 3 illustrates the taxonomy of cyberattacks on microgrids with MAS-based adaptive protection.

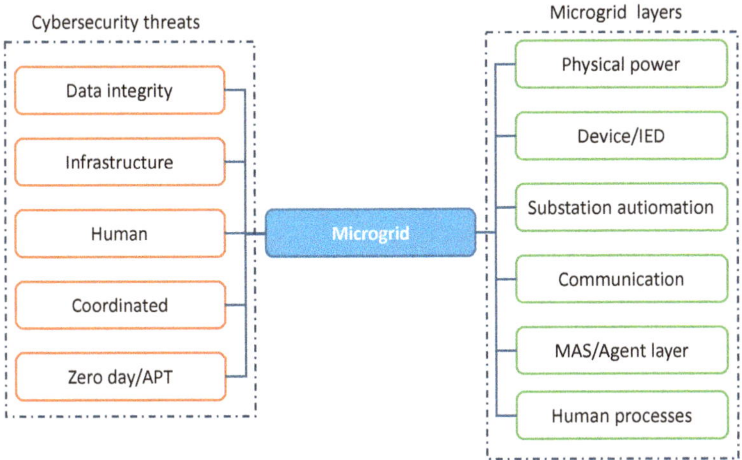

Figure 3. Taxonomy of cyberattacks in microgrids with MAS-based adaptive protection.

4.1.1. Data Integrity Attacks

In data integrity attacks, there is a malicious attempt to alter the data used for critical functions within the power system, such as false data injection (FDI) and data manipulation attacks. These attacks attempt to inject misleading data into the system, or modify it, which can corrupt the decision-making processes of control systems. FDI can lead to incorrect operational responses, potentially causing instability in power delivery and system failures [60,74].

4.1.2. Infrastructure and Communication Attacks

Infrastructure attacks try to exploit weaknesses in the hardware, software, and communication infrastructure connecting digital measuring devices and communication protocols, which can lead to unauthorized access and control over the power systems [15,75], including central control and monitoring systems, and often take the form of denial-of-service (DoS) attacks. In DoS attacks, the aim is to disrupt the availability of services by overwhelming the communication channels or control systems. This can prevent legitimate users from accessing critical services, leading to operational disruptions in power systems [60,75]. Another form is in which an attacker discreetly intercepts and manipulates communication between two legitimate parties or systems while remaining unnoticed, called a man-in-the-middle (MitM) attack [76]. These cyber threats can also target DERs, compromising their control systems and leading to cascading failures and power outages. These attacks exploit vulnerabilities in the communication networks that connect the DER to the microgrid [15,60].

4.1.3. Human-Focused Threats

Ransomware attacks can target critical infrastructure, such as pipelines and power grids, locking operators out of their systems until a ransom is paid. This type of attack, in which the human factor is involved, can halt operations and cause significant economic losses [16,75]. Additionally, employees or contractors with access to sensitive systems can intentionally or unintentionally cause harm. Insider threats can lead to data breaches or sabotage, making them a significant concern for power system security [75].

4.1.4. Coordinated and Multi-Stage Attacks

Coordinated attacks are simultaneous, sophisticated, and potentially stealthy cyberattacks implemented at multiple nodes in a networked MG system [77]. Coordinated attacks on all nodes present a significant challenge for traditional detection and isolation-based countermeasures [68]. These may come in the form of replicating device readings (replay attack) or constructing an FDI attack vector (optimized attack) to mask a physical attack on the power system, designed to neutralize the physical attack's impact on measurement residuals and go undetected, or a series of coordinated and interconnected attack stages (multi-stage attack), often combining cyber and physical components [73].

4.1.5. Zero-Day and Advanced Persistent Threats (APTs)

Cyber-physical systems are often found to be vulnerable to unseen or unknown threats and exploited by what is called a zero-day attack. This is a type of cyberattack for which knowledge-based detection strategies are ineffective [58]. The primary reason for this ineffectiveness is that such strategies rely on an up-to-date database of each known attack vector. Since a zero-day attack is novel and previously unseen, an existing database would not contain information about its specific attack vector, rendering these traditional detection methods useless.

By means of advanced tactics, stealthy techniques, and persistent focus on specific targets, advanced persistent threats (APT) can infiltrate MASs over time, gathering intelligence and eventually disrupting the coordination and decision-making processes critical for adaptive protection in power systems [76].

4.2. Impact of Cyberattacks on MAS-Based Adaptive Protection in Microgrids

MAS-based schemes rely on communication networks to coordinate and execute protection strategies. Cyberattacks can target these networks to introduce delays or alter data, leading to incorrect fault detection and isolation, thereby compromising the adaptive protection mechanism [39]. Moreover, the interconnected nature of hybrid AC-DC microgrids means that an attack on one sub-grid could have repercussions on the entire system. For instance, an attack affecting the AC sub-grid's frequency stability might harm the DC side's battery voltage stability, leading to system-wide instability [1].

By manipulating data integrity, FDI attacks work by data injections that can manipulate the data exchanged between agents in MASs, leading to false fault detection or mis-operation of protection devices. This can disrupt the adaptive protection strategies designed to respond to real-time grid conditions [60]. Subsequently, this results in voltage and frequency instability, potentially causing cascading failures and power outages in AC microgrids [16,78]. Additionally, denial-of-service (DoS) attacks overwhelm the communication channels used by MASs. These DoS attacks can prevent timely data exchange and decision-making, hindering the system's ability to adapt to changing conditions and effectively protect the grid [79].

Attacks targeting communication networks can delay or block data exchange between components, such as those used in multi-agent systems (MASs), which are crucial for adaptive protection [17]. This disruption can prevent timely decision-making and coordination, affecting the overall stability and reliability of the system. Individuals with access to these MAS components can intentionally disrupt operations by altering configurations or injecting malicious code, thereby affecting the system's adaptive protection capabilities [79,80].

Additionally, MASs in power systems rely on software for developing algorithms for decision-making and coordination. Taking advantage of this fact, cyberattacks targeting software vulnerabilities can compromise agent functionality, leading to incorrect protection actions or failure to respond to faults [77,79].

Table 3 summarizes the impact of cyberattacks on the MAS-based protection of microgrids, categorizing these attacks across system layers and highlighting representative techniques and their objectives.

Table 3. Crosswalk of cyberattacks and their impact on MAS adaptive protection.

Ref.	Attack Type	Target Layer	Example	Impact on MAS
[60,74,78]	Data Integrity (FDI)	MAS/Agent, Comm.	False measurement injection	False fault detection, mis-tripping
[60,75,79]	DoS	Comm. Network, MAS	Flooding IEC 61850 GOOSE	Delayed/failed adaptive response
[76,80]	MitM	Comm. Network	Altered relay settings	Hidden miscoordination
[75,80]	Insider Threat	MAS Software, Ops.	Config. sabotage	Loss of coordination
[58,76,77]	APT/Zero-Day	All layers	Persistent infiltration	Long-term disruption of MAS

5. Existing Solutions

MGs' increasing integration of digital control systems, communication networks, and DERs makes them highly susceptible to various cyber threats. Existing solutions, encompassing prevention, detection, isolation, and resilient control, are essential for the effective management and secure operation of microgrids. The most noteworthy strategies found in the literature are presented in the subsequent subsections.

5.1. *Cybersecurity Standards in Power Systems*

Embracing structured cybersecurity standards is at the forefront of a secured microgrid, providing a general framework for developing, deploying, and managing cybersecurity best practices fitted to the unique needs of microgrids [81,82]. On this note, several standards and frameworks are highlighted in the literature as being relevant to microgrid cybersecurity.

The NERC CIP standards provide a foundational and broad framework for the overall cybersecurity and resilience of the bulk power system against various forms of cyber threats and attacks by focusing on robust security practices and incident management [76]. The NERC CIP standards, while mandatory for bulk electric system entities, are often criticized for their compliance-heavy focus, extensive documentation requirements, and limited agility in adapting to emerging threats [12,83].

IEC 62351 is a primary security-related international standard that specifies measures for ensuring end-to-end security in power systems, enhancing IEC 61850 and IEC 60870 communication standards. It is applicable in defending against breaches in confidentiality, system availability, data integrity attacks, unauthorized access, and network vulnerabilities [84]. The IEC 62443 [85] framework focuses on risk assessment and mitigation, providing guidelines covering security levels and system requirements crucial for maintaining operational continuity in smart grids [81,86]. The primary challenge identified with IEC 62443 relates to justifying its adoption when other standards are already in place [83].

ISO/IEC 27000 [87] outlines the requirements for establishing, employing, and maintaining an information security management system (ISMS) [81,88]. This standard is particularly relevant for ensuring security during potential data integrity attacks. However, these standards are often IT-centric, requiring significant tailoring to accommodate the real-time, safety-critical constraints of OT environments [83,88].

The NIST standard provides technical guidance to develop effective wide cybersecurity strategies in the United States (U.S.), including for Smart Grid applications regarding information systems, SCADA and industrial automation systems, IoT devices, and cryptographic security with the NIST SP 800 Series [76]. Additionally, NISTIR 7628 provides high-level requirements for cybersecurity and definitions of the logical interfaces of DER systems. The main challenge with NIST is that its IT-centric security controls, such as automatic account disabling, often cannot be directly applied to Operational Technology (OT) devices due to their lack of inherent automatic functionality, necessitating manual processes or compensating controls.

The IEEE is a leading professional organization dedicated to advancing technology, and it develops numerous standards for power and energy systems. The IEEE 2030 [89] Series covers MG, smart grid (SG), and battery energy storage system (BESS) interoperability and cybersecurity [84,90]. IEEE 1686 [91] defines the cybersecurity capabilities of intelligent electronic devices (IEDs) used in electric power systems [84]. Cybersecurity in DER is managed by IEEE 1547.3 [92], while the security of electric power substations in smart grids is addressed by the IEEE 1402 and IEEE C37.240 standards [93,94]. The challenges associated with IEEE standards for microgrids include fragmented coverage, where existing regional codes may have apparent gaps in technical specifications for interconnection and interoperability with distributed generation, leading to reliance on multiple IEEE standards with varying relevance that can quickly become outdated by emerging control technologies [12]. Additionally, minor regional electrical parameter differences and distinct environmental conditions further complicate the effective implementation of these universal standards, requiring careful adaptation for local microgrid protection and operation [95].

Table 4 presents a summary with the primary focus of each standard, as well as its main limitations in the context of microgrids.

Table 4. Summary of cybersecurity-related standards for power systems.

Ref.	Standard	Focus/Use	Key Limitations
[12]	NERC CIP	Widely used in power utilities to identify cyber vulnerabilities and recommend protections	Manual, costly compliance; not tailored for microgrids or modern communication tech like Wireless Sensor Networks (WSNs)
[96]	IEC 62351	Enhances IEC 61850 with security for communication/data transfer	Optional use; introduces latency; lacks key management; vulnerable to replay, DoS, Simple Network Time Protocol (SNTP) attacks
[88]	ISO/IEC 27000-series	ISMS framework, widely used in information security governance	Limited technical controls; less adaptable to resource-constrained devices; variable implementation
[81,97,98]	NIST Cybersecurity Framework	Flexible framework for critical infrastructure protection	IT centric; voluntary adoption; incomplete mapping to other standards; limited smart grid specificity
[12]	IEEE Standards	Technical protocols and cybersecurity for DER, IEDs, and substation systems	Fragmented coverage; multiple standards with varying relevance to microgrid-scale protection

The implementation of these standards in power systems faces persistent challenges that span regulatory, technical, and operational domains. These challenges are amplified by the complex and evolving nature of standards, the unique requirements of operational technology, and the dynamic threat landscape confronting modern grids [14,83]. Table 5 consolidates the primary barriers identified in recent literature, highlighting issues such as the proliferation of overlapping standards, interoperability limitations, insufficient domain-specific guidance, and the difficulty of integrating legacy infrastructure. By summarizing these challenges, the table provides a structured foundation for understanding where current standards fall short and where future efforts should focus to ensure robust, effective cybersecurity for critical energy systems.

Table 5. Key challenges in implementing cybersecurity standards in power systems.

Challenge	Description
Complexity of Standards	- Proliferation of evolving, and sometimes conflicting, standards across jurisdictions complicates implementation.
Interoperability Issues	- Inconsistent requirements and interpretations hinder seamless integration of multi-vendor systems.
Domain-Specific Gaps	- Generic guidelines often fail to address the unique operational constraints of OT environments.
Resource Constraints	- Limited technical expertise, funding, and time restrict comprehensive deployment, particularly in smaller utilities.
Compliance vs. Effectiveness	- Emphasis on regulatory compliance may overshadow the pursuit of genuine security robustness.
Evolving Threat Landscape	- The rapid evolution of attack techniques frequently outpaces the update cycles of standards.
Testing and Validation	- Absence of standardized testbeds and validation methodologies limits assurance of effectiveness.
Legacy System Integration	- Inability of aging infrastructure to meet modern security requirements leads to partial or insecure adaptations.

5.2. Encryption in MAS-Based Microgrids

Cryptography is one of the most popular and widely used security mechanisms, with a history dating back to the history of written language itself. The approach is to develop mathematical methods for encoding information into ciphers to protect it from unauthorized access, which could be potentially hostile. Therefore, encryption of sensitive data is one of the first and most effective steps toward countering threats such as MitM and FDI attacks, unauthorized access to malicious agents, agent log modification, and provenance attacks. In this last one, a mobile agent's itinerary details are disclosed to an adversary [58].

In multi-agent-based microgrids, researchers have explored a range of cryptographic and trust-enhancing mechanisms to secure agent-to-agent communications and ensure privacy-preserving consensus and optimization. Current approaches commonly integrate cryptographic primitives (e.g., homomorphic encryption, secret sharing, symmetric/asymmetric encryption), programmable cryptographic controllers, and emerging technologies such as distributed ledgers and quantum key distribution (QKD). While these methods demonstrate effectiveness in laboratory testbeds and simulation environments, they exhibit significant variations in computational cost, latency, and deployment assumptions.

The researchers in [99] propose a cryptography-based programmable (crypto-control) method, such as the Dynamic Encrypted Weight Addition (DEWA) scheme, which combines partial homomorphic encryption with secret sharing to enable secure distributed control. This method has already been validated in RTDS simulator testbeds that integrate software-defined network (SDN) and IoT components. However, this approach would require further work to adapt DEWA to other communication-based controls. Similarly, fully homomorphic encryption has been used to solve distributed optimal power flow and energy management systems (EMS) problems, showing that privacy-preserving optimization is feasible, although the computational burden remains significant [100]. Additionally, partial homomorphic encryption paired with event-triggered communication has also proven promising, as it reduces unnecessary data exchanges while preserving the confidentiality of local agent information [101,102]. Other scholars have applied lightweight elliptic-curve-based homomorphic hybrids to improve efficiency further, for example, in AMI data falsification detection, offering a more practical alternative to computationally heavy Cheon–Kim–Kim–Song (CKKS) schemes [103]. Despite these advances, homomorphic encryption imposes significant latency and computational overheads on the low-power controllers typically used in distributed energy resources [6].

The studies summarized in Table 6 emphasize that advanced cryptographic primitives can impose prohibitive computational burdens on resource-constrained smart grid devices, underscoring the need for careful architectural design and optimization [104]. Therefore, the various approaches in existing studies demonstrate technical feasibility, but reveal gaps in scalability, standardization, and operational guidelines.

Table 6. Performance impact, threats, and key management.

Ref.	Encryption Type	Performance	MAS Use Cases	Notes and Mitigations
[104]	Symmetric authenticated encryption	Low latency, suitable for fast control	Frequent control and telemetry	Use for time-critical channels; combine with mutual authentication and rolling keys
[100–102]	Homomorphic encryption (partial/fully)	High CPU utilization and latency on-device	Privacy-preserving aggregation, distributed optimal power flow (OPF)	Useful for sums/optimization; mitigate by offloading, event-triggering, or encrypting only aggregates

Mohamed et al. [105] use a private proof-of-work blockchain to safeguard control and measurement data against denial-of-service (DoS) attacks while employing H-infinity controllers to mitigate uncertainty. Awais et al. [106] propose a blockchain-secured peer-to-peer energy marketplace enhanced with shielded execution environments to prevent data tampering by third parties. Sharma and Sarojwal [107] extend this to adaptive smart contracts and identity-based microgrid schemes, utilizing hashing and blockchain to enhance point-to-point (P2P) trading against DDoS attacks. These approaches collectively highlight blockchain's capacity to ensure integrity and secure transactions, though scalability, latency, and the absence of AI-driven detection remain open challenges.

5.3. Authentication

Authentication is the process of recognizing an agent's identity in the MAS by associating an incoming request with a set of identifying credentials. Since a request may originate on a remote host and may traverse several machines and network channels that are secured in different ways (and are not equally trusted), it is non-trivial to authenticate the source of communication in a distributed system [58]. In the literature, multiple techniques have been identified. Table 7 below presents a summary.

Table 7. Summary of authentication solutions.

Ref.	Technique	Authentication	Description/Use	Drawbacks
[108,109]	Public Key Infrastructure (PKI)	Digital Certificates	Authenticates agents and ensures message integrity and non-repudiation	Poor scalability and high management and maintenance costs
[110,111]		JSON Web Token (JWT)	Used with PKI for stateless, secure communication, and frequent remote calls	Highly centralized is limiting for MG
[112,113]	One-Pass Authentication	Lightweight Mechanisms	Reduces communication and computational overhead, especially in large agent systems	Challenges in maintaining user anonymity and session key secrecy Limitations in balancing efficiency, performance, and security
[114,115]	Group Key Management	Group Re-key Protocols	Ensures that all agents share the same encryption key for secure group communication	Most are insecure and susceptible to MitM and impersonation attacks, or are not suitable for autonomous MG
[104,116]	Blockchain + ZKP	-	Audit, settlements, reputation, market transactions; off-chain techniques and permissioned ledgers reduce overhead	High storage and computation, throughput limits; best for non-real-time market functions

Various researchers have expressed that traditional identity authentication based on PKI digital certificates has poor scalability and high management and maintenance costs, making it unsuitable for large-scale power end-side devices [111,117]. Furthermore, there are inherent security issues with certificate authorities that manage digital certificates, which can be a significant drawback in complex power systems in terms of managing the lifecycle of certificates, such as issuance, distribution, and revocation [110,117].

JWT-based authentication solutions are highly centralized, which poses challenges in meeting legal obligations on privacy. This centralization can be a significant limitation in decentralized environments like microgrids [110,111,118]. Other scientific articles identified that blockchain-based authentication mechanisms that use JWTs face issues such as not fully protecting prosumer privacy, susceptibility to multiple security attacks, and high computational and communication resource demands [119].

Lightweight authentication mechanisms, while reducing computational costs, may still face challenges in maintaining user anonymity and session key secrecy. For example, the SE-LAKAF framework did not initially preserve user anonymity and session key confidentiality, which are critical for secure communications [113]. Furthermore, lightweight group authentication methods need to balance efficiency with security, and there are limitations in overseeing the performance and security requirements of smart grid applications [120]. Subsequently, group re-key protocols must efficiently manage the complexity of key distribution and re-keying processes.

Permissioned blockchains with zero-know proof (ZKP) and blockchain-backed trust management schemes have been shown to provide accountability, auditability, and agent credibility assessments in MAS-based energy markets [116]. Additionally, quantum key distribution (QKD) authentication schemes have been prototyped for SCADA/machine-to-machine (M2M) communication channels in DER environments, demonstrating their feasibility for quantum-secured authentication [121]. However, these strategies impose high storage and computation requirements and are not well suited for real-time performance functionality [104,116]. Despite these advances, most of the schemes are insecure and susceptible to MitM and impersonation attacks or are not suitable for autonomous MGs [115].

Mezquita et al. [122] present a multi-agent architecture where producers and consumers trade through blockchain-backed smart contracts that ensure non-repudiation. Cheng and Chow [123] propose a Bayesian reputation metric that helps distributed agents evaluate one another's trustworthiness during consensus, complementing cryptographic methods without requiring blockchain ledgering. Rath et al. [74] offer a blockchain-based peer-to-peer validation layer that authenticates nodes, paired with anomaly detection and recovery. These studies reveal how blockchain and probabilistic trust metrics can enhance MAS authentication, though comprehensive identity management at scale is still an unresolved bottleneck. Consequently, the proposed methods need to ensure that they do not introduce significant computational overhead or additional security vulnerabilities [117,120].

5.4. Distributed Detection, Mitigation, and Prevention

Multi-agent systems (MASs) are particularly effective in distributed environments due to their ability to operate autonomously and collaboratively. This is crucial for detecting and mitigating cyber threats in real time across large networks [124–126]. Consequently, agent-based architecture offers capabilities such as autonomy, reactivity, proactivity, and mobility, which are desirable for intrusion detection systems (IDSs) [125]. These features enable the system to adapt to new threats dynamically and efficiently. Accordingly, a hybrid approach that combines different detection methodologies, such as signature-based, anomaly-based, and stateful protocol analysis, has been found to enhance the detection capabilities of IDPSs in MAS environments [127,128]. This hybrid approach helps in identifying both known and unknown threats.

In this context, the integration of machine learning algorithms, such as convolutional and recurrent neural networks, improves accuracy and reduces the false favorable rates of IDPSs [129,130]. These algorithms can analyze large datasets and detect patterns indicative of cyber threats. Other researchers have established a decentralized multi-agent reinforcement learning (MARL) scheme that is used to enhance the detection process by allowing agents to learn from their environment and improve their detection strategies over time [131,132]. Such a case would be the work of Roy et al. [62], which can potentially provide a solution for unexpected threats such as zero-day attacks.

MAS-based IDPSs can process and analyze data in real time, ensuring timely detection and response to cyber threats. This is achieved through continuous monitoring and the use of advanced data processing techniques [126,133]. Some authors argue that the use of MASs enables scalable solutions that can effectively manage the vast amounts of data generated in large networks. This is particularly important in environments like cloud computing and big data networks [132,134]. Wang et al. [68] present the ability to maintain frequency/voltage regulation and real/reactive power sharing under both single and multiple node attacks, including coordinated node attacks on all control agents, which are difficult for detection and isolation methods to manage. Subsequently, MAS-based IDPSs can provide proactive responses to detected threats, such as automatic incident response and endpoint quarantine, thereby mitigating the impact of cyberattacks [128]. Bougueroua et al. [124] argue that the high messaging rates for collaboration are the primary performance bottleneck in multi-agent IDPS.

Zhang et al. [135] propose an event-triggered distributed detection and recovery method using stationarity and consistency features for attack identification, complemented by neural networks that reconstruct normal states and controls in DC microgrids. Instead of building a blockchain-enabled framework that validates transactions, Rath et al. [74] detect compromised nodes through physics-informed analysis and use predictive controllers for multi-hop recovery. Cheng and Chow [123] advance this line by incorporating a Bayesian reputation metric to enhance distributed trust and expose stealthy attacks in a real-time testbed—meanwhile, Zhang et al. [136] model false data injection within load-sharing protocols and analyze stability regions under attack. Together, these works demonstrate the promise of AI and MAS-based distributed detection, though they often lack integration with blockchain consensus or broader end-to-end resilience frameworks.

5.5. Resilience Strategies

The decentralized coordination feature of MAS-based adaptive protection, combined with cybersecurity best practices, is crucial for enhancing resilience in microgrids. For instance, in the event of a fault, agents negotiate to determine the optimal protection strategy, considering factors such as fault currents and communication latencies [137]. This decentralized approach ensures that the system remains functional even if some components fail due to cyberattacks or other sources, thereby improving resilience [36,52]. However, these approaches require real-time verification and are computation intensive.

Another approach to building resilience is the integration of protection strategies for both pre- and post-contingency conditions [63]. As mentioned by the authors, during pre-contingency, relays autonomously adapt their settings to maintain coordination with other relays. In post-contingency scenarios, the microgrid central controller (MGCC) plays a crucial role in adapting protection settings to manage subsequent faults. This unified approach would ensure that the system is prepared for potential cascading failures, thereby enhancing overall resilience.

Moreover, to address the challenges of communication failures, researchers have proposed minimal communication protection schemes [138]. These schemes utilize voltage and current measurements to detect faults and isolate only the faulted feeder, thereby reducing reliance on extensive communication networks. This approach not only improves resilience but also minimizes power outages by limiting the impact of faults caused by communication issues.

Another development is the integration of AI and machine learning (ML) with MAS to enhance fault detection, classification, and coordination [52]. Subsequently, deep learning algorithms can be trained to detect faults with high accuracy and determine the optimal protection strategy, as presented by [48]. Additionally, recurrent neural networks (RNNs) have been used to estimate state variables and detect faults in real time [139]. Federated reinforcement learning (Fed-RL) has been proposed to improve the resilience of networked microgrids. This approach allows multiple agents to learn optimal control policies while preserving data privacy. The learned policies are then transferred to hardware-in-the-loop testbeds, bridging the gap between simulation and real-world implementation [140]. This method is particularly effective in handling model complexities and unknown dynamic behaviors of inverter-based resources. Therefore, the combination of AI and MAS enables the system to adapt to various operational modes and improve resilience.

The literature shows that energy storage devices, such as supercapacitors and lithium-ion batteries, have been integrated into MAS-based protection systems to enhance resilience. These devices can provide the necessary fault current during islanded mode or communication failures, ensuring that circuit breakers can trip correctly [141,142]. This approach would not incur additional costs, as energy storage devices are already part of the microgrid infrastructure.

Hierarchical protection strategies leveraging MASs have been developed to address dynamic operational variations in microgrids. These strategies integrate a dual-tier system, where higher tiers make operational decisions and lower tiers oversee event analysis and relay configuration updates [143]. This hierarchical approach ensures swift and autonomous protection coordination, even during prolonged adjustments of the main relay. State-observer-based protection schemes have been proposed to enhance fault detection and classification. These schemes use particle filters to estimate the measured current and voltage signals, enabling the detection of both high- and low-impedance faults with high accuracy [139]. The use of state observers ensures that the system can operate effectively in both grid-connected and islanded modes.

A three-stage emergency approach has been developed to improve the resilience of hybrid networked microgrids. This approach involves offline analysis to determine the resilient operation zone, real-time monitoring of the operating point, and the implementation of corrective countermeasures during contingencies [144]. This method ensures that the microgrid operates within a secure zone, preventing unnecessary tripping of the DER.

MAS-based fault localization and restoration methods have been proposed to improve resilience. These methods use phase angle comparison of current signals to detect faults and isolate them without requiring voltage transformers or relays [145]. Additionally, power restoration processes are implemented to restore the microgrid to its normal state after the fault is cleared.

Ahmad et al. [146] propose a blockchain-integrated distributed energy resources management system (DERMS) that substitutes for central control during outages, thus ensuring DER coordination. Mohamed et al. [105] couple blockchain replication with H-infinity control to maintain frequency stability under DoS, while Babahajiani and Zhang [147] introduce push-sum synchronization and quantum-inspired methods to strengthen distributed control against communication failures and cyber threats. Zhang et al. [135] also contribute here by pairing distributed anomaly detection with neural network recovery to restore microgrid operations. Collectively, these works highlight creative pathways toward resilient MAS-based microgrids, but often lack a unified integration of blockchain, AI, and MAS within a single pipeline.

Figure 4 provides a summary of the various resiliency and redundancy strategies found in recent relevant literature. Here, each of the nine blocks represents a strategy, along with the features listed under each one. The blocks, as depicted in the literature, are identified as follows: MAS coordination [36,137], Pre- and post-contingency [36], Minimum communication [138], Integration of AI and ML [52], BESS [141,142], Hierarchical protection [143], State observer-based schemes [139], Three-stage emergency approach [144], and Localization and restoration with MAS [145].

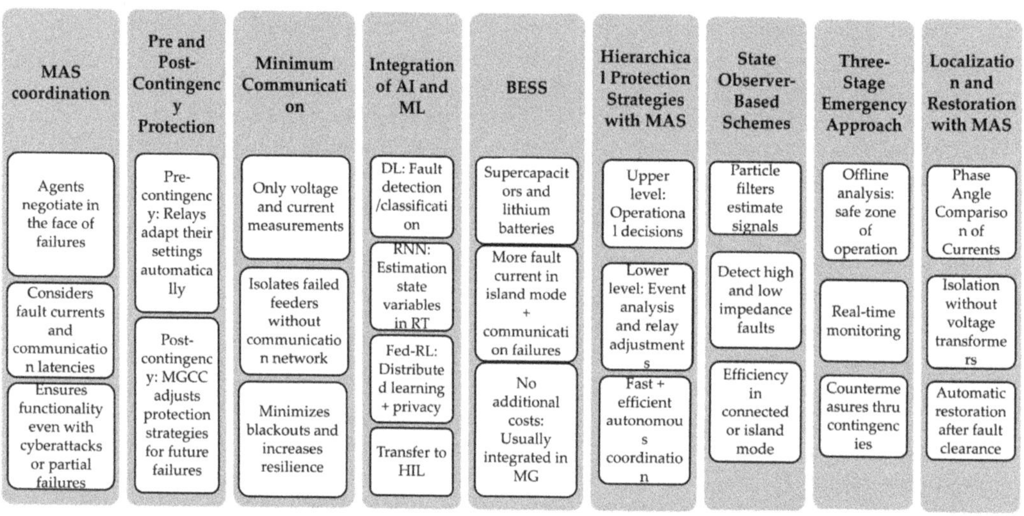

Figure 4. Resiliency strategies.

5.6. Scalability and Performance

Scaling cybersecurity measures in microgrids within the MAS presents several challenges, particularly in maintaining system performance. The integration of DER and advanced communication technologies increases system complexity and uncertainties, making microgrid management more challenging [148,149]. Moreover, the need for real-time responsiveness to cyberattacks further complicates the technical landscape, requiring innovative frameworks to manage disturbances without complete recalculation of power flow equations [150].

Ensuring optimal control performance while addressing cybersecurity threats is a challenging task. For instance, the informatics artificial neural network (I-ANN) designed for DERs in weak microgrids in [151] aims to enhance robustness and damping while mitigating rapid fluctuations in voltage and frequency. However, replacing conventional proportional integral (PI) controllers with I-ANN introduces new complexities in system optimization.

Maintaining real-time adaptability and computational efficiency is crucial for effective operation. The proposed frameworks must optimize operational points, including resource generation and network reconfiguration, while considering technical, economic, and reliability parameters [150]. This requires balancing the need for immediate updates with the computational load. Khatana et al. [59] propose an attack detection solution with a low computational and communication footprint for attack detection. This would be very much in line with the highly distributed resources of MGs.

Multi-agent systems must coordinate effectively to manage controller components within individual microgrids and neighboring grids. This coordination is essential for maintaining system performance during cyberattacks [152,153]. Abianeh et al. [69] propose a distributed, multi-agent malicious DER detection and isolation scheme that offers modular integration and is less impacted by communication delays compared to centralized methods. Nonetheless, it does require extensive data and high-performance computation.

Ensuring scalability and resilience in large-scale systems is a significant challenge. Innovative methods, like the considerable change sensitivity (LCS) method and hierarchical distributed control systems, are proposed to enhance real-time adaptability and resilience [150,153]. Al-dulaimi et al. [63] provide an approach that reduces the need for constant control updates and adapts to limited bandwidth. However, due to the need for communication and coordination among numerous agents, it has become increasingly complex.

Babahajiani and Zhang [147] explore scalable synchronization protocols, such as push-sum consensus, which remain robust in unbalanced networks, while Mezquita et al. [122] demonstrate how MAS negotiation and blockchain settlement can reduce transaction costs while increasing throughput and posing challenges in identity management. Mohamed [154] broaden the view with a review of AI and blockchain in renewables, highlighting the computational and environmental costs associated with scaling these technologies. These contributions underscore that while MAS, blockchain, and AI offer compelling security and control benefits, performance constraints and integration trade-offs remain central to future research.

Table 8 summarizes the main challenges of scaling cybersecurity measures in multi-agent microgrids, along with corresponding approaches proposed in the literature. The table highlights the trade-offs between system complexity, control performance, real-time adaptability, coordination, and scalability.

Table 8. Summary of the scalability and performance challenges of MAS-based MG cybersecurity.

Ref.	Challenge Area	Key Issues	Proposed Approaches	Limitations/Trade-Offs
[148–150]	System Complexity	DER integration and advanced communication technologies increase uncertainties and complicated management.	Innovative frameworks to manage disturbances without recalculating full power flow equations.	Added complexity in microgrid operations.
[151]	Control vs. Cybersecurity	Difficult to ensure optimal control performance while addressing cyber threats.	I-ANN enhances robustness and damping, mitigating rapid voltage/frequency fluctuations.	Replacement of PI controllers with I-ANN introduces new optimization complexities.
[59,150]	Real-Time Adaptability and Efficiency	Balancing immediate updates with computational efficiency.	Frameworks optimizing generation/reconfiguration under technical, economic, and reliability constraints; low-footprint attack detection.	High trade-offs between updating speed and computational load.
[69,152,153]	Multi-Agent Coordination	Agents must coordinate within and across microgrids during cyberattacks.	Distributed malicious DER detection and isolation, modular and less delay-sensitive than centralized methods.	Requires extensive data and high-performance computing.
[63,150,153]	Scalability and Resilience	Large-scale adaptability and resilience remain challenging.	Large change sensitivity (LCS) method, hierarchical distributed control, bandwidth-efficient solutions.	Increased complexity in communication and coordination with scale.

6. Discussion

A review of the tables and figures shows that research on MAS-based adaptive protection for microgrids has clearly matured in terms of technical sophistication, yet the attention given to cybersecurity is far less consistent. In fact, the treatment of security often feels partial or secondary. A few patterns stand out quite firmly: adaptive protection itself has advanced more quickly than its secure implementation, most studies lean heavily toward detection rather than full resilience, and many of the same structural issues—scalability, interoperability, and real-time performance—continue to surface.

One prominent example is provided in Table 1, where most schemes aim to improve the speed of fault detection, isolation, and selectivity. That is significant progress, but what is evident is how rarely these same studies consider encryption, authentication, or standards compliance. The result is a curious mismatch: technically advanced controllers that, in practice, remain vulnerable to even basic cyber threats.

A second pattern is evident in Table 2. Much of the cybersecurity research focuses on two well-known categories of attack: false data injection (FDI) and denial-of-service (DoS) attacks. While these are serious risks, the narrow focus is problematic. Real microgrids are likely to face more complex conditions, including coordinated or long-term intrusions. Some recent work attempts to address this by combining detection and mitigation within the same framework, which is a welcome move in the direction of resilience rather than relying solely on anomaly flagging.

The trade-offs involved are most evident in Figures 2 and 4, as well as in the synthesis presented in Table 8. Pushing more intelligence to local nodes or reducing communication needs can indeed improve responsiveness; however, this often comes with a heavier computational burden and greater design complexity. On the other hand, hierarchical or distributed consensus methods can scale well and improve resilience, but they introduce additional latency and communication overhead. In short, each solution solves one problem while creating another, and this balancing act is at the heart of MAS-based cybersecurity. Metrics, therefore, need to go beyond detection accuracy and consider delays, communication footprint, and the system's ability to remain stable under less-than-ideal conditions.

Finally, Table 4 highlights the underutilization of existing standards, including NERC CIP, IEC 62351, ISO/IEC 27000, and several IEEE guidelines. The reasons are not surprising: implementation is costly, many standards were designed for IT rather than OT environments, and DER controllers are often too resource-constrained to support the cryptographic load. Standards provide an essential starting point, but they cannot be taken as ready-made solutions for microgrids.

Taken together, these results suggest a field that is both creative and fast-paced, yet also fragmented. Researchers are exploring diverse approaches, yet there is little sign of convergence on shared practices, datasets, or benchmarks. The tables and figures provide a valuable overview of current progress, but they also highlight the significant work still needed before MAS-based cybersecurity for microgrids becomes a practical reality.

7. Potential Future Directions

Future research could focus on developing security mechanisms and architectures that adhere to established industry standards. This would make them more practical and trustworthy for real-world applications. Validating cybersecurity strategies, especially those using distributed control or multi-agent systems (MASs) in cyber-physical testbeds that realistically emulate power system communication networks and protocols (such as IEC 61850 GOOSE/MMS or DNP3), could provide valuable insights. Evaluating how these strategies impact performance and their effectiveness in realistic protocol environments is essential. Designing solutions in alignment with critical infrastructure standards, such as IEC 62443 or NERC CIP (where applicable), and evaluating their compliance and interoperability should also be a priority.

Another critical area is the integration of lightweight, real-time cryptographic techniques into microgrid communication and control systems. These methods must account for the computational limitations of devices while still providing robust protection against data manipulation and eavesdropping. Researchers should aim to move beyond isolated detection or mitigation tools and develop comprehensive security frameworks. These frameworks would ideally cover intrusion prevention, detection, identification, mitigation, and recovery and be validated in complex, multi-domain cyber-physical testbeds with realistic attack scenarios and diverse grid operations. At the same time, authentication and access control methods tailored to the unique operational technology (OT) environment of microgrids are needed to prevent unauthorized access and control actions.

There is also an opportunity to focus on secure-by-design MAS-based adaptive protection frameworks for microgrids. This involves embedding cryptographic protocols and trust management directly into the communication and decision-making processes of distributed agents. Achieving this requires developing novel MAS architectures and control/security algorithms that include lightweight encryption, strong authentication, and secure key management, all of which are designed for resource-constrained power system devices.

Research could investigate how machine learning (ML) and AI-based detection and mitigation systems can be effectively integrated with other cybersecurity functions, such as secure configuration, vulnerability scanning, and automated incident response. This integration is crucial for MAS-based adaptive protection in microgrids, where reliable and safe communication is critical.

8. Conclusions

A significant gap exists in developing and validating microgrid cybersecurity solutions that explicitly adhere to or implement relevant cybersecurity standards (e.g., NIST, IEC 62443, IEC 62351 beyond NSM). Furthermore, the absence of applied encryption, digital signatures, and secure key management is a critical gap. These techniques are crucial for ensuring the confidentiality, integrity, and authenticity of data and control signals exchanged within microgrids, particularly in distributed architectures that rely on communication networks. While some studies address secure communication channels, there is a lack of applied research on robust authentication mechanisms, especially those suitable for diverse and potentially resource-constrained microgrid components and agents (e.g., behavioral authentication, mutual authentication protocols). Although an MAS is used for distributed control, the applied research does not explicitly detail how communication between agents is secured using techniques like encryption or digital signatures, or how trust among agents is cryptographically established and maintained. While detection and mitigation are active areas of research, a comprehensive approach encompassing identification, protection, detection, response, and recovery, as outlined in frameworks like NIST, is less evident in the applied techniques.

The lack of standard security practices was a limiting factor when developing this literature review; therefore, a great deal of insight could have been derived regarding the performance of the research articles discussed under close-to-real-world conditions. A summarized overview and a critical analysis of the existing cybersecurity landscape for MAS-based adaptive protection in microgrids were presented. Furthermore, this review identified multiple research gaps and various future research directions.

Author Contributions: Conceptualization, A.J.T.C. and M.A.-M.; methodology, A.J.T.C., M.A.-M. and J.C.H.; writing—original draft preparation, A.J.T.C. and M.A.-M.; writing—review and editing, D.M.-H., C.G.C.-G., J.C.H. and F.A.-R.; supervision, J.C.H., L.H.-C. and M.A.-M. All authors have read and agreed to the published version of the manuscript.

Funding: This research was funded by Fondocyt Grant No. FONDOCYT-2023-1-3C1-0547 and FONDOCYT-2023-1-1C3-0732 in the Dominican Republic.

Data Availability Statement: Not applicable.

Acknowledgments: The authors acknowledge the support provided by the Thematic Network 723RT0150, "Red para la integración a gran escala de energías renovables en sistemas eléctricos (RIBIERSE-CYTED)". The authors acknowledge the support provided by the Sustainability Energy Center, ECE Department, Research & Development building, University of Puerto Rico at Mayagüez.

Conflicts of Interest: The authors declare that they have no conflicts of interest.

Abbreviations

The following abbreviations are used in this manuscript:

AC	Alternating Current
ACL	Agent Communication Language
AES	Advanced Encryption Standard
AGC	Automatic Generation Control
AI	Artificial Intelligence
APT	Advanced Persistent Threat
AsyE	Asymmetric Encryption
BESS	Battery Energy Storage System
BIC	Bidirectional Interlinking Converter
BLK	Blockchain
CB	Circuit Breaker
CHIL	Controller Hardware-in-the-Loop
CIP	Critical Infrastructure Protection
CKKS	Cheon–Kim–Kim–Song

DAS	Distribution Automation System
DC	Direct Current
DCDM	Decentralized Consensus Decision-Making
DER	Distributed Energy Resource
DES	Data Encryption Standard
DG	Distributed Generation
DL	Deep Learning
DM	Detection And Mitigation
DNP3	Distributed Network Protocol 3
DoS	Denial-of-Service
DT	Detection
ECC	Elliptic Curve Cryptography
EIoT	Electrical Internet of Things
ELK	Elasticsearch, Logstash, and Kibana
EMS	Energy Management System
ESS	Energy Storage System
FDI	False Data Injection
Fed-RL	Federated Reinforcement Learning
FIPA	Foundation for Intelligent Physical Agent
FNR	False Negative Rate
FPR	False Positive Rate
GOOSE	Generic Object-Oriented Substation Event
HIL	Hardware-in-the-Loop
HMI	Human–Machine Interface
HVDC	High-Voltage Direct Current
I-ANN	Informatics Artificial Neural Network
ICT	Information and Communication Technology
IDPS	Intrusion Detection Protection System
IDS	Intrusion Detection System
IED	Intelligent Electronic Device
IoT	Internet of Things
IP	Internet Protocol
IPS	Interface Protection System
ISE	Integral Squared Error
ISMS	Information Security Management System
JADE	Java Agent Development Framework
JWT	Json Web Token
LAN	Local Area Network
LC	Inductance Capacitance
LCS	Large Change Sensitivity
M2M	Machine-to-Machine
MA	Master Agent
MAC	Media Access Control
MARL	Multi-Agent Reinforcement Learning
MAS	Multi-Agent System
MG	Microgrid
MGCC	Microgrid Central Controller
MHDM	MAS-Based Detection and Mitigation
MitM	Man-in-the-Middle
ML	Machine Learning
MMG	Multi-Microgrid
MT	Mitigation
NERC	North American Electric Reliability Corporation

NSM	Network and System Management
OCC	One-Class Classifier
OPF	Optimal Power Flow
OT	Operational Technology
PCC	Point of Common Coupling
PCS	Paillier Cryptosystem
PF	Performance Function
PI	Proportional-Integral
PKI	Public Key Infrastructure
RI	Resilience Index
RL	Reinforcement Learning
RNN	Recurrent Neural Network
RSA	Rivest–Shamir–Adleman
RT	Real-Time
RTAC	Real-Time Automation Controller
SA	Slave Agent
SCADA	Supervisory Control and Data Acquisition
SG	Smart Grid
SMV	Sampled Measured Value
SNTP	Simple Network Time Protocol
SPoF	Single Point of Failure
SVR	Support Vector Regression
SyE	Symmetric Encryption
TADR	True Attack Detection Rate
TCP/IP	Transport Control Protocol/Internet Protocol
TNR	True-Negative Rate
TPR	True-Positive Rate
UDP	User Datagram Protocol
U.S.	United States
VLAN	Virtual Local Area Network
W-MSR	Weighted Mean Subsequence Reduced
WMSR	Weighted Mean Subsequence Reduced
WSN	Wireless Sensor Network

References

1. Taher, M.A.; Tariq, M.; Sarwat, A.I. Enhancing Security in Islanded AC Microgrid: Detecting and Mitigating FDI Attacks in Secondary Consensus Control through AI-Based Method. In Proceedings of the 2023 IEEE International Conference on Energy Technologies for Future Grids, ETFG 2023, Wollongong, Australia, 3–6 December 2023. [CrossRef]
2. Rahmoune, F.; Ait Aali, N.; El Bouzekri El Idrissi, Y. Enhancing Green Energy Security: Exploring Multi-Agent Systems Perspectives for a Secure Smart Grid Information System. *Procedia Comput. Sci.* **2024**, *236*, 485–492. [CrossRef]
3. Patarroyo-Montenegro, J.F.; Andrade, F.; Guerrero, J.M.; Vasquez, J.C. A Linear Quadratic Regulator with Optimal Reference Tracking for Three-Phase Inverter-Based Islanded Microgrids. *IEEE Trans. Power Electron.* **2021**, *36*, 7112–7122. [CrossRef]
4. Patarroyo-Montenegro, J.F.; Vasquez-Plaza, J.D.; Andrade, F. A State-Space Model of an Inverter-Based Microgrid for Multivariable Feedback Control Analysis and Design. *Energies* **2020**, *13*, 3279. [CrossRef]
5. Satpathy, P.R.; Ramachandaramurthy, V.K.; Padmanaban, S. Advanced Protection Technologies for Microgrids: Evolution, Challenges, and Future Trends. *Energy Strategy Rev.* **2025**, *58*, 101670. [CrossRef]
6. Do Nascimento, L.L.; Rolim, J.G. Multi-Agent System for Adaptive Protection in Microgrids. In Proceedings of the 2013 IEEE PES Conference on Innovative Smart Grid Technologies, ISGT LA 2013, Washington, DC, USA, 24–27 February 2013.
7. Khazaei, A.A.; Mahmoudi, A. Decentralized Adaptive Protection Structure for Microgrids Based on Multi–Agent Systems. In Proceedings of the 2019 Iranian Conference on Renewable Energy & Distributed Generation (ICREDG), Tehran, Iran, 11–12 June 2019; pp. 1–8.
8. Senarathna, T.S.S.; Udayanga Hemapala, K.T.M. Review of Adaptive Protection Methods for Microgrids. *AIMS Energy* **2019**, *7*, 557–578. [CrossRef]

9. *IEC 61850-8-1:2011+AMD1:2020 CSV*; Communication Networks and Systems for Power Utility Automation—Part 8-1: Specific Communication Service Mapping (SCSM). International Electrotechnical Commission: Geneva, Switzerland, 2020.
10. Abbaspour, E.; Fani, B.; Sadeghkhani, I.; Alhelou, H.H. Multi-Agent System-Based Hierarchical Protection Scheme for Distribution Networks with High Penetration of Electronically-Coupled DGs. *IEEE Access* **2021**, *9*, 102998–103018. [CrossRef]
11. Gutierrez-Rojas, D.; Nardelli, P.H.J.; Mendes, G.; Popovski, P. Review of the State of the Art on Adaptive Protection for Microgrids Based on Communications. *IEEE Trans. Industr Inform.* **2021**, *17*, 1539–1552. [CrossRef]
12. Alvarez-Alvarado, M.S.; Apolo-Tinoco, C.; Ramirez-Prado, M.J.; Alban-Chacón, F.E.; Pico, N.; Aviles-Cedeno, J.; Recalde, A.A.; Moncayo-Rea, F.; Velasquez, W.; Rengifo, J. Cyber-Physical Power Systems: A Comprehensive Review about Technologies Drivers, Standards, and Future Perspectives. *Comput. Electr. Eng.* **2024**, *116*, 109149. [CrossRef]
13. Shobole, A.A.; Abafogi, M. Adaptive Protection in Smart Distribution Networks: Coordination Demonstration of Multi-Agent Systems. In Proceedings of the 2023 5th Global Power, Energy and Communication Conference (GPECOM), Cappadocia, Turkiye, 14–16 June 2023; pp. 483–488.
14. North American Electric Reliability Corporation. *2025 ERO Reliability Risk Priorities Report*; North American Electric Reliability Corporation (NERC): Atlanta, GA, USA; Washington, DC, USA, 2025. Available online: https://www.nerc.com/comm/RISC/Related%20Files%20DL/2025_RISC_ERO_Priorities_Report.pdf (accessed on 25 August 2025).
15. Gurina, L.; Zoryna, T. Distributed Energy: Benefits of Use and Threats to Cybersecurity. In Proceedings of the 2023 International Conference on Industrial Engineering, Applications and Manufacturing, ICIEAM 2023, Sochi, Russia, 15–19 May 2023; pp. 87–92. [CrossRef]
16. Bidram, A.; Poudel, B.; Damodaran, L.; Fierro, R.; Guerrero, J.M. Resilient and Cybersecure Distributed Control of Inverter-Based Islanded Microgrids. *IEEE Trans. Industr Inform.* **2020**, *16*, 3881–3894. [CrossRef]
17. Zhou, Q.; Shahidehpour, M.; Alabdulwahab, A.; Abusorrah, A.; Che, L.; Liu, X. Cross-Layer Distributed Control Strategy for Cyber Resilient Microgrids. *IEEE Trans. Smart Grid* **2021**, *12*, 3705–3717. [CrossRef]
18. Gehbauer, C.; Black, D.R. Secured Microgrid Operation for Military Facilities—Findings from a Crowd Hacking Event. In Proceedings of the 2024 IEEE Power and Energy Society Innovative Smart Grid Technologies Conference, ISGT 2024, Washington, DC, USA, 19–22 February 2024. [CrossRef]
19. Anand, A.; Nagu, B.; Gudur, K. Transition Control and Operation of Hybrid Energy System. In Proceedings of the 2024 IEEE International Conference on Smart Power Control and Renewable Energy, ICSPCRE 2024, Rourkela, India, 19–21 July 2024.
20. Rouhani, A.; Abasi, M.; Joorabian, M.; Davatgaran, V. Designing a Novel Optimal Energy Management to Determine the Contribution and Effect of Primary Sources on the Structure and Components Size of Smart Microgrids and Smart Buildings. In Proceedings of the 2024 9th International Conference on Technology and Energy Management, ICTEM 2024, Behshar, Iran, 14–15 February 2024.
21. Guarnieri, M.; Bovo, A.; Giovannelli, A.; Mattavelli, P. A Real Multitechnology Microgrid in Venice: A Design Review. *IEEE Ind. Electron. Mag.* **2018**, *12*, 19–31. [CrossRef]
22. Chae, W.-K.; Won, J.-N.; Lee, H.-J.; Kim, J.-E.; Kim, J. Comparative Analysis of Voltage Control in Battery Power Converters for Inverter-Based AC Microgrids. *Energies* **2016**, *9*, 596. [CrossRef]
23. Eyimaya, S.E.; Altin, N. *Microgrids: Definitions, Architecture, and Control Strategies*; Academic Press: Cambridge, MA, USA, 2023.
24. Sinha, R.R.; Kanwar, N. *Hybrid Microgrids: Architecture, Modeling, Limitations, and Solutions*; Academic Press: Cambridge, MA, USA, 2023.
25. Weber, L.; Nasiri, A.; Hyypio, D.; Dittman, W. Modeling and Control of a Synchronous Generator in an AC Microgrid Environment. In Proceedings of the ECCE 2016—IEEE Energy Conversion Congress and Exposition, Milwaukee, WI, USA, 18–22 September 2016.
26. Udoha, E.; Das, S.; Abusara, M. A Power Management System for Interconnected Ac Islanded Microgrids Using Back-To-Back Converter. *IET Conf. Proc.* **2023**, *2023*, 177–183. [CrossRef]
27. De la Cruz, J.; Wu, Y.; Candelo-Becerra, J.E.; Vásquez, J.C.; Guerrero, J.M. Review of Networked Microgrid Protection: Architectures, Challenges, Solutions, and Future Trends. *CSEE J. Power Energy Syst.* **2024**, *10*, 448–467. [CrossRef]
28. Rouhani, S.H.; Su, C.L.; Mobayen, S.; Razmjooy, N.; Elsisi, M. Cyber Resilience in Renewable Microgrids: A Review of Standards, Challenges, and Solutions. *Energy* **2024**, *309*, 133081. [CrossRef]
29. Hussain, N.; Nasir, M.; Vasquez, J.C.; Guerrero, J.M. Recent Developments and Challenges on AC Microgrids Fault Detection and Protection Systems–A Review. *Energies* **2020**, *13*, 2149. [CrossRef]
30. Vegunta, S.C.; Higginson, M.J.; Kenarangui, Y.E.; Li, G.T.; Zabel, D.W.; Tasdighi, M.; Shadman, A. AC Microgrid Protection System Design Challenges—A Practical Experience. *Energies* **2021**, *14*, 2016. [CrossRef]
31. Gaggero, G.B.; Rossi, M.; Girdinio, P.; Marchese, M. Cybersecurity Issues in Communication-Based Electrical Protections. In Proceedings of the International Conference on Electrical, Computer, and Energy Technologies, ICECET 2022, Prague, Czech Republic, 20–22 July 2022. [CrossRef]

32. Uzair, M.; Li, L.; Eskandari, M.; Hossain, J.; Zhu, J.G. Challenges, Advances and Future Trends in AC Microgrid Protection: With a Focus on Intelligent Learning Methods. *Renew. Sustain. Energy Rev.* **2023**, *178*, 113228. [CrossRef]
33. Kaur, G.; Prakash, A.; Rao, K.U. A Critical Review of Microgrid Adaptive Protection Techniques with Distributed Generation. *Renew. Energy Focus* **2021**, *39*, 99–109. [CrossRef]
34. Shobole, A.A.; Wadi, M. Multiagent Systems Application for the Smart Grid Protection. *Renew. Sustain. Energy Rev.* **2021**, *149*, 111352. [CrossRef]
35. Patnaik, B.; Mishra, M.; Bansal, R.C.; Jena, R.K. AC Microgrid Protection—A Review: Current and Future Prospective. *Appl. Energy* **2020**, *271*, 115210. [CrossRef]
36. dos Reis, F.B.; Pinto, J.O.C.P.; dos Reis, F.S.; Issicaba, D.; Rolim, J.G. Multi-Agent Dual Strategy Based Adaptive Protection for Microgrids. *Sustain. Energy Grids Netw.* **2021**, *27*, 100501. [CrossRef]
37. Khalid, H.; Shobole, A. Existing Developments in Adaptive Smart Grid Protection: A Review. *Electr. Power Syst. Res.* **2021**, *191*, 106901. [CrossRef]
38. Fawzy, N.; Habib, H.F.; Mohammed, O.; Brahma, S. Protection of Microgrids with Distributed Generation Based on Multiagent System. In Proceedings of the 2020 IEEE International Conference on Environment and Electrical Engineering and 2020 IEEE Industrial and Commercial Power Systems Europe (EEEIC/I&CPS Europe), Madrid, Spain, 9–12 June 2020; pp. 1–5.
39. Albarakati, J.A.; Azeroual, M.; Boujoudar, Y.; EL Iysaouy, L.; Aljarbouh, A.; Tassaddiq, A.; EL Markhi, H. Multi-Agent-Based Fault Location and Cyber-Attack Detection in Distribution System. *Energies* **2023**, *16*, 224. [CrossRef]
40. Aazami, R.; Esmaeilbeigi, S.; Valizadeh, M.; Javadi, M.S. Novel Intelligent Multi-Agents System for Hybrid Adaptive Protection of Micro-Grid. *Sustain. Energy Grids Netw.* **2022**, *30*, 100682. [CrossRef]
41. De La Cruz, J.; Vasquez, J.C.; Guerrero, J.M.; Luna, E.G.; Candelo-Becerra, J.E. Adaptive Multi-Agent-Zonal Protection Scheme for AC Microgrids. In Proceedings of the 2023 25th European Conference on Power Electronics and Applications (EPE'23 ECCE Europe), Aalborg, Denmark, 4–8 September 2023; pp. 1–9.
42. Ataei, M.A.; Gitizadeh, M. A Distributed Adaptive Protection Scheme Based on Multi-agent System for Distribution Networks in the Presence of Distributed Generations. *IET Gener. Transm. Distrib.* **2022**, *16*, 1521–1540. [CrossRef]
43. Memon, A.A.; Kauhaniemi, K. Real-Time Hardware-in-the-Loop Testing of IEC 61850 GOOSE-Based Logically Selective Adaptive Protection of AC Microgrid. *IEEE Access* **2021**, *9*, 154612–154639. [CrossRef]
44. Satuyeva, B.; Sultankulov, B.; Nunna, H.S.V.S.K.; Kalakova, A.; Doolla, S. Q-Learning Based Protection Scheme for Microgrid Using Multi-Agent System. In Proceedings of the 2019 International Conference on Smart Energy Systems and Technologies (SEST), Porto, Portugal, 9–11 September 2019; pp. 1–6.
45. Rahman, M.S.; Isherwood, N.; Oo, A.M.T. Multi-Agent Based Coordinated Protection Systems for Distribution Feeder Fault Diagnosis and Reconfiguration. *Int. J. Electr. Power Energy Syst.* **2018**, *97*, 106–119. [CrossRef]
46. Tripathi, J.M.; Yadav, N.; Mallik, S.K.; Chandel, A. A Multi-Agent Approach for Protection Coordination in a Microgrid Using JADE Platform. In Proceedings of the 2025 IEEE 1st International Conference on Smart and Sustainable Developments in Electrical Engineering, SSDEE 2025, Dhanbad, India, 28 February–2 March 2025; pp. 1–6. [CrossRef]
47. Alzahrani, S.; Sinjari, K.; Mitra, J. Multi-Agent and State Observer-Based Technique for Microgrid Protection. *IEEE Trans. Ind. Appl.* **2024**, *60*, 2697–2705. [CrossRef]
48. Najar, A.; Kazemi Karegar, H.; Esmaeilbeigi, S. Multi-agent Protection Scheme for Microgrid Using Deep Learning. *IET Renew. Power Gener.* **2024**, *18*, 663–678. [CrossRef]
49. Dizioli, F.A.S.; Barra, P.H.A.; Menezes, T.S.; Lacerda, V.A.; Coury, D.V.; Fernandes, R.A.S. Multi-Agent System-Based Microgrid Protection Using Angular Variation: An Embedded Approach. *Electr. Power Syst. Res.* **2023**, *220*, 109324. [CrossRef]
50. *IEC 60870-5-104:2006/AMD1:2016/COR1:2023*; Telecontrol Equipment and Systems—Part 5-104: Transmission Protocols—Network Access for IEC 60870-5-101 Using Standard Transport Profiles (Corrigendum 1 to Amendment 1). International Electrotechnical Commission: Geneva, Switzerland, 2023.
51. Abbaspour, E.; Fani, B.; Karami-Horestani, A. Adaptive Scheme Protecting Renewable-Dominated Micro-Grids against Usual Topology-Change Events. *IET Renew. Power Gener.* **2021**, *15*, 2686–2698. [CrossRef]
52. Uzair, M.; Li, L.; Zhu, J.G.; Eskandari, M. A Protection Scheme for AC Microgrids Based on Multi-Agent System Combined with Machine Learning. In Proceedings of the 2019 29th Australasian Universities Power Engineering Conference (AUPEC), Nadi, Fiji, 26–29 November 2019; pp. 1–6.
53. Abbaspour, E.; Fani, B.; Heydarian-Forushani, E. A Bi-Level Multi Agent Based Protection Scheme for Distribution Networks with Distributed Generation. *Int. J. Electr. Power Energy Syst.* **2019**, *112*, 209–220. [CrossRef]
54. Faria, I.M.; Furlan, R.H.; Martins, P.E.T.; Menezes, T.S.; Oleskovicz, M.; Coury, D.V. The Proposition of a Multiagent System for Adaptive Protection of a Distribution System. In Proceedings of the 2018 Simposio Brasileiro de Sistemas Eletricos (SBSE), Niteroi, Brazil, 12–16 May 2018; pp. 1–6.

55. Daryani, M.J.; Karkevandi, A.E. Decentralized Cooperative Protection Strategy for Smart Distribution Grid Using Multi-Agent System. In Proceedings of the 2018 6th International Istanbul Smart Grids and Cities Congress and Fair (ICSG), Istanbul, Turkey, 25–26 April 2018; pp. 134–138.
56. Daryani, M.J.; Karkevandi, A.E.; Usta, O. Multi-Agent Approach to Wide-Area Integrated Adaptive Protection System of Microgrid for Pre- and Post-Contingency Conditions. In Proceedings of the 2018 IEEE PES Innovative Smart Grid Technologies Conference Europe (ISGT-Europe), Sarajevo, Bosnia and Herzegovina, 21–25 October 2018; pp. 1–6.
57. Pullaguram, D.; Sahoo, S. Cyber Security Threats in Multi-Agent Microgrids. In *Cyber Security for Microgrids*; Institution of Engineering and Technology: Stevenage, UK, 2022; pp. 105–125.
58. Owoputi, R.; Ray, S. Security of Multi-Agent Cyber-Physical Systems: A Survey. *IEEE Access* **2022**, *10*, 121465–121479. [CrossRef]
59. Khatana, V.; Chakraborty, S.; Saraswat, G.; Patel, S.; Salapaka, M.V. A Distributed Malicious Agent Detection Scheme for Resilient Power Apportioning in Microgrids. In Proceedings of the IECON 2024—50th Annual Conference of the IEEE Industrial Electronics Society, Chicago, IL, USA, 3–6 November 2024; pp. 1–7. [CrossRef]
60. Karanfil, M.; Rebbah, D.E.; Debbabi, M.; Kassouf, M.; Ghafouri, M.; Youssef, E.N.S.; Hanna, A. Detection of Microgrid Cyberattacks Using Network and System Management. *IEEE Trans. Smart Grid* **2023**, *14*, 2390–2405. [CrossRef]
61. *IEC 62351-7:2017*; Power Systems Management and Associated Information Exchange—Data and Communications Security—Part 7: Network and System Management (NSM) Data Object Models. International Electrotechnical Commission: Geneva, Switzerland, 2017.
62. Roy, S.D.; Debbarma, S.; Guerrero, J.M. Machine Learning Based Multi-Agent System for Detecting and Neutralizing Unseen Cyber-Attacks in AGC and HVDC Systems. *IEEE J. Emerg. Sel. Top. Circuits Syst.* **2022**, *12*, 182–193. [CrossRef]
63. Al-dulaimi, F.N.S.; Kurnaz, S. Enhancing Cybersecurity and Frequency Control Efficiency in AC Islanded Microgrids: A Distributed Approach with a Stochastic Models. In Proceedings of the 2024 8th International Conference on Green Energy and Applications (ICGEA), Singapore, 14–16 March 2024. [CrossRef]
64. Hu, B.; Zhou, C.; Tian, Y.-C.; Hu, X.; Junping, X. Decentralized Consensus Decision-Making for Cybersecurity Protection in Multimicrogrid Systems. *IEEE Trans. Syst. Man. Cybern. Syst.* **2021**, *51*, 2187–2198. [CrossRef]
65. Zhou, T.L.; Xiahou, K.S.; Zhang, L.L.; Wu, Q.H. Multi-Agent-Based Hierarchical Detection and Mitigation of Cyber Attacks in Power Systems. *Int. J. Electr. Power Energy Syst.* **2021**, *125*, 106516. [CrossRef]
66. Choi, I.-S.; Hong, J.; Kim, T.-W. Multi-Agent Based Cyber Attack Detection and Mitigation for Distribution Automation System. *IEEE Access* **2020**, *8*, 183495–183504. [CrossRef]
67. Boakye-Boateng, K.; Ghorbani, A.A.; Lashkari, A.H. Securing Substations with Trust, Risk Posture, and Multi-Agent Systems: A Comprehensive Approach. In Proceedings of the 2023 20th Annual International Conference on Privacy, Security and Trust, PST 2023, Copenhagen, Denmark, 21–23 August 2023. [CrossRef]
68. Wang, Y.; Mondal, S.; Deng, C.; Satpathi, K.; Xu, Y.; Dasgupta, S. Cyber-Resilient Cooperative Control of Bidirectional Interlinking Converters in Networked AC/DC Microgrids. *IEEE Trans. Ind. Electron.* **2021**, *68*, 9707–9718. [CrossRef]
69. Abianeh, A.J.; Wan, Y.; Ferdowsi, F.; Mijatovic, N.; Dragicevic, T. Vulnerability Identification and Remediation of FDI Attacks in Islanded DC Microgrids Using Multiagent Reinforcement Learning. *IEEE Trans. Power Electron.* **2022**, *37*, 6359–6370. [CrossRef]
70. Stout, W.M.S. Toward a Multi-Agent System Architecture for Insight & Cybersecurity in Cyber-Physical Networks. In Proceedings of the 2018 International Carnahan Conference on Security Technology (ICCST), Montreal, QC, Canada, 22–25 October 2018; pp. 1–5.
71. Ogbogu, C.E.; Thornburg, J.; Okozi, S.O. Smart Grid Fault Mitigation and Cybersecurity with Wide-Area Measurement Systems: A Review. *Energies* **2025**, *18*, 994. [CrossRef]
72. Liu, M.; Teng, F.; Zhang, Z.; Ge, P.; Sun, M.; Deng, R.; Cheng, P.; Chen, J. Enhancing Cyber-Resiliency of DER-Based Smart Grid: A Survey. *IEEE Trans. Smart Grid* **2024**, *15*, 4998–5030. [CrossRef]
73. Jimada-Ojuolape, B.; Teh, J.; Lai, C.M. Securing the Grid: A Comprehensive Analysis of Cybersecurity Challenges in PMU-Based Cyber-Physical Power Networks. *Electr. Power Syst. Res.* **2024**, *233*, 110509. [CrossRef]
74. Rath, S.; Nguyen, L.D.; Sahoo, S.; Popovski, P. Self-Healing Secure Blockchain Framework in Microgrids. *IEEE Trans. Smart Grid* **2023**, *14*, 4729–4740. [CrossRef]
75. Rath, S.; Pal, D.; Sharma, P.S.; Panigrahi, B.K. A Cyber-Secure Distributed Control Architecture for Autonomous AC Microgrid. *IEEE Syst. J.* **2020**, *15*, 3324–3335. [CrossRef]
76. Abdelkader, S.; Amissah, J.; Kinga, S.; Mugerwa, G.; Emmanuel, E.; Mansour, D.E.A.; Bajaj, M.; Blazek, V.; Prokop, L. Securing Modern Power Systems: Implementing Comprehensive Strategies to Enhance Resilience and Reliability against Cyber-Attacks. *Results Eng.* **2024**, *23*, 102647. [CrossRef]
77. Yaacoub, J.P.; Noura, H.; Azar, J.; Salman, O.; Chahine, K. Cybersecurity in Smart Renewable Energy Systems. In Proceedings of the 2024 International Wireless Communications and Mobile Computing (IWCMC), Ayia Napa, Cyprus, 27–31 May 2024; pp. 1534–1540.

78. Nair, P.S.; Mandal, S.K.; Sharma, N. Securing Smart Microgrids: A Cybersecurity Survey. In Proceedings of the 2023 International Conference on Power Energy, Environment & Intelligent Control (PEEIC), Greater Noida, India, 19–23 December 2023; pp. 1318–1322.
79. Rath, S.; Das, T.; Sengupta, S. Improvise, Adapt, Overcome: Dynamic Resiliency Against Unknown Attack Vectors in Microgrid Cybersecurity Games. *IEEE Trans. Smart Grid* **2024**, *15*, 4245–4258. [CrossRef]
80. Tripathy, M.; Niyogi, R.; Kumar, P.S.; Kumbhar, G.B.; Singh, R.; Thakur, V. A Novel Approach for Detection of Cyber Attacks in Microgrid SCADA System. In Proceedings of the 2023 IEEE 3rd International Conference on Sustainable Energy and Future Electric Transportation (SEFET), Bhubaneswar, India, 9–12 August 2023; pp. 1–6.
81. Abraham, D.; Toftegaard, Ø.; Retnam, B.B.J.D.; Gebremedhin, A.; Yayilgan, S.Y. Consequence Simulation of Cyber Attacks on Key Smart Grid Business Cases. *Front. Energy Res.* **2024**, *12*, 1395954. [CrossRef]
82. Khalaf, M.; Ayad, A.; Tushar, M.H.K.; Kassouf, M.; Kundur, D. A Survey on Cyber-Physical Security of Active Distribution Networks in Smart Grids. *IEEE Access* **2024**, *12*, 29414–29444. [CrossRef]
83. Anderson, J.; Bougie, J.; Dood, M.; Falk, H.; Formea, J.; Haveron, S.; Holstein, D.; Lacroix, M.; Laughner, T.; Mix, S.; et al. Task Force on Utility & Municipality Challenges on Analyzing and Implementing Cybersecurity Standards and Best Practices. Available online: https://resourcecenter.ieee-pes.org/publications/technical-reports/pes_tr_tr116_psccc_100824 (accessed on 9 September 2025).
84. Hasan, M.K.; Abdulkadir, R.A.; Islam, S.; Gadekallu, T.R.; Safie, N. A Review on Machine Learning Techniques for Secured Cyber-Physical Systems in Smart Grid Networks. *Energy Rep.* **2024**, *11*, 1268–1290. [CrossRef]
85. *IEC 62443-2-4:2015*; Security for Industrial Automation and Control Systems—Part 2-4: Security Program Requirements for IACS Service Providers. International Electrotechnical Commission: Geneva, Switzerland, 2015.
86. Téglásy, B.Z.; Gran, B.A.; Katsikas, S.; Gkioulos, V.; Lundteigen, M.A. Clarification of the Cybersecurity and Functional Safety Interrelationship in Industrial Control Systems: Barrier Concepts and Essential Functions. In Proceedings of the 30th European Safety and Reliability Conference and the 15th Probabilistic Safety Assessment and Management Conference, Venice, Italy, 1–5 November 2020; pp. 1980–1987.
87. *ISO/IEC 27000:2018*; Information Technology—Security Techniques—Information Security Management Systems—Overview and Vocabulary. International Organization for Standardization/International Electrotechnical Commission: Geneva, Switzerland, 2018.
88. Culot, G.; Nassimbeni, G.; Podrecca, M.; Sartor, M. The ISO/IEC 27001 Information Security Management Standard: Literature Review and Theory-Based Research Agenda. *TQM J.* **2021**, *33*, 76–105. [CrossRef]
89. *IEEE 2030-2011*; IEEE Guide for Smart Grid Interoperability of Energy Technology and Information Technology Operation with the Electric Power System (EPS), and End-Use Applications, and Loads. Institute of Electrical and Electronics Engineers (IEEE): New York, NY, USA, 2011.
90. Hasan, M.K.; Habib, A.A.; Shukur, Z.; Ibrahim, F.; Islam, S.; Razzaque, M.A. Review on Cyber-Physical and Cyber-Security System in Smart Grid: Standards, Protocols, Constraints, and Recommendations. *J. Netw. Comput. Appl.* **2023**, *209*, 103540. [CrossRef]
91. *IEEE 1686-2022*; Standard for Intelligent Electronic Devices Cybersecurity Capabilities. IEEE (Institute of Electrical and Electronics Engineers): New York, NY, USA, 2022.
92. *IEEE 1547.3-2007*; Guide for Monitoring, Information Exchange, and Control of Distributed Resources Interconnected with Electric Power Systems. IEEE: New York, NY, USA, 2007.
93. *IEEE 1402-2021*; Guide for Physical Security of Electric Power Substations. IEEE: New York, NY, USA, 2021.
94. *IEEE C37.240-2014*; Standard Cybersecurity Requirements for Substation Automation, Protection, and Control Systems. IEEE: New York, NY, USA, 2014.
95. Alsafran, A.S. A Feasibility Study of Implementing IEEE 1547 and IEEE 2030 Standards for Microgrid in the Kingdom of Saudi Arabia. *Energies* **2023**, *16*, 1777. [CrossRef]
96. Kim, Y.-S.; Lee, G.; Kang, J. Lightweight IEC 61850 Secure Communication Module for Microgrids. In *Lecture Notes on Data Engineering and Communications Technologies*; Springer: Cham, Switzerland, 2017; Volume 2, pp. 443–451. [CrossRef]
97. Cusimano, J. Overview of ISA 62443 and Its Relationship to the NIST Framework. In Proceedings of the ISA Process Control and Safety Symposium 2014, PCS 2014, Houston, TX, USA, 7–9 October 2014; pp. 859–872.
98. Goodwin, S. The Need for a Financial Sector Legal Standard to Support the NIST Cybersecurity Framework. In Proceedings of the IEEE Southeast Con, Mobile, AL, USA, 26 March–3 April 2022; pp. 89–95.
99. Wang, L.; Zhang, P.; Tang, Z. Programmable Crypto-Control for Networked Microgrids. In *Microgrids*; Wiley: Hoboken, NJ, USA, 2024; pp. 335–357.
100. Cheng, Z.; Ye, F.; Cao, X.; Chow, M.-Y. A Homomorphic Encryption-Based Private Collaborative Distributed Energy Management System. *IEEE Trans. Smart Grid* **2021**, *12*, 5233–5243. [CrossRef]

101. Zhuo, Q.; Zhang, H.; Hu, X. A Homomorphic Cryptography Based Privacy-Preserving Consensus Algorithm with Event-Triggering in Distributed Energy Management. In Proceedings of the 2021 40th Chinese Control Conference (CCC), Shanghai, China, 26–28 July 2021; pp. 6772–6777.
102. Liu, B.; Wu, J.; Chai, L. Distributed Privacy-Preserving Algorithm for Economic Dispatch and Demand Response of Smart Grid with Homomorphic Encryption. *IEEE Trans. Smart Grid* **2025**, *16*, 173–182. [CrossRef]
103. Joshi, S.; Li, R.; Bhattacharjee, S.; Das, S.K.; Yamana, H. Privacy-Preserving Data Falsification Detection in Smart Grids Using Elliptic Curve Cryptography and Homomorphic Encryption. In Proceedings of the 2022 IEEE International Conference on Smart Computing (SMARTCOMP), Helsinki, Finland, 20–24 June 2022; pp. 229–234.
104. Raso, E.; Bracciale, L.; Gallo, P.; Bernardinetti, G.; Bianchi, G.; Sanseverino, E.R.; Loreti, P. Performance Evaluation of Cryptographic Schemes for Blockchain Security of Smart Grids. In Proceedings of the 2022 Workshop on Blockchain for Renewables Integration (BLORIN), Palermo, Italy, 2–3 September 2022; pp. 113–117.
105. Mohamed, M.O.; Abdelaziz, A.Y.; Abo-Elyousr, F.K. Blockchain-Based Approach for Load Frequency Control of Smart Grids under Denial-of-Service Attacks. *Comput. Electr. Eng.* **2024**, *116*, 109150. [CrossRef]
106. Awais, M.; Abbas, Q.; Tariq, S.; Warraich, S.H. Blockchain Based Secure Energy Marketplace Scheme to Motivate P2P Microgrids. *Int. J. Inform. Commun. Technol. (IJ-ICT)* **2022**, *11*, 177. [CrossRef]
107. Sharma, D.D.; Lin, J.; Sarojwal, A.; Sharma, A.; Sharma, A. Blockchain Based Adaptive Non-Cooperative Game Strategy for Smart Power Contracts. In Proceedings of the 2023 IEEE 8th International Conference for Convergence in Technology (I2CT), Lonavla, India, 7–9 April 2023; pp. 1–6.
108. Sabir, B.E.; Youssfi, M.; Bouattane, O.; Allali, H. Authentication Model Based on JWT and Local PKI for Communication Security in Multi-Agent Systems. In *Learning and Analytics in Intelligent Systems*; Springer: Cham, Switzerland, 2020; Volume 7, pp. 469–479. [CrossRef]
109. Feng, L.; Mei, H. Research and Design of Security in Multi-Agent System. In Proceedings of the IET International Conference on Wireless Mobile and Multimedia Networks Proceedings (ICWMMN 2006), Online, 6–9 November 2006; p. 417. [CrossRef]
110. Jiang, Y.; Du, S.; Xu, W.; Dong, Y. V2G Charging and Discharging Information Authentication Based on Blockchain Technology. In Proceedings of the 2024 IEEE 4th International Conference on Digital Twins and Parallel Intelligence, DTPI 2024, Wuhan, China, 18–20 October 2024; pp. 732–736. [CrossRef]
111. Mahmood, S.; Gohar, M.; Choi, J.G.; Koh, S.J.; Alquhayz, H.; Khan, M. Digital Certificate Verification Scheme for Smart Grid Using Fog Computing (Fonica). *Sustainability* **2021**, *13*, 2549. [CrossRef]
112. Ogunnusi, O.S.; Razak, S.A.; Abdullah, A.H. A Lightweight One-Pass Authentication Mechanism for Agent Communication in Multi-Agent System Based Applications. *J. Teknol.* **2015**, *77*, 1–9. [CrossRef]
113. Mehta, P.J.; Parne, B.L.; Patel, S.J. SE-LAKAF: Security Enhanced Lightweight Authentication and Key Agreement Framework for Smart Grid Network. *Peer Peer Netw. Appl.* **2023**, *16*, 1513–1535. [CrossRef]
114. Zhu, L.; Cao, Y.; Liao, L.; Tan, Y.; Durad, M.H.; Wang, D. Secure Group Communication in Multi-Agent Systems. *WSEAS Trans. Commun.* **2006**, *5*, 781–787.
115. Bolgouras, V.; Ntantogian, C.; Panaousis, E.; Xenakis, C. Distributed Key Management in Microgrids. *IEEE Trans. Industr. Inform.* **2020**, *16*, 2125–2133. [CrossRef]
116. Samuel, O.; Javaid, N.; Khalid, A.; Imrarn, M.; Nasser, N. A Trust Management System for Multi-Agent System in Smart Grids Using Blockchain Technology. In Proceedings of the GLOBECOM 2020—2020 IEEE Global Communications Conference, Taipei, Taiwan, 7–11 December 2020; pp. 1–6.
117. Xu, M.; Qin, Y.; Mei, W.; Lin, C.; Shen, L. Lightweight Access Authentication Management Method for Complex Power End-Side Devices. In Proceedings of the Sixth International Conference on Information Science, Electrical, and Automation Engineering (ISEAE 2024), Wuhan, China, 19–21 April 2024; Volume 13275, p. 57.
118. Boi, B.; Esposito, C. Decentralized Authentication in Microservice Architectures with SSI and DID in Blockchain. In Proceedings of the International Conference on Cloud Computing Technology and Science, CloudCom, Naples, Italy, 4–6 December 2023; pp. 216–223.
119. Pathak, A.; Al-Anbagi, I.; Hamilton, H.J. Privacy-Preserving Authentication Mechanism for P2P Energy Trading in Smart Grid Networks. In Proceedings of the IEEE International Conference on Communications, Denver, CO, USA, 9–13 June 2024; pp. 3085–3090.
120. Daubry, W.; Dricot, J.-M.; Henneaux, P. Decentralized Group Authentication with Membership Verification in Islanded Smart Grids. In Proceedings of the 2023 12th Mediterranean Conference on Embedded Computing (MECO), Budva, Montenegro, 6–10 June 2023; pp. 1–6.
121. Alshowkan, M.; Evans, P.; Starke, M.; Earl, D.; Peters, N. Authentication of Smart Grid Communications Using Quantum Key Distribution. *Sci. Rep.* **2022**, *12*, 12731. [CrossRef]

122. Mezquita, Y.; Gazafroudi, A.S.; Corchado, J.M.; Shafie-Khah, M.; Laaksonen, H.; Kamišalić, A. Multi-Agent Architecture for Peer-to-Peer Electricity Trading Based on Blockchain Technology. In Proceedings of the 2019 XXVII International Conference on Information, Communication and Automation Technologies (ICAT), Sarajevo, Bosnia and Herzegovina, 20–23 October 2019; pp. 1–6.
123. Cheng, Z.; Chow, M.-Y. An Augmented Bayesian Reputation Metric for Trustworthiness Evaluation in Consensus-Based Distributed Microgrid Energy Management Systems with Energy Storage. In Proceedings of the 2020 2nd IEEE International Conference on Industrial Electronics for Sustainable Energy Systems (IESES), Cagliari, Italy, 1–3 September 2020; pp. 215–220.
124. Bougueroua, N.; Mazouzi, S.; Belaoued, M.; Seddari, N.; Derhab, A.; Bouras, A. A Survey on Multi-Agent Based Collaborative Intrusion Detection Systems. *J. Artif. Intell. Soft Comput. Res.* **2021**, *11*, 111–142. [CrossRef]
125. Isaza, G.A.; Castillo, A.G.; Duque, N.D. An Intrusion Detection and Prevention Model Based on Intelligent Multi-Agent Systems, Signatures and Reaction Rules Ontologies. In *Advances in Intelligent and Soft Computing*; Springer: Berlin/Heidelberg, Germany, 2009; Volume 55, pp. 237–245. [CrossRef]
126. Ouiazzane, S.; Addou, M.; Barramou, F. A Multi-Agent Model for Network Intrusion Detection. In Proceedings of the ICSSD 2019—International Conference on Smart Systems and Data Science, Rabat, Morocco, 3–4 October 2019.
127. Mudzingwa, D.; Agrawal, R. A Study of Methodologies Used in Intrusion Detection and Prevention Systems (IDPS). In Proceedings of the IEEE Southeastcon, Orlando, FL, USA, 15–18 March 2012.
128. Devi, V.A.; Bhuvaneswari, E.; Tummala, R.K. Decentralized Hybrid Intrusion Detection System for Cyber Attack Identification Using Machine Learning. In Proceedings of the 2023 International Conference on Data Science, Agents and Artificial Intelligence, ICDSAAI 2023, Chennai, India, 21–23 December 2023.
129. Mezghani, S.; Ktata, F.B. A Distributed Intelligent Agent Based Intrusion Detection System Using Deep Learning Algorithms. In Proceedings of the DTUC '20: Proceedings of the 2nd International Conference on Digital Tools & Uses Congress, Virtual, 7–15 October 2020.
130. Ali, A.; Zia, A.; Razzaque, A.; Shahid, H.; Sheikh, H.T.; Saleem, M.; Yousaf, F.; Muneer, S. Enhancing Cybersecurity with Artificial Neural Networks: A Study on Threat Detection and Mitigation Strategies. In Proceedings of the 2nd International Conference on Cyber Resilience, ICCR 2024, Dubai, United Arab Emirates, 26–28 February 2024.
131. Louati, F.; Barika Ktata, F.; Amous, I. An Intelligent Security System Using Enhanced Anomaly-Based Detection Scheme. *Comput. J.* **2024**, *67*, 2317–2330. [CrossRef]
132. Louati, F.; Ktata, F.B.; Amous, I. Big-IDS: A Decentralized Multi Agent Reinforcement Learning Approach for Distributed Intrusion Detection in Big Data Networks. *Cluster Comput.* **2024**, *27*, 6823–6841. [CrossRef]
133. Tesnim, Y.; Farah, J. A Multi-Agent-Based System for Intrusion Detection. In *Smart Innovation, Systems and Technologies*; Springer: Singapore, 2021; Volume 241, pp. 177–191. [CrossRef]
134. Achbarou, O.; El Kiram, M.A.; Bourkoukou, O.; Elbouanani, S. A Multi-Agent System-Based Distributed Intrusion Detection System for a Cloud Computing. In *Communications in Computer and Information Science*; Springer: Cham, Switzerland, 2018; Volume 929, pp. 98–107. [CrossRef]
135. Zhang, X.; Zhang, Z.; Zhang, R.; Liu, W.; Li, H.; Peng, J. Event-Triggered Resilient Recovery Learning Control Protocol for Interconnected DC Microgrids with Distributed Attack Detection. *Sustain. Energy Grids Netw.* **2024**, *38*, 101364. [CrossRef]
136. Zhang, H.; Meng, W.; Qi, J.; Wang, X.; Zheng, W.X. Distributed Load Sharing Under False Data Injection Attack in an Inverter-Based Microgrid. *IEEE Trans. Ind. Electron.* **2019**, *66*, 1543–1551. [CrossRef]
137. Hosseini, S.A.; Sadeghi, S.H.H.; Nasiri, A. Decentralized Adaptive Protection Coordination Based on Agents Social Activities for Microgrids with Topological and Operational Uncertainties. *IEEE Trans. Ind. Appl.* **2021**, *57*, 702–713. [CrossRef]
138. Samkari, H.S.; Johnson, B.K. Multi-Agent Protection Scheme for Resilient Microgrid Systems with Aggregated Electronically Coupled Distributed Energy Resources. In Proceedings of the IECON 2018—44th Annual Conference of the IEEE Industrial Electronics Society, Washington, DC, USA, 21–23 October 2018; pp. 752–757.
139. Mumtaz, F.; Khan, H.H.; Zafar, A.; Ali, M.U.; Imran, K. A State-Observer-Based Protection Scheme for AC Microgrids with Recurrent Neural Network Assistance. *Energies* **2022**, *15*, 8512. [CrossRef]
140. Mukherjee, S.; Hossain, R.R.; Mohiuddin, S.M.; Liu, Y.; Du, W.; Adetola, V.; Jinsiwale, R.A.; Huang, Q.; Yin, T.; Singhal, A. Resilient Control of Networked Microgrids Using Vertical Federated Reinforcement Learning: Designs and Real-Time Test-Bed Validations. *arXiv* **2023**, arXiv:2311.12264. [CrossRef]
141. Habib, H.F.; Esfahani, M.M.; Mohammed, O. Improvement of Protection Scheme for Microgrids Using Lithium- Ion Battery during Islanding. In Proceedings of the 2018 IEEE Industry Applications Society Annual Meeting (IAS), Portland, OR, USA, 23–27 September 2018; pp. 1–8.
142. Habib, H.F.; Esfahani, M.M.; Mohammed, O.A. Investigation of Protection Strategy for Microgrid System Using Lithium-Ion Battery During Islanding. *IEEE Trans. Ind. Appl.* **2019**, *55*, 3411–3420. [CrossRef]

143. Diaz Caicedo, A.M.; Gómez-Luna, E.; Franco Mejia, E. Revolutionizing Protection Dynamics in Microgrids: Local Validation Environment and a Novel Global Management Control Through Multi-Agent Systems. *Comput. Electr. Eng.* **2024**, *120*, 109748. [CrossRef]
144. Shaker, A.; Bozorg, M.; Safari, A.; Najafi-Ravadanegh, S. An Adaptive Emergency Approach for Hybrid Networked Microgrids Resilience. *IEEE Access* **2022**, *10*, 103164–103175. [CrossRef]
145. Habib, H.F.; Mohammed, O. Decentralized Multi-Agent System for Protection and the Power Restoration Process in Microgrids. In Proceedings of the 2017 Ninth Annual IEEE Green Technologies Conference (GreenTech), Denver, CO, USA, 29–31 March 2017; pp. 358–364.
146. Ahmad, S.; Ahn, B.; Kim, T.; Choi, J.; Chae, M.; Han, D.; Won, D. Blockchain-Integrated Resilient Distributed Energy Resources Management System. In Proceedings of the 2022 IEEE International Conference on Communications, Control, and Computing Technologies for Smart Grids (SmartGridComm), Singapore, 25–28 October 2022; pp. 59–64.
147. Babahajiani, P.; Zhang, P. Cyber-Resilient Distributed Microgrid Control. In *Microgrids*; Wiley: Hoboken, NJ, USA, 2024; pp. 307–334.
148. Chhor, J.; Sourkounis, C. Networked Control Approach for Voltage Regulation with Optimal Reactive Power-Sharing. In Proceedings of the IECON 2018—44th Annual Conference of the IEEE Industrial Electronics Society, Washington, DC, USA, 21–23 October 2018; pp. 225–230.
149. Yan, C.; Han, Y.; Yang, P.; Wang, C. Microgrid Cybersecurity: Addressing Challenges and Ensuring Resilience. In Proceedings of the 2023 IEEE 4th China International Youth Conference on Electrical Engineering, CIYCEE 2023, Chengdu, China, 8–10 December 2023. [CrossRef]
150. Jahromi, M.Z.; Yaghoubi, E.; Yaghoubi, E.; Maghami, M.R.; Chamorro, H.R. An Innovative Real-Time Recursive Framework for Techno-Economical Self-Healing in Large Power Microgrids Against Cyber–Physical Attacks Using Large Change Sensitivity Analysis. *Energies* **2025**, *18*, 190. [CrossRef]
151. Surinkaew, T.; Kerdphol, T. Informatics-Centric Neural Network for Distributed Energy Resources Against Diverse Cyber Threats. *IEEE Trans. Industr Inform.* **2024**, *20*, 14029–14041. [CrossRef]
152. Nguyen, T.L.; Wang, Y.; Nguyen, H.T.; Hoang, T.T. Interconnected Microgrid Systems: Architecture, Hierarchical Control, and Implementation. In *Smart Cyber-Physical Power Systems: Fundamental Concepts, Challenges, and Solutions*; Wiley: Hoboken, NJ, USA, 2025.
153. Luu, N.A.; Nguyen, T.L. Secondary Control for Cyber-Physical Interconnected Microgrid Systems. *Eng. Technol. Appl. Sci. Res.* **2025**, *15*, 21944–21950. [CrossRef]
154. Mohamed, N. Renewable Energy in the Age of AI: Cybersecurity Challenges and Opportunities. In Proceedings of the 2024 15th International Conference on Computing Communication and Networking Technologies (ICCCNT), Kamand, India, 24–28 June 2024; pp. 1–6.

electronics

MDPI

Article

Comprehensive Power Regulation of a Novel Shared Energy Storage Considering Demand-Side Response for Multi-Scenario Bipolar DC Microgrid

Gongqiang Li [1], Bin Zhao [1], Xiaoqiang Ma [1], Xiaofan Ji [2] and Hanqing Yang [3,*]

[1] CHN Energy (Gonghe) Renewable Energy Development Co., Ltd., Hainan Tibetan Autonomous Prefecture 813000, China; 12090165@ceic.com (G.L.); 20087055@ceic.com (B.Z.); 20084521@ceic.com (X.M.)
[2] National Institute of Clean-and-Low-Carbon Energy, Beijing 102211, China; 20024057@ceic.com
[3] School of Automation Engineering, University of Electronic Science and Technology of China, Chengdu 611756, China
* Correspondence: hqyang5517@uestc.edu.cn

Abstract: In order to improve the ability to suppress unbalanced voltage in bipolar DC microgrids, a comprehensive power regulation control of a novel shared energy storage system is proposed for a multi-scenario bipolar DC microgrid. The novel shared energy storage system is composed of an electric spring (ES) with a full-bridge DC/DC converter and non-critical load (NCL) in series, considering demand-side response. The proposed comprehensive power regulation control can enable the bipolar DC microgrid to deal with various scenarios. When operating in stand-alone mode, the unbalanced voltage caused by greater unbalanced power can still be suppressed under the proposed control of the shared energy storage. In case of distributed energy storage (DES) failure on the source side, the shared energy storage can realize DC voltage regulation and maintain system operation by reducing NCL power. In grid-connected operation, the shared energy storage can actively cooperate with the power dispatching of the utility grid for storage reduction of DES on the source side. Thus, the reliability and resilience of the bipolar microgrid have been improved. Finally, to verify the effectiveness of the proposed control strategy, hardware-in-the-loop experimental results are presented in this paper.

Keywords: novel shared energy storage; bipolar DC microgrid; electric spring; power dispatching; demand-side response

1. Introduction

Recently, DC loads such as electronic equipment, electric vehicles, and DC motors, have been widely used in daily life. In addition, due to not suffering from the issues of harmonics, reactive power loss, and phase synchronization, DC microgrids are of great significance to renewable energy consumption, and have received much attention [1–3]. According to the form of power distribution and the number of power supply buses, DC microgrids are classified into unipolar and bipolar. Bipolar DC microgrids with three-wire structures (positive bus, negative bus, and neutral bus) not only inherit the general advantages of unipolar DC microgrids, but also are superior in terms of having different voltage levels, further improving the flexibility of the DC power supply [4–6].

During the operation of a bipolar DC microgrid, unbalanced voltage between the positive and negative poles is a problem worthy of attention. Unbalanced current caused by the power difference will increase the power loss of the network. Moreover, the current

https://doi.org/10.3390/electronics14091866

in the neutral bus will also bring about load voltage deviation, deteriorating the quality of the power supply [7,8]. Therefore, it is necessary to suppress unbalanced voltage to ensure the efficient and reliable operation of bipolar DC microgrids.

In previous studies on bipolar DC microgrids, unbalanced voltage and power regulation have usually been accomplished by adopting a splitting capacitor-based converter with balanced voltage control for distributed energy storage (DES) on the source side [9,10]. Aiming at splitting capacitor-based bipolar DC microgrids, scholars mainly focus on the converter topology and control method. For the unbalanced voltage suppression control method of a traditional three-level converter, Refs. [11–13] studied a three-level buck converter and boost converter, and proposed corresponding control methods for voltage regulation; while [14] focused on model predictive control (MPC) of a three-level bidirectional converter (T-LBC) to realize bidirectional power flow and voltage balance in a bipolar DC microgrid. However, the topologies and control methods mentioned above use DES to achieve power regulation (e.g., suppress unbalanced voltage and power dispatching) of a bipolar DC microgrid only from the perspective of the source side, without considering demand-side response.

In fact, distributed resources such as household energy storage and electric vehicles near the demand side have bidirectional regulation functions, which can flexibly switch roles between being power producers and consumers to achieve clean, low-carbon, safe, and efficient development of power systems [15–17]. In recent years, the concept of shared energy storage has been proposed and rapidly developed. Shared energy storage can effectively reduce operating costs, promote on-site consumption of renewable energy, and achieve power/frequency regulation by utilizing the complementarity of DES among different users; Refs. [18–22] proposed shared energy storage operation management frameworks based on capacity allocation and energy interaction. Based on shared energy storage devices, Ref. [23] proposed an energy capacity trading strategy to minimize the operating costs for microgrids; Ref. [21] studied a day-ahead scheduling model of a regional integrated energy system considering sharing energy storage equipment utilizing cooperative game theory. However, the above studies only involve optimizing the operation and capacity configuration of shared energy storage to achieve long-term economic benefits, with little consideration given to the participation of shared energy storage in short-term power regulation under changeable demand-side response.

Aiming at the control of shared energy storage considering demand-side response, using a DC electric spring (ES) to eliminate voltage imbalance has emerged as a potential method [24]. By dividing the load into critical load (CL) and non-critical load (NCL), a DC ES can take advantage of the loose voltage or power adjustment range of the NCL to obtain flexible regulation ability. Meanwhile the CL and NCL are both in operation. Up till now, DC ESs have been studied in the fields of stabilizing DC bus voltage [25,26], restraining harmonics [27,28], and improving system inertia; Ref. [29] proposed a distributed control method based on MPC for both improving system inertia and reducing voltage deviation; Ref. [30] studied a DC ES and its functionalities of DC bus regulation and fault-ride through support; Ref. [31] utilized a DC ES to reduce the power loss on the line resistance in a DC microgrid. However, the above studies only aimed to improve the performance of DC microgrids in islanded operation.

Motivated by the above, this paper establishes a novel shared energy storage system based on a DC ES configuration with the combination of economic control in grid-connected operation and performance improvement in islanded operation. The proposed comprehensive power regulation method can not only improve the ability of unbalanced voltage suppression and realize voltage support in the case of DES failure, but also adjust power consumption according to the power dispatching command.

2. Analysis of Studied Bipolar DC Microgrid Equipped with a Novel Shared Storage System

2.1. Overview and Operation Principle

Figure 1 shows the configuration of the studied bipolar DC microgrid with three DC buses, namely, the positive bus (P), negative bus (N), and neutral bus (0), which is equipped with a novel shared storage system. A cluster of distributed generators (DGs) is installed on the power supply side. The generated power can be fed with $\pm 0.5\ V_{DC}$ (P−0 and 0-N) and V_{DC}, where V_{DC} is the nominal value of the P-N bus voltage. DES is normally responsible for both the bus voltage regulation and unbalanced voltage suppression through a T-LBC. On the demand side, except for the conventional loads composed of the constant power load (CPL) and resistive load, ESs with the structure of NCLs in series are combined to inform the shared energy storage system for both the positive and negative poles, allowing flexible power consumption. This novel shared energy storage system consisting of an ES and series-connected NCLs can achieve charging and discharging by making the NCLs bear voltage deviations with full consideration of users' demands.

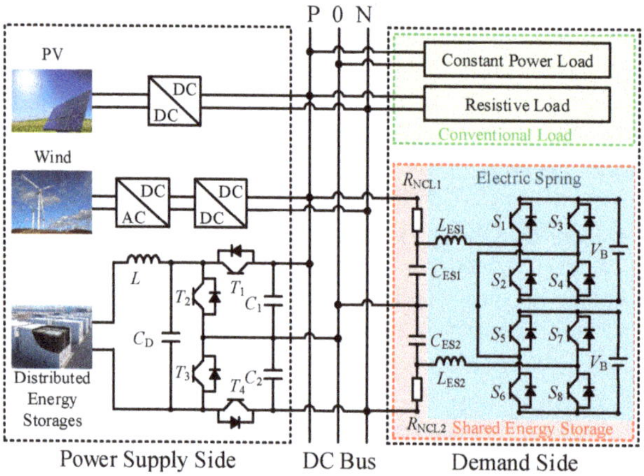

Figure 1. Configuration of studied bipolar DC microgrid with novel shared energy storage.

The bipolar DC microgrid with novel shared energy storage is generally able to operate in stand-alone or grid-connected mode with the disconnection or connection of the utility grid. For the DES control system, a droop control method with voltage balance of the neutral bus is adopted to manage the parallel operations of the DGs, which has been well explored in previous research work. Thus, the proposed control strategy is targeted at the shared energy storage on the demand side, which can adapt to both stand-alone and grid-connected operation modes. Based on the system structure in Figure 1, the proposed comprehensive power-regulation-controlled shared energy storage system can consequently work in three operation modes in response to the specific scenarios illustrated in Figure 2.

(1) Secondary compensation mode: In scenario 1, when DES exists, the shared energy storage is controlled to realize secondary compensation for eliminating the bus voltage deviation caused by the droop characteristic and improving the ability to suppress unbalanced voltage.
(2) Voltage regulation mode: In scenario 2, when DES fails, the shared energy storage is controlled as two DC voltage sources, regulating the bus voltage and maintaining the power balance between the positive and negative poles by adjusting the power consumption of the NCL.

(3) Power dispatching mode: In scenario 3, when the bipolar DC microgrid operates in grid-connected mode, the shared energy storage is controlled as the current source, managing the part of the power flow between the microgrid and the utility grid.

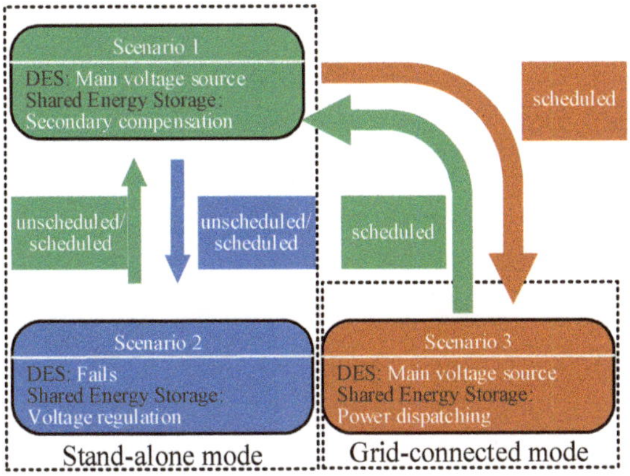

Figure 2. Main control objectives of bipolar DC microgrid mode switch in different scenarios.

2.2. Realization of Secondary Compensation

In order to further analyze the voltage imbalance suppression ability, the bipolar DC microgrid with the novel shared energy storage shown in Figure 1 was simplified, as demonstrated in Figure 3. Since a DG controlled by MPPT can be regarded as a power source with changing power in real time, it can be represented by the negative of the CPL power. Define $P_{1,2} = P_{\mathrm{CPL1,2}} - P_{\mathrm{RES1,2}}$, then $P_{1,2} > 0$ indicates that the power generated by the DGs is less than the power consumed by the loads. On the contrary, $P_{1,2} < 0$ means that the supplied power is more than load power. The subscripts 1 and 2 represent the power between P-0 and 0-N, respectively.

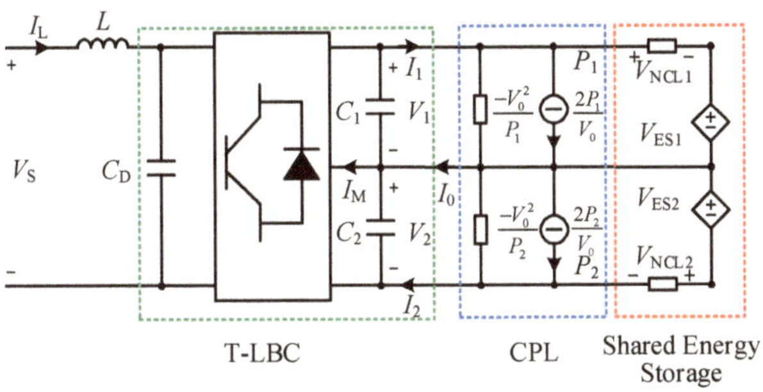

Figure 3. Simplified configuration of studied bipolar DC microgrid with novel shared energy storage.

A CPL can be simulated by the controlled current source and resistance connected in parallel expressed in (1).

$$I_{\mathrm{CPL},i} = 2\frac{P_{\mathrm{CPL},i}}{V_0} - \frac{P_{\mathrm{CPL},i}}{V_0^2}V_{\mathrm{CPL},i} \quad (i = 1,2) \tag{1}$$

where $I_{\mathrm{CPL},i}$ is the current absorbed by CPL_i, $V_{\mathrm{CPL},i}$ is the voltage across CPL_i, $P_{\mathrm{CPL},i}$ is the power consumed by CPL_i, V_0 is the voltage of the normal operation point, and $V_0 = 0.5\ V_{\mathrm{DC}}$.

In a shared energy storage system, an ES with the structure of an NCL in series can be equivalent to a controlled voltage source and resistance connected in series. The relationship of voltage, current, and power is described in (2).

$$\begin{cases} I_{\mathrm{NCL},i} = \frac{V_{\mathrm{NCL},i}}{R_{\mathrm{NCL},i}} \\ V_i = V_{\mathrm{NCL},i} + V_{\mathrm{ES},i} \qquad (i = 1,2) \\ P_{\mathrm{SL},i} = \frac{V_{\mathrm{NCL},i}^2 + V_{\mathrm{NCL},i} V_{\mathrm{ES},i}}{R_{\mathrm{NCL},i}} \end{cases} \tag{2}$$

where $V_{\mathrm{NCL},i}$, $I_{\mathrm{NCL},i}$, and $R_{\mathrm{NCL},i}$ are the voltage, current, and resistance of NCL_i. $P_{\mathrm{NCL},i}$ is the sum of the power of NCL_i and ES_i.

According to the KCL law in Figure 3 and the CPL equivalent model in (1), the relationship between the currents in the positive bus I_1, negative bus I_2, and neutral bus I_0 can be obtained as in (3) without considering shared energy storage.

$$I_0 = I_1 - I_2 = \frac{2P_1}{V_0} - \frac{P_1}{V_0^2} V_1 - \frac{2P_2}{V_0} + \frac{P_2}{V_0^2} V_2 \tag{3}$$

where $P_{1,2}$ is the net power consumed after equivalence of the RES and CPL. V_1 and V_2 are the voltages of the positive and negative poles. The subscripts 1, 2, and 0 denote the positive, negative, and neutral buses, respectively.

The neutral bus current of the T-LBC can be calculated by (4):

$$I_{\mathrm{M}} = I_0 + C_1 \frac{dV_1}{dt} - C_2 \frac{dV_2}{dt} \tag{4}$$

where I_{M} is the output current in the neutral line of the T-LBC. C_1 and C_2 are the output capacitance.

When C_1 and C_2 are designed to be the same and equal to C, (5) can be deduced by integrating (4).

$$V_1 - V_2 = \frac{1}{C} \int_0^{T_s} (I_{\mathrm{M}} - I_0) dt \tag{5}$$

Because the value of I_0 is related to the difference of P_1 and P_2, $V_1 = V_2$ can only be realized by controlling I_{M}. According to (4), $V_1 = V_2$ will cause $I_{\mathrm{M}} = I_0$. Using the relationship of $V_1 = V_2 = V_0 = 0.5\ V_{\mathrm{DC}}$, the following can be obtained.

$$I_{\mathrm{M}} = I_0 = \frac{(P_1 - P_2)}{V_0} = \frac{2(P_1 - P_2)}{V_{\mathrm{DC}}} \tag{6}$$

Substituting the relationship between the input and output currents of the T-LBC, shown as $I_{\mathrm{M}} \leq (1 - D)I_{\mathrm{L}}$, into (6), the imbalance in power can be further obtained.

$$P_1 - P_2 \leq \frac{(1 - D) I_{\mathrm{L}} V_{\mathrm{DC}}}{2} \tag{7}$$

where D and I_{L} are the duty cycle and input current of the T-LBC.

Therefore, in the bipolar DC microgrid without shared energy storage, only when the imbalance power meets (7) can neutral-point voltage balance control be achieved by the T-LBC.

After considering shared energy storage, combining (2) and (3), I_0, I_1, and I_2 are changed into the following.

$$\begin{cases} I_0 = I_1 - I_2 \\ I_1 = \frac{2P_1}{V_0} - \frac{P_1}{V_0^2}V_1 + \frac{V_1 - V_{ES1}}{R_{NCL1}} \\ I_2 = \frac{2P_2}{V_0} - \frac{P_2}{V_0^2}V_2 + \frac{V_2 - V_{ES2}}{R_{NCL2}} \end{cases} \quad (8)$$

According to (8), and further assuming $R_{NCL,1} = R_{NCL,2}$, (4) and (7) can be rewritten as (9) and (10).

$$I_M = \frac{2(P_1 - P_2)}{V_{DC}} + \frac{V_{ES2} - V_{ES1}}{R_{NCL}} \quad (9)$$

$$P_1 - P_2 \leq \frac{(1-D)I_L V_{DC}}{2} + \frac{(V_{ES1} - V_{ES2})V_{DC}}{2R_{NCL}} \quad (10)$$

where $V_{ES1} - V_{ES2}$ can be controlled to be larger than 0.

Therefore, after equipping shared energy storage in the bipolar DC microgrid, the value of unbalanced power can be greater than that in (7). That is, the system has a stronger ability to suppress unbalanced voltage.

2.3. Regulation of Bus Voltage as Voltage Source

When DES in the system fails and exits the operation, shared energy storage is controlled as follows.

(1) ES_1 and ES_2 can keep V_1 and V_2 running at the rated voltage point V_0 by controlling V_{ES1} and V_{ES2}. Therefore, the stability of the bus voltage within a certain range is maintained.

(2) Coordinating P_{SL1} and P_{SL2} can ensure $P_1 + P_{SL1} = P_2 + P_{SL2}$. Thus, unbalanced voltage control is achieved.

2.4. Ability of Power Dispatching

Define that P_G is the interactive power between the bipolar DC microgrid and the grid. It is worth noting that it is assumed that P_G is a known parameter obtained by upper-level optimization. How factors such as user behavior and electricity prices affect P_G is not considered in this paper. $P_G > 0$ means there is power transmission from the bipolar DC microgrid to the grid, while $P_G < 0$ indicates that the bipolar DC microgrid absorbs power from the grid. Then, the load power consumption of the whole system can be calculated by $P_1 + P_{SL1} + P_2 + P_{SL2}$. Assuming that the power transmitted by DES through the T-LBC is P_S, $P_S > 0$ means discharging of DES, otherwise the DES is charging. Thus, according to the V-I relationship of the series ESs in (2), P_G can be described by (11).

$$\begin{aligned} P_G &= P_S - (P_1 + P_2 + P_{SL1} + P_{SL2}) \\ &= P_S - P_1 - P_2 - \frac{V_{DC}}{2R_{NCL}}(2V_{DC} - V_{ES1} - V_{ES2}) \end{aligned} \quad (11)$$

It can be seen from (11) that P_G can be adjusted by controlling V_{ES1} and V_{ES2}, which means power consumed by NCLs can be dispatched by controlling the ES according to the dispatching command P_G. Although the efficiency of the NCLs will be affected, the bus voltage of the system can be compensated and economic power dispatching can be achieved by increasing or decreasing the power consumption of the NCLs.

3. Proposed Comprehensive Power Regulation Control Strategy

The comprehensive power regulation control strategy proposed in this paper is divided into two parts: the voltage support of shared energy storage in stand-alone operation mode, and the power dispatching in grid-connected operation mode. In this section, these two parts are described in detail. The topology of shared energy storage is composed of an ES with full-bridge DC/DC converter and NCLs connected in series, as shown in Figure 4.

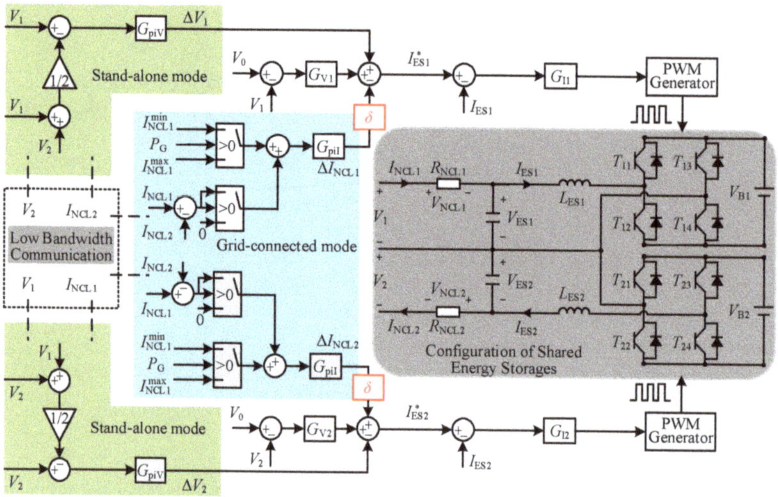

Figure 4. Topology and control diagram of shared energy storages.

We take one shared energy storage system as an example to explain the control strategy in detail; the control system of another one can be designed in the same way. Since DES adopts droop control with voltage balance control for the neutral bus, the load voltage V_{DC} has a deviation due to the droop characteristics. Thus, there is also a voltage drop in V_1 and V_2. In order to compensate for the voltage deviation, the ES with full-bridge DC/DC converter adopts voltage–current double-loop control, as shown in (12).

$$\begin{cases} V_1^{\text{ref}} = V_0 \\ I_{\text{ES1}}^{\text{ref}} = (V_1^{\text{ref}} - V_1) * G_{\text{V1}} \\ d_1 = (I_{\text{ES1}}^{\text{ref}} - I_{\text{ES1}}) * G_{\text{I1}} \end{cases} \tag{12}$$

where G_{V1} and G_{I1} are the transfer functions of the PI controller in the voltage loop and current loop, respectively. d_1 is the duty cycle of IGBT (Insulated Gate Bipolar Transistor); T_{11}/T_{14}, and T_{12}/T_{14} are controlled by $(1 - d_1)$. V_1^{ref} and $I_{\text{ES1}}^{\text{ref}}$ are the reference values of the voltage outer loop and current inner loop.

3.1. Design of Unbalanced Voltage Suppression Control with Voltage Support in Stand-Alone Mode

When the bipolar DC microgrid operates in stand-alone mode, shared energy storage needs to further improve the ability of unbalanced voltage suppression on the basis of bus voltage deviation compensation. Therefore, by adding the unbalanced voltage correction to (12), the ES can compensate the voltage imbalance, as shown in (13).

$$\Delta V_1 = (V_1 - \frac{V_1 + V_2}{2}) * G'_{\text{piV}} \tag{13}$$

where G'_{V1} is the transfer function of the PI controller with the objective of enforcing zero $(V_1 - V_2)$, achieving voltage balance between V_1 and V_2 by adding the imbalance correction ΔV_1 to the bus voltage reference.

On the other hand, (13) is also critical for enabling the bus voltage regulation capability in the event of DES failure. At this moment, due to the constant voltage control of the ES in (12), as long as the power balance is maintained, the system will not collapse when DES fails. The NCL power can be either decreased or increased by controlling the ES voltage, thus the power provided/absorbed by the failed DES before can be balanced by the NCL.

3.2. Design of Unbalanced Voltage Suppression Control with Power Dispatching in Grid-Connected Mode

In grid-connected operation mode, the power command P_G on the inverter is sent from the dispatching center. In order to cooperate with the power dispatching of the utility grid, power consumed by NCLs should be adjusted according to P_G. Different from the condition in stand-alone operation mode, load power should be reduced when $P_G > 0$, which indicates the utility grid requests power from the microgrid. When the utility grid needs to transmit power to the microgrid, that is, $P_G < 0$, the NCL needs to increase the consumed power, improving the power transmission capacity of the microgrid. Considering that the ES adopts constant voltage control, the change in power can only be obtained by correcting the reference current of the NCL. Therefore, the current correction can be calculated by the following.

$$\begin{cases} \Delta I_{1d} = (V_{NCL1}^{min} - V_{NCL1}) * G'_{piI} \; P_G > 0 \\ \Delta I_{1i} = (V_{NCL1}^{max} - V_{NCL1}) * G'_{piI} \; P_G < 0 \end{cases} \tag{14}$$

where V_{NCL1}^{min} and V_{NCL1}^{max} are the allowable minimum and maximum voltages of NCL_1. G'_{piI} is the transfer function of the PI controller with the objective of enforcing zero (V_{NCL1}^{min} − V_{NCL1}) or (V_{NCL1}^{max} − V_{NCL1}).

In order to further deal with the case of using shared energy storage to improve the unbalanced voltage suppression capability, a power imbalance correction is carried out for the current correction according to the different consumed powers of P_{SL1} and P_{SL2}. Therefore, the current correction shown in (14) is rewritten as (15).

$$\begin{cases} \Delta I_{1d} = G'_{piI} * (V_{NCL1}^{min} - V_{NCL1} + \max\{(V_{NCL1} - V_{NCL2}), 0\}) \; P_G > 0 \\ \Delta I_{1i} = G'_{piI} * (V_{NCL1}^{max} - V_{NCL1} + \min\{(V_{NCL1} - V_{NCL2}), 0\}) \; P_G < 0 \end{cases} \tag{15}$$

where max {(V_{NCL1} − V_{NCL2}), 0} stands for the larger value between (V_{NCL1} − V_{NCL2}) and 0, and min {(V_{NCL1} − V_{NCL2}), 0} represents the smaller value between (V_{NCL1} − V_{NCL2}) and 0.

Equation (15) means that when $P_G > 0$, the shared energy storage with less total power between P_{SL1} and P_{SL2} is limited to the allowable minimum power value of the NCL, and the other one is adjusted according to the voltage balance. Similarly, when $P_G < 0$, the shared energy storage with larger power between P_{SL1} and P_{SL2} is set to the allowable NCL maximum power. According to the above analysis, the current reference value of NCL_1 in power dispatching mode is calculated as follows:

$$\Delta I_{NCL1} = \begin{cases} \Delta I_{1d}, \; P_G > 0 \\ 0, \quad P_G = 0 \\ \Delta I_{1i}, \; P_G < 0 \end{cases} \tag{16}$$

$$I_{ES1}^{ref} = (V_1^{ref} - V_1) * G_{V1} + \Delta V_1 - \Delta I_{NCL1} \tag{17}$$

Thus, power dispatching can be achieved by using (16) and (17). It is worth mentioning that the power dispatching function in grid-connected mode can be enabled by a variable δ.

When P_G is equal to 0, δ is equal to 0. Otherwise, δ is equal to 1. As shown in Figure 4, the blue box is controlled by δ, marked in red.

4. Stability Analysis

Since the proposed control strategy is aimed at the full-bridge DC/DC converter of the ES in the shared energy storage, the dynamics of the voltage outer loop and current inner loop can be decoupled for stability analysis. A stability analysis of a conventional voltage-controlled full-bridge DC/DC has been presented in [31]. The analysis for the proposed unbalanced voltage suppression-controlled ESs in the bipolar DC microgrid can be carried out based on it. Figure 5 shows the equivalent control diagram of the autonomous controlled ES_1, where the output voltage and current can be obtained from (18), and the current relationship between ES_1 and ES_2 is described in (19).

$$\begin{cases} \dot{V}_{ES1} = -\frac{V_{ES1}}{R_{NCL1}C_{ES1}} - \frac{I_{ES1}}{C_{ES1}} + \frac{V_1}{R_{NCL1}C_{ES1}} \\ \dot{I}_{ES1} = \frac{V_{ES1}}{2L_{ES1}} - \frac{r}{L_{ES1}}I_{ES1} - \frac{\Delta d}{2L_{ES1}}V_{B1} \end{cases} \tag{18}$$

where r is the equivalent series resistor of inductor L_{ES1}. Δd is the difference in the duty ratios between T_{11} and T_{13}, calculated as $\Delta d = 2d_1 - 1$.

$$\begin{cases} C_D\dot{V}_1 = \frac{V_{S1}-V_1}{r_d} - \frac{P_1}{V_1} - I_{NCL1} \\ C_D\dot{V}_2 = \frac{V_{S2}-V_2}{r_d} - \frac{P_2}{V_2} - I_{NCL2} \end{cases} \tag{19}$$

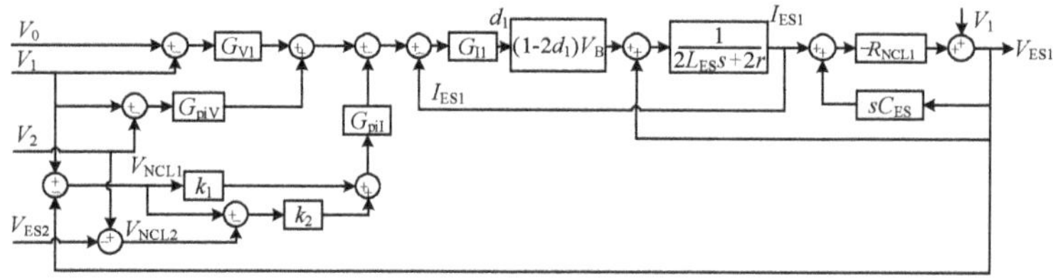

Figure 5. Equivalent control diagram of one shared energy storage.

The state-space model of unbalanced voltage correction and NCL current correction can be obtained by introducing intermediate variables φ_{piV} and

$$\begin{cases} k_{iiV1}\dot{\varphi}_{piV1} = \frac{V_1-V_2}{2} \\ k_{iiI1}\dot{\varphi}_{piI1} = k_1V_{NCL1} + k_2(V_{NCL1} - V_{NCL2}) \end{cases} \tag{20}$$

The state-space model of the voltage loop and current loop can also be obtained in a similar way.

$$\begin{cases} k_{iV1}\dot{\varphi}_{V1} = V_0 - V \\ k_{iI1}\dot{\varphi}_{I1} = k_{pV1}(V_0 - V_1) + \varphi_{V1} + k_{piV1}(V_1 - V_2)/2 + \varphi_{piV1} + k_{piI1}[(k_1 - k_2)V_{NCL1} - k_2V_{NCL2}] + \varphi_{piI1} \end{cases} \tag{21}$$

The state space model of ES_1 and its control system can be expressed in the following by integrating (18)–(21) and combining the voltage current relationship shown in (2).

$$\begin{cases} \dot{X}_1 = A_1X_1 \\ X_1 = [V_1, V_{ES1}, I_{ES1}, \varphi_{piV}, \varphi_{piI}, \varphi_{V1}, \varphi_{I1}]^T \end{cases} \tag{22}$$

Thus, duty cycle d_1 can be obtained through (23).

$$d_1 = k_{pI1}k_{iI1}\dot{\varphi}_{I1} + \varphi_{I1} \tag{23}$$

In the same way, the state-space model of ES_2 can be obtained, which will not be repeated here. Therefore, based on the current relationship shown in (19), the state-space model of Figure 4 can be expressed as $\dot{X} = AX$, where $X = [V_{1,2}, V_{ES1,2}, I_{ES1,2}, \varphi_{piV1,2}, \varphi_{piI1,2}, \varphi_{V1,2}, \varphi_{I1,2}]^T$.

With the state-space model of the whole system presented in (4), the stability analysis of unbalanced voltage suppression-controlled ESs can be developed with the parameters listed in Table 1. In order to analyze the stability under different unbalanced load power and dispatching power from the utility grid, the root locus plot considering changes in $\Delta P = |P_1 - P_2|$ and P_G is shown in Figure 6.

Table 1. Parameters of controller.

Description	Symbol	Nominal Value
ES output capacitance	$C_{ES1,2}$	500×10^{-6} F
ES input inductance	$L_{ES1,2}$	3 mH
Equivalent series resistance	r	0.01 Ω
Proportional gain in G_V	$k_{pV1,2}$	3
Integral gain in G_V	$k_{iV1,2}$	10
Proportional gain in G_I	$k_{pI1,2}$	5
Integral gain in G_I	$k_{iI1,2}$	50
Proportional gain in G_{piV}	$k_{piV1,2}$	0.5
Integral gain in G_{piV}	$k_{iiV1,2}$	25
Proportional gain in G_{piI}	$k_{piI1,2}$	0.01
Integral gain in G_{piI}	$k_{iiI1,2}$	3
Equivalent control parameters	$k_{1,2,3,4}$	0.65/0/0.65/1
Resistance of NCL	$R_{NCL1,2}$	4.5 Ω/7.5 Ω
Power of CPL	$P_{1,2}$	3 kW/1 kW

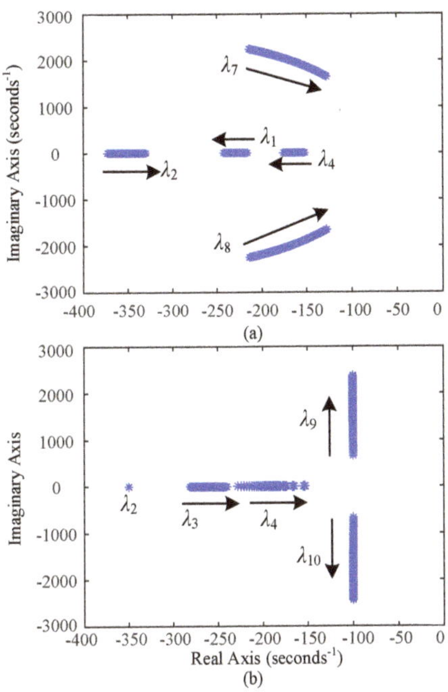

Figure 6. Root locus diagram of the overall system for stability analysis. (**a**) $0 < \Delta P < 500$. (**b**) $-3000 < P_G < 3000$.

It can be seen from Figure 6a, with the increase in ΔP, the eigenvalues $\lambda_{1,2}$, λ_4, and $\lambda_{7,8}$ are vulnerable to being influenced. However, although $\lambda_{7,8}$ move towards the unstable area, all the eigenvalues are located in the left half-plane with the range of 0~5000, which indicates the system is stable.

When P_G increases, the small-signal stability is also changed, especially $\lambda_{3,4}$ and $\lambda_{9,10}$, which, respectively, move towards the right half-plane and move far away from the real axis, as shown in Figure 6b. A stability analysis of the proposed control in the condition where $P_G = 0$, k_1, k_2, k_3, and k_4 are set to 0 in stand-alone operation is also presented.

5. Results and Discussion

To verify the effectiveness of the proposed control strategy, hardware-in-the-loop experiments based on the OPAL-RT real-time simulator and MATLAB/Simulink 2020a simulation environment were developed, as presented in Figure 7. The control systems of the shared energy storage were implemented by a DSP(TMS320F28335) controller, while the mathematic model of the system, including the loads shown in Equation (1), were established by OPAL-RT based on the MATLAB/Simulink modeling environment. The DSP controller and OPAL-RT were connected through built-in input/output modules, while the host computer and OPAL-RT were connected through TCP/IP. The effect of the communication delay is not considered in this paper. The basic parameters of the tested bipolar DC microgrid shown in Figure 3 can be seen in Table 2, which shows the power scale of the studied system and the limit values of both power generation and power consumption.

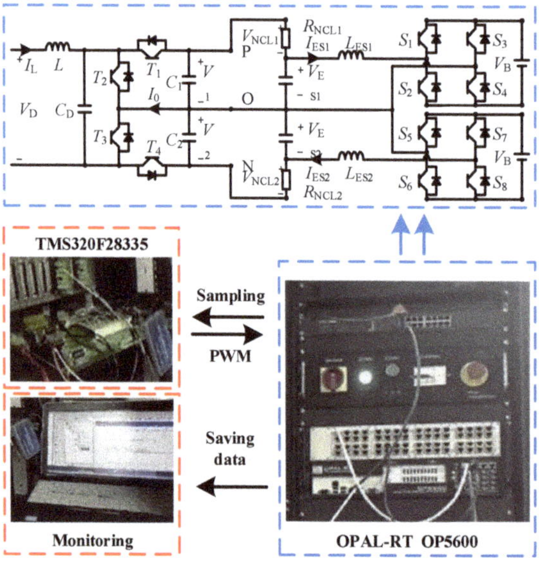

Figure 7. HIL experimental setup.

Three cases with different scenarios for changes in stand-alone mode and grid-connected mode were tested. With the controller parameters given in Table 1, the power of the NCLs, CPLs, and DES of the studied system are observed to explain the control performance, and also the voltages of the positive and negative poles.

Table 2. Parameters of the studied bipolar DC microgrid.

Descriptions	Value
Nominal DC voltage of positive and negative pole	150 V
Maximum power of DES	3 kW
Range of power generation by DGs	0~7 kW
Range of power consumption by CPLs	2~10 kW
Range of P_1 and P_2	−10~10 kW
Resistance of NCL in positive pole	4.5 Ω
Resistance of NCL in negative pole	7.5 Ω
Allowable voltage deviation of NCL	−20~20%
Droop coefficient of T-LBC	7×10^{-3}

5.1. Case 1: Secondary Compensation Mode

In case 1, the performance of shared energy storage operating in secondary compensation mode with the change in CPLs is studied and the experimental results are presented in Figure 8. The results without shared energy storage are also presented. Initially, P_1 and P_2 are −2.5 kW and −4 kW, respectively. Here, the negatives of P_1 and P_2 mean that the power generated by the DGs is greater than that consumed by the CPLs. The power difference between the positive and negative poles can be calculated by $\Delta P = |P_1 - P_2|$. In this condition, the bipolar DC microgrid without shared energy storage can also regulate the DC bus voltage and suppress unbalanced voltage through the control of the T-LBC, as shown in Figure 8b before t = 2 s. However, there are still voltage deviations in V_1 and V_2. Fortunately, this drawback is solved when the bipolar DC microgrid is equipped with shared energy storage, as shown in Figure 8a. It can be concluded that with the proposed control of shared energy storage, the DC bus voltage deviation caused by droop control of the T-LBC has been reduced from 2.67% to 0.01%.

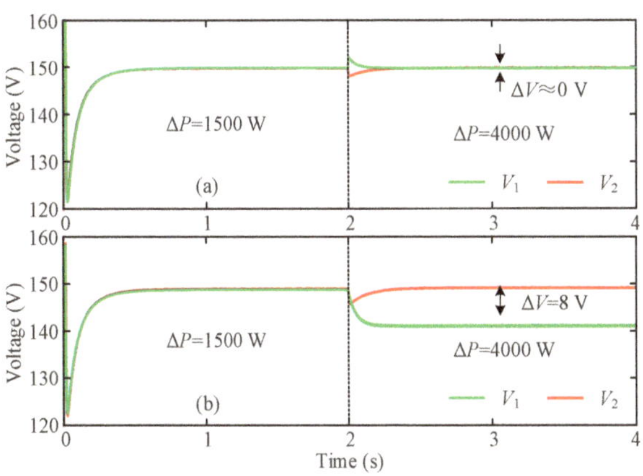

Figure 8. Experimental results of DC voltage in case 1. (**a**) With proposed control of shared energy storage in bipolar DC microgrid. (**b**) Without shared energy storage in bipolar DC microgrid.

Power change happens at t = 2 s. P_1 suddenly changes from −2.5 kW to 0 kW. ΔP subsequently increases from 1.5 kW to 4 kW, which has gone beyond the unbalanced voltage suppression ability of the T-LBC. Thus, in Figure 8b, the voltage difference occurs after t = 2 s, where V_1 = 138 V and V_2 = 146 V. While in Figure 8a, with the proposed control of shared energy storage, the voltage imbalance has been well eliminated for the different power consumed in NCL_1 and NCL_2. The difference between V_1 and V_2 has been reduced from 5.3% to around 0%.

5.2. Case 2: Secondary Compensation Mode to Voltage Regulation Mode

Case 2 demonstrates the mode switch from secondary compensation to voltage regulation mode in stand-alone operation. The experimental results are shown in Figure 9. The initial conditions are the same as in case 1. At t = 2 s, the T-LBC fails and disconnects from the DC bus. As shown in Figure 10, the shortage power caused by the failing of the T-LBC is compensated by further reducing the power consumed by NCL_1, from 4.5 kW to 3.2 kW, and by NCL_2, from 2.75 kW to 2 kW. Consequently, the DC bus voltages of the positive and negative poles are regulated and balanced by the shared energy storage, as shown in Figure 10, where $V_1 = V_2 \approx 150$ V. Thus, the effectiveness of the proposed control strategy from secondary compensation mode to voltage regulation mode in stand-alone mode has been verified.

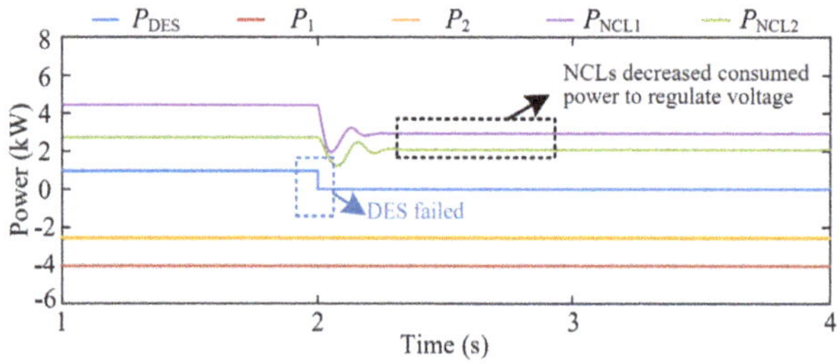

Figure 9. Experimental results of power distribution in case 2.

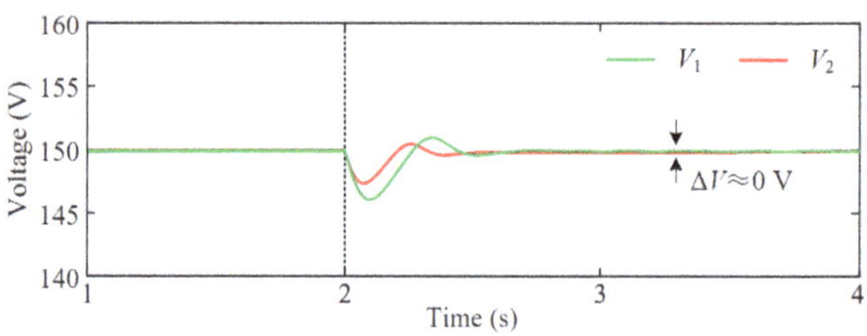

Figure 10. Experimental results of DC voltage in case 2.

5.3. Case 3: Secondary Compensation Mode to Power Dispatching Mode

In case 3, the performance of the studied system operating from secondary compensation mode to power dispatching mode is explored, and Figures 11 and 12 show the experimental results. The initial conditions are the same as in case 1 before t = 2 s, as shown in Figure 11. The bipolar DC microgrid is connected with the utility grid at t = 2 s. With the condition of P_G = 3 kW, NCLs reduce its power consumption to enable the microgrid output a larger P_G. Thus, DES only needs to output 1.9 kW to smooth the power imbalance. However, if the NCL does not reduce its power consumption, DES will output a power greater than 3 kW, which has already exceeded the maximum output power of DES. Therefore, the bipolar DC microgrid cannot meet the power dispatching of the utility grid without reducing NCL power in some conditions.

Furthermore, the performance of the power dispatching mode with PG change in grid-connected operation is also presented between t = 2 and 6 s. When t = 4 s, the dispatching power P_G changes from 3 kW to −4 kW, which means the utility grid transmits power to the bipolar DC microgrid. The consumed power of the NCLs increases to the maximum load power P_{NCL1}^{max} and P_{NCL2}^{max}. With the condition of $P_{NCL,1}$ = 7.2 kW and $P_{NCL,2}$ = 4.3 kW, DES only needs to output power of 1 kW through the T-LBC.

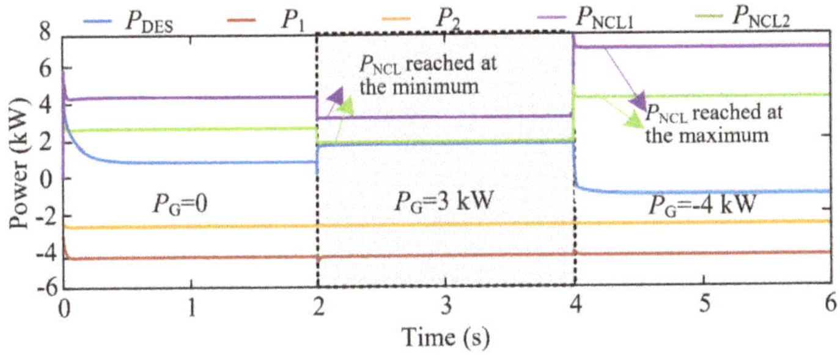

Figure 11. Experimental results of power distribution in case 3.

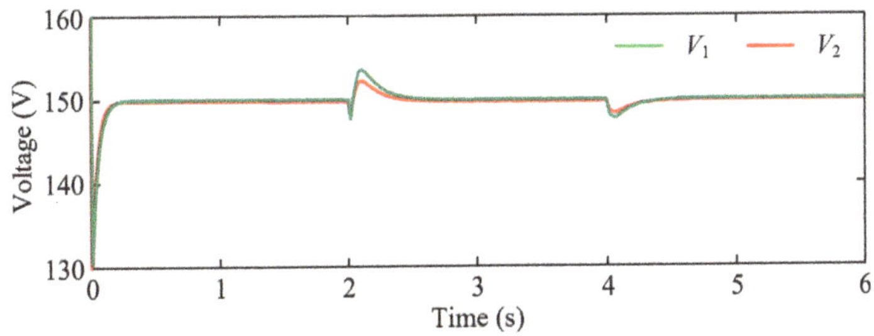

Figure 12. Experimental results of DC voltage in case 3.

During the operation of mode transferring and the change in P_G, the DC voltage of V_1 and V_2 can be well maintained to be balanced and stable, as shown in Figure 12. The effectiveness of the proposed control strategy from secondary compensation mode in stand-alone operation to power dispatching mode in grid-connected operation has been verified.

In order to reflect the advantages of the power dispatching mode of the ESs, Table 3 shows the NCL consumption power and DES output power under the conditions of actively adjusting power and non-regulation power, respectively.

Table 3. Comparison of results with/without power dispatching control.

	With Power Dispatching Control	Without Power Dispatching Control
$P_G = 0$ kW	$P_{NCL,1} = 4.5$ kW $P_{NCL,2} = 2.75$ kW $P_{DES} = 1.12$ kW	$P_{NCL,1} = 4.5$ kW $P_{NCL,2} = 2.75$ kW $P_{DES} = 1.12$ kW
$P_G = 3$ kW	$P_{NCL,1} = 3.2$ kW $P_{NCL,2} = 1.92$ kW $P_{DES} = 1.9$ kW	$P_{NCL,1} = 4.2$ kW $P_{NCL,2} = 2.1$ kW $P_{DES} = 3$ kW $= P_{DES}^{max}$
$P_G = -4$ kW	$P_{NCL,1} = 7.2$ kW $P_{NCL,2} = 4.32$ kW $P_{DES} = -1$ kW	$P_{NCL,1} = 5.2$ kW $P_{NCL,2} = 3.5$ kW $P_{DES} = -2.0$ kW

It can be seen from Table 3 that when $P_G > 0$, NCLs can reduce the power consumption to enable DES to output less power. So, the capacity of DES can be reduced and the bipolar microgrid can output more P_G. Similarly, when $P_G < 0$, NCLs can reduce the input power of DES by increasing the power consumption, so as to make the microgrid have the ability to absorb greater P_G.

6. Conclusions

In this paper, comprehensive power regulation control of shared energy storage with both voltage support and power dispatching for a multi-scenario bipolar DC microgrid is proposed. The proposed control is targeted at ensuring the performance of the bipolar DC microgrid in different conditions, which are stand-alone operation, DES failure condition, and grid-connected operation. In this way, the voltage deviation is reduced by about 2.67% in stand-alone operation, and DES achieves a lower power in power dispatching mode than in grid-connected operation. Furthermore, when DES fails, DC voltage can still be regulated by decreasing the power consumption of the shared energy storage. With the proposed control, the reliability and resilience of the bipolar DC microgrid are both improved. Then, the system's stability under proposed control is analyzed based on small-signal modeling, and the HIL experimental results for different operating conditions are carried out to verify the effectiveness of the proposed control.

Author Contributions: Conceptualization, G.L. and B.Z.; methodology, X.M.; Validation, X.J. and H.Y.; writing-original draft preparation, H.Y.; writing-review and editing, G.L., B.Z., X.M. and X.J. All authors have read and agreed to the published version of the manuscript.

Funding: This paper was supported by Research and Technology Project for 100 MW High Voltage Energy Storage System (Grant Nos. E565400011).

Data Availability Statement: The original contributions presented in this study are included in the article. Further inquiries can be directed to the corresponding author.

Conflicts of Interest: Author Gongqiang Li, Bin Zhao and Xiaoqiang Ma were employed by the company CHN Energy (Gonghe) Renewable Energy Development Co., Ltd., Author Xiaofan Ji were employed by the company National Institute of Clean-and-Low-Carbon Energy. The remaining authors declare that the research was conducted in the absence of any commercial or financial relationships that could be construed as a potential conflict of interest.

References

1. Lotfi, H.; Khodaei, A. Ac versus dc microgrid planning. *IEEE Trans. Smart Grid* **2015**, *8*, 296–304. [CrossRef]
2. Han, Y.; Zhang, G.; Li, Q.; You, Z.; Chen, W.; Liu, H. Hierarchical energy management for pv/hydrogen/battery island dc microgrid. *Int. J. Hydrogen Energy* **2019**, *44*, 5507–5516. [CrossRef]
3. Li, X.; Jiang, W.; Wang, J.; Wang, P.; Wu, X. An autonomous control scheme of global smooth transitions for bidirectional dc-dc converter in dc microgrid. *IEEE Trans. Energy Convers.* **2020**, *36*, 950–960. [CrossRef]
4. Lee, J.-O.; Kim, Y.-S.; Moon, S.-I. Current injection power flow analysis and optimal generation dispatch for bipolar dc microgrids. *IEEE Trans. Smart Grid* **2020**, *12*, 1918–1928. [CrossRef]
5. Tavakoli, S.D.; Zhang, P.; Lu, X.; Hamzeh, M. Mutual interactions and stability analysis of bipolar dc microgrids. *CSEE J. Power Energy Syst.* **2019**, *5*, 444–453.
6. Carvalho, E.L.; Meneghetti, L.H.; Bellinaso, L.; Cardoso, R.; Michels, L. Bidirectional interlink converter for bipolar dc microgrids. In Proceedings of the 2019 IEEE PES Innovative Smart Grid Technologies Conference-Latin America (ISGT Latin America), Gramado, Brazil, 15–18 September 2019; pp. 1–6.
7. Liao, J.; Zhou, N.; Huang, Y.; Wang, Q. Unbalanced voltage analysis and suppression method in a radial bipolar dc distribution network. *IEEE J. Emerg. Sel. Top. Power Electron.* **2021**, *9*, 5687–5702. [CrossRef]
8. Khodabakhsh, J.; Moschopoulos, G. Distributed unbalanced voltage suppression in bipolar dc microgrids with smart loads. In Proceedings of the 2021 IEEE Applied Power Electronics Conference and Exposition (APEC), Phoenix, AZ, USA, 14–17 June 2021; pp. 2692–2697.
9. Li, B.; Fu, Q.; Mao, S.; Zhao, X.; Xu, D.; Gong, X.; Wang, Q. DC/DC converter for bipolar LVDC system with integrated voltage balance capability. *IEEE Trans. Power Electron.* **2020**, *36*, 5415–5424. [CrossRef]
10. Lin, B.-R.; Zhang, S.-Z. Analysis and implementation of a three-level hybrid dc–dc converter with the balanced capacitor voltages. *IET Power Electron.* **2016**, *9*, 457–465. [CrossRef]
11. Zhang, X.; Gong, C.; Yao, Z. Three-level dc converter for balancing dc 800-v voltage. *IEEE Trans. Power Electron.* **2014**, *30*, 3499–3507. [CrossRef]

12. Carstensen, C.; Biela, J. Novel 3 level bidirectional buck converter with wide operating range for hardware-in-the-loop test systems. In Proceedings of the 2012 15th International Power Electronics and Motion Control Conference (EPE/PEMC), Novi Sad, Serbia, 4–6 September 2012; p. DS2b-2.
13. Chen, H.-C.; Liao, J.-Y. Modified interleaved current sensorless control for three-level boost pfc converter with considering voltage imbalance and zero-crossing current distortion. *IEEE Trans. Ind. Electron.* **2015**, *62*, 6896–6904. [CrossRef]
14. Nisha, K.; Gaonkar, D.N. Model predictive controlled three-level bidirectional converter with voltage balancing capability for setting up EV fast charging stations in bipolar dc microgrid. *Electr. Eng.* **2022**, *104*, 2653–2665. [CrossRef]
15. Wang, W.; Kang, K.; Sun, G.; Xiao, L. Configuration optimization of energy storage and economic improvement for household photovoltaic system considering multiple scenarios. *J. Energy Storage* **2023**, *67*, 107631. [CrossRef]
16. Sun, Y.; Gao, J.; Wang, J.; Huang, Z.; Li, G.; Zhou, M. Evaluating the reliability of distributed photovoltaic energy system and storage against household blackout. *Glob. Energy Interconnect.* **2021**, *4*, 18–27. [CrossRef]
17. Bai, J.; Ding, T.; Jia, W.; Zhu, S.; Bai, L.; Li, F. Online rectangle packing algorithm for swapped battery charging dispatch model considering continuous charging power. *IEEE Trans. Autom. Sci. Eng.* **2024**, *21*, 320–331. [CrossRef]
18. Zhai, X.; Li, Z.; Li, Z.; Xue, Y.; Chang, X.; Su, J.; Jin, X.; Wang, P.; Sun, H. Risk-averse energy management for integrated electricity and heat systems considering building heating vertical imbalance: An asynchronous decentralized approach. *Appl. Energy* **2025**, *383*, 125271. [CrossRef]
19. Zhang, W.; Wei, W.; Chen, L.; Zheng, B.; Mei, S. Service pricing and load dispatch of residential shared energy storage unit. *Energy* **2020**, *202*, 117543. [CrossRef]
20. Zhang, H.; Li, Z.; Xue, Y.; Chang, X.; Su, J.; Wang, P.; Guo, Q.; Sun, H. A stochastic bi-level optimal allocation approach of intelligent buildings considering energy storage sharing services. *IEEE Trans. Consum. Electron.* **2024**, *70*, 5142–5153. [CrossRef]
21. Jo, J.; Park, J. Demand-side management with shared energy storage system in smart grid. *IEEE Trans. Smart Grid* **2020**, *11*, 4466–4476. [CrossRef]
22. Zhang, W.-Y.; Zheng, B.; Wei, W.; Chen, L.; Mei, S. Peer-to-peer transactive mechanism for residential shared energy storage. *Energy* **2020**, *246*, 23204. [CrossRef]
23. Doroudchi, E.; Khajeh, H.; Laaksonen, H. Increasing self-sufficiency of energy community by common thermal energy storage. *IEEE Access* **2022**, *10*, 85106–85113. [CrossRef]
24. Mok, K.-T.; Wang, M.-H.; Tan, S.-C.; Hui, S.Y.R. Dc electric springs-a technology for stabilizing dc power distribution systems. *IEEE Trans. Power Electron.* **2016**, *32*, 1088–1105. [CrossRef]
25. Wang, M.-H.; Yan, S.; Tan, S.-C.; Xu, Z.; Hui, S.Y. Decentralized control of dc electric springs for storage reduction in dc microgrids. *IEEE Trans. Power Electron.* **2019**, *35*, 4634–4646. [CrossRef]
26. Chen, X.; Shi, M.; Sun, H.; Li, Y.; He, H. Distributed cooperative control and stability analysis of multiple dc electric springs in a dc microgrid. *IEEE Trans. Ind. Electron.* **2017**, *65*, 5611–5622. [CrossRef]
27. Chen, T.; Zheng, Y.; Chaudhuri, B.; Hui, S.-Y.R. Distributed electric spring based smart thermal loads for overvoltage prevention in lv distributed network using dynamic consensus approach. *IEEE Trans. Sustain. Energy* **2019**, *11*, 2098–2108. [CrossRef]
28. Wang, M.-H.; Yan, S.; Tan, S.-C.; Hui, S.Y. Hybrid-dc electric springs for dc voltage regulation and harmonic cancellation in dc microgrids. *IEEE Trans. Power Electron.* **2017**, *33*, 1167–1177. [CrossRef]
29. Yang, H.; Li, T.; Long, Y.; Chen, C.L.P.; Xiao, Y. Distributed virtual inertia implementation of multiple electric springs based on model predictive control in dc microgrids. *IEEE Trans. Ind. Electron.* **2021**, *69*, 13439–13450. [CrossRef]
30. Wang, M.-H.; Mok, K.-T.; Tan, S.-C.; Hui, S.Y.R. Multifunctional dc electric springs for improving voltage quality of dc grids. *IEEE Trans. Smart Grid* **2018**, *9*, 2248–2258. [CrossRef]
31. Yang, Y.; Tan, S.-C.; Hui, S.Y.R. Mitigating distribution power loss of dc microgrids with dc electric springs. *IEEE Trans. Smart Grid* **2017**, *9*, 5897–5906. [CrossRef]

electronics

MDPI

Article

Autonomous Decentralized Cooperative Control DC Microgrids Realized by Directly Connecting Batteries to the Baseline

Hirohito Yamada

International Research Institute of Disaster Science, Tohoku University, Sendai 980-8572, Japan; hirohito.yamada.c1@tohoku.ac.jp; Tel.: +81-022-752-2149

Abstract: Recent years have seen increasing attention paid to autonomous decentralized microgrids that are disaster-resistant and suitable for local consumption of locally generated renewable energy power. Although various methods have been discussed for achieving microgrids through autonomous decentralized cooperative control, there are few reports that have reached the stage of field testing. In this study, I propose a novel method for configuring the baseline of DC microgrids, where storage batteries are distributed and directly connected to the DC bus. I have built a testbed to demonstrate the operation of the DC microgrid through autonomous decentralized cooperative control. My method simply employs the droop characteristics inherent in batteries, and I introduce the new concept of a 'weakly coupled grid'. This approach allows the realization of microgrids with autonomous decentralized cooperative control without the need for advanced and complex grid control technologies using DC/DC converters, and with a simple configuration. Additionally, by directly connecting batteries to the baseline, I introduce a grid stabilization method achieved by imparting electrical inertia to the baseline. This paper describes the construction method, the operation principle, and safe and stable operational methods for autonomous decentralized microgrids using this approach, aiming to serve as a guide for those who wish to build autonomous decentralized controlled microgrids in practice.

Keywords: DC microgrid; autonomous decentralized cooperative control; directly connecting batteries to the baseline; droop control; electrical inertia; weakly coupled grid

1. Introduction

In recent years, large-scale and catastrophic natural disasters have become increasingly frequent in many countries around the world. The primary cause of these disasters is believed to be global warming, and the reduction in greenhouse gas emissions, such as CO_2, has become an urgent issue. Therefore, efforts are being made to shift from fossil fuels to renewable energy sources in power generation. However, there are several challenges to overcome. Power generation by renewable energy, such as solar or wind power, is intermittent and unstable, and as the ratio of power generated by renewable energy increases, it can destabilize power grid systems. Furthermore, there is insufficient capacity in the existing power grids to accommodate additional renewable energy power generation. Additionally, the existing power grids, which are based on large-scale centralized power generation and long-distance transmission, are vulnerable to major disasters and can also be targets of terrorist attacks.

Against this backdrop, regional microgrids based on the principle of local production and consumption of electricity are gaining attention [1–5]. Particularly in residential areas, direct current (DC) microgrids that connect small power generators using renewable energies, such as solar panels and wind turbines, and storage batteries through DC buses are drawing keen interest [6–11]. Although DC microgrids require the installation of dedicated power lines, they are simpler to control and offer greater flexibility in grid configuration compared to alternating current (AC) grids. Field testing and social implementations have already been conducted in various parts of the world, although control methods and baseline voltages vary [12–14]. However, even if a microgrid is constructed, maintaining and operating it over the long term is not easy.

Microgrid control methods include centralized control, decentralized control, hierarchical control, and various other approaches [15–21]. The centralized control method, illustrated in Figure 1a, is employed in existing power grids. This approach relies on centralized controllers situated in the central control rooms of power companies, which monitor and manage the entire grid. Several devices within the grid are connected to the centralized controller via communication lines, allowing for unified management of the grid. This method enables optimization across the entire grid. However, if the centralized controller fails or the communication lines are disrupted, the entire grid becomes uncontrollable, making it vulnerable to large-scale disasters [22–25].

In contrast, the autonomous decentralized cooperative control (ADCC) method, as depicted in Figure 1b, is a unique control method that does not involve a centralized controller and communication lines that connect devices. Each device connected to the grid independently makes decisions and controls itself [26–32]. This method lacks the framework to monitor the entire grid, making overall grid optimization challenging and limiting it to local or partial optimization of grids. Nevertheless, even if part of the grid is damaged, the remaining parts can still function to some extent, making this method effective during disasters. Additionally, this approach allows for relatively easy modifications and expansions of the grid, enabling the creation of a flexible and scalable grid. However, since the ADCC does not have a mechanism for monitoring the entire grid, each device does not know the status of the entire grid. Under such circumstances, the issue is how each device can autonomously control itself.

The key is for each device to understand the situation around it as accurately as possible. Although ADCC does not know the situation of the entire grid, it is possible for each device to understand the situation of the grid around it to a certain extent. (That is the subject of this paper, which I will explain in detail later.) On that basis, some rules regarding grid use are established so as not to have a negative impact on the grid, and the system is such that all devices follow those rules and achieve their own purpose. For example, the purpose of each device is to feed renewable-energy-generated electricity into the grid in the case of a renewable energy generator such as PV or wind power. Also, in the case of a power load, it is to obtain power from the grid and start up when it is needed. However, in that case, it is necessary to avoid having a negative impact on the grid, so some kind of rule (Grid Code) regarding grid use is necessary. (That is also the subject of this paper, as I will explain in detail later.)

By introducing such a system, it is possible to maintain and operate the grid without centralized controllers and communication lines. In essence, this system establishes rules for proper grid utilization, with each device adhering to those rules while exercising self-restraint. This approach creates a democratic grid that can be effectively maintained and managed.

This paper focuses on the operating principles, construction methods, control methods, and safety measures of DC microgrids with ADCC, which enable practical use. Specifically, I propose an ADCC DC microgrid created by directly connecting distributed small batteries to the baseline. I built an experimental testbed for this type of DC microgrid to verify the feasibility of the method and have successfully operated it stably for over three years.

Herein, I share the knowledge and challenges gained from this experience, hoping to provide useful insights for those planning to construct microgrids in the future.

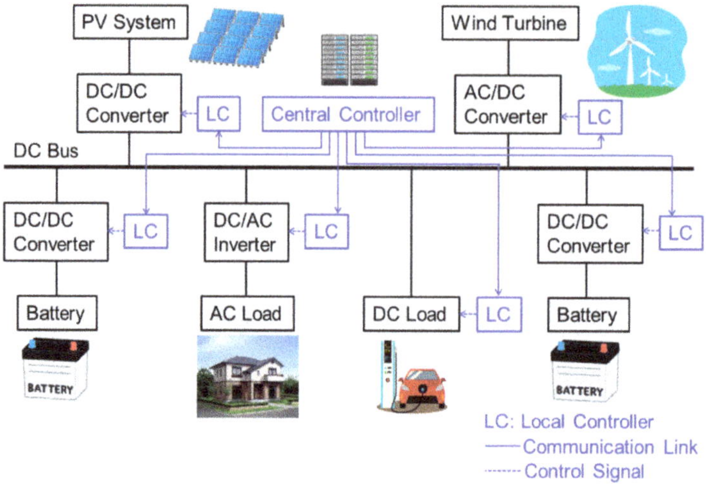

(a)

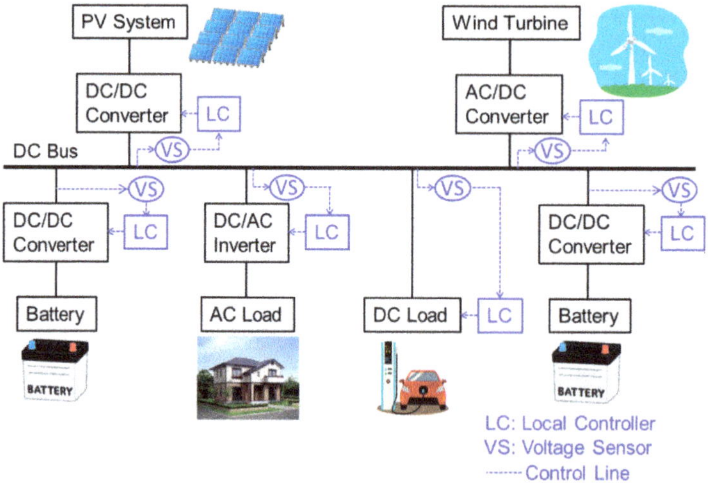

(b)

Figure 1. Control methods of power grids: (**a**) centralized control method; (**b**) autonomous decentralized cooperative control (ADCC) method.

2. Key Elements in Grid Control

Regardless of ADCC, there are two important factors in grid control: (1) droop control, which allows the coordinated operation of multiple generators within the grid autonomously, and (2) electrical inertia, which is essential for stabilizing the grid. Droop control is used in existing power grid systems to enable multiple synchronous generators to cooperate while sharing the load and supplying power to the grid. Specifically, each synchronous generator has droop characteristics, as shown in Figure 2a, where the generator's rotational speed decreases as the load rate increases, causing the output voltage or frequency to decrease. This helps counteract frequency changes in the system by increasing (decreasing) the load rate when the system frequency decreases (increases) due to some factor. Additionally, because each synchronous generator has this characteristic, neighboring synchronous generators can share the load and operate in coordination.

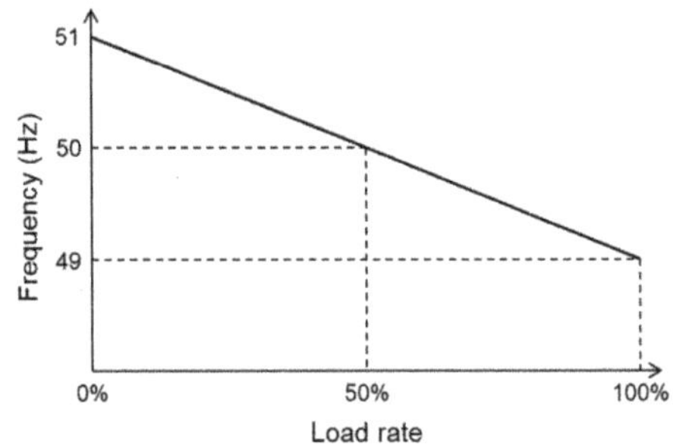

(**a**)

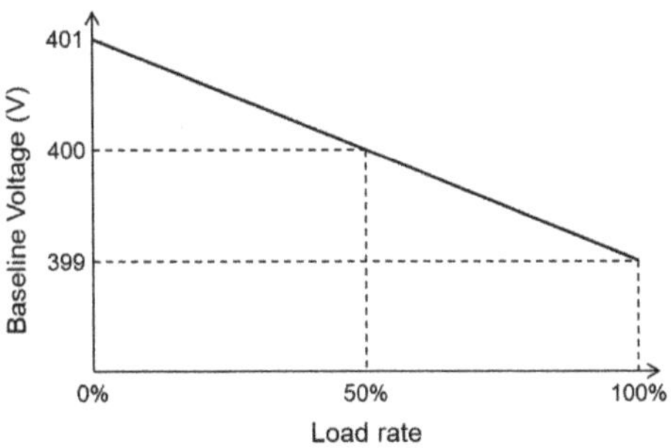

(**b**)

Figure 2. Droop characteristics of (**a**) synchronous generators; (**b**) DC/DC converters.

On the other hand, electrical inertia is very important for stabilizing the grid. When using electrical appliances at home, some appliances require several times their rated power consumption for a very short time when the switch is turned on. This power consumed at the start of the appliance is called inrush power or starting power, and the reason why even large electrical appliances can start up normally is because of the significant electrical inertia delivered to the household power plug. In existing power systems, there is essentially no energy storage, so there is always a need to balance generation and consumption (the principle of simultaneous power supply and demand). Therefore, a mechanism to cope with instantaneous increases in power load is necessary, and in power systems, power plant generators handle this. In other words, the source of electrical inertia is the rotational kinetic energy stored in the heavy turbines of thermal and hydro powerplant generators. This energy is instantaneously (within a few microseconds) converted into electrical energy and released into the grid system in response to sudden increases in power consumption. Since wind turbines in wind power generation do not have heavy turbines or synchronization mechanisms to store kinetic energy, there is a concern that the introduction of wind power generation will reduce the electrical inertia of the existing power system.

The two factors mentioned above are also important in the case of DC grids, and these mechanisms must also be present in ADCC DC microgrids. In DC microgrids, distributed power sources such as generators and batteries are generally connected to the baseline via DC/DC converters. Therefore, the DC/DC converters have droop characteristics with respect to the baseline voltage, as shown in Figure 2b, to enable coordinated operation of the distributed power sources [33–35]. However, as the number of distributed power sources increases, the number of droop parameters to be handled also increases, making control more complex. Therefore, I devised a method to achieve coordinated operation by focusing on the droop characteristics inherent in the battery itself. This will be discussed in Section 4.

Electrical inertia in DC grids is also crucial for stabilizing the grid and coping with sudden changes in power load. For instance, when rapid charging of electric vehicles (EVs) begins, the baseline voltage drops due to a sudden increase in power load. If the baseline has significant electrical inertia, this voltage drop will be small; however, if the electrical inertia is low, a large voltage drop will occur. In general DC microgrids, DC/DC converters connected to distributed power sources handle this change [36,37]. When the baseline voltage starts to drop, control works to supply more power from distributed power sources such as generators or batteries to the baseline. This can handle some increases in power load, but if the power supplied exceeds the rated power of the DC/DC converters, protection circuits will activate and stop the output. Additionally, this control requires a few milliseconds, during which the power load may give up on starting, or in the worst-case scenario, the grid may go down.

From this perspective, electrical inertia in DC grids can be considered the 'maximum power' that can be instantaneously extracted from a certain point on the baseline or power line. Here, 'instantaneous' specifically refers to a period ranging from a few seconds to at most a few tens of seconds, differing from power supply sustainability over several hours and referring solely to instantaneous force. This is a very short time sufficient to start household appliances requiring high inrush power. The power sustainability for long-term operation of appliances is related to battery capacity and is distinct from electrical inertia.

3. DC Grid with Distributed Batteries Directly Connected to the Baseline

Based on that background, I propose a DC microgrid, as shown in Figure 3, where multiple small batteries are directly connected to the baseline. In a typical DC microgrid, distributed power sources such as batteries are connected to the baseline via DC/DC converters. This makes it possible to connect batteries with terminal voltages different from the baseline voltage, as DC/DC converters perform voltage conversion. Additionally, DC/DC converters allow free power transfer between the batteries and the baseline. However, when directly connecting batteries to the baseline, the terminal voltage of the batteries must match the baseline voltage. This can be achieved by connecting multiple battery cells in series to form the bulk battery, but some constraints may arise if different types of battery cells with varying cell voltages are mixed. Nonetheless, the following significant advantages outweigh these drawbacks:

1. Utilizing the droop characteristics inherent in the batteries themselves for coordinated operation among distributed batteries.
2. Directly delivering the large electrical inertia of the batteries to the baseline without passing through the DC/DC converters, resulting in substantial electrical inertia for the baseline.
3. Since each distributed battery is directly connected to the baseline, the state of charge (SoC) of each battery spontaneously equalizes over time.

These aspects will be discussed in detail in the following sections.

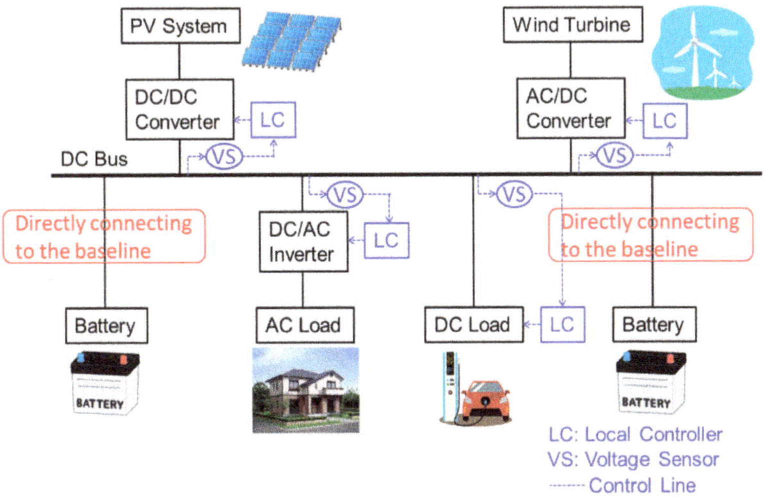

Figure 3. My method of directly connecting distributed batteries to the baseline.

4. Inherent Droop Characteristics of Batteries

The droop characteristics inherent to batteries can be formulated from the equivalent circuit of batteries. The basic equivalent circuit of a battery consists of an electromotive force E in series with an internal resistance r_i, as shown in Figure 4a. If a load resistance is connected, and the value of the load resistance R_L is equal to r_i, the battery can deliver the maximum power, defining a load rate of 100%. When a load R_L of any value is connected, the load rate α is given by

$$\alpha = \frac{4r_i R_L}{(r_i + R_L)^2}, \tag{1}$$

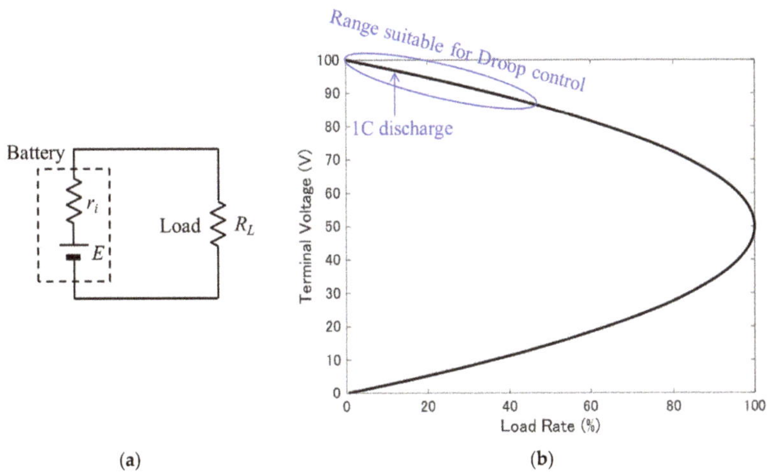

Figure 4. Droop characteristics of a battery: (**a**) basic equivalent circuit of a battery; (**b**) relationship between the load rate α and the terminal voltage V_t.

Meanwhile, the terminal voltage V_t of the battery, given the electromotive force E, is

$$V_t = \frac{R_L}{r_i + R_L} E, \tag{2}$$

The relationship between the load rate α and the terminal voltage V_t is thus

$$\alpha = 4\left(\frac{V_t}{E} - \frac{V_t^2}{E^2}\right), \quad (3)$$

This relationship is depicted in Figure 4b, representing the droop characteristics of inherent batteries. Generally, as the load rate increases (i.e., as the output current increases), the terminal voltage drops due to the voltage drop across the internal resistance. However, the droop coefficient changes its sign depending on whether the value of the load resistance is greater than or less than the internal resistance of the battery.

To achieve ADCC, it is necessary to use the region where the droop coefficient is negative, i.e., the upper half of the graph, when the load resistance is greater than the internal resistance of the battery. Moreover, where the load rate is 50% or less, a nearly linear relationship is observed between the load rate and the terminal voltage, enabling stable droop operation.

Equation (3) indicates that the droop characteristics of the battery do not depend on the value of internal resistance. Consequently, even if the battery deteriorates and its internal resistance increases, the load rate will be maintained at the same value as when it was new, as long as the value of the electromotive force (equal to the terminal voltage without loading), does not change. Furthermore, even with a mix of different types of batteries, if their electromotive force is uniform, the load rate can be equalized across the batteries. While the electromotive force of batteries naturally depends on the SoC, as will be explained in Section 9, the electromotive force of lithium-ion batteries can be regarded as nearly constant within a specific range of SoC values. Thus, by directly connecting batteries of the same type and equal terminal voltage to the baseline, the inherent droop characteristics of the batteries can be leveraged for cooperative operation.

Measurement of the internal resistance of a lithium iron phosphate ($LiFePO_4$) battery rated at 51.2 V, 40 Ah showed that the internal resistance was approximately 75 mΩ, depending on the state of charge (SoC) and the magnitude of the discharge current. Therefore, the load rate for a 1C (40 A) discharge of this battery is approximately 12%, and under normal usage conditions, the load rate remains relatively low, which allows for effective droop control.

Thus, by directly connecting batteries with a matched terminal voltage to the baseline, ADCC can be achieved using the batteries' inherent droop characteristics.

5. Electrical Inertia in DC Grids

In Section 2, I defined electrical inertia in DC grids as 'the maximum power that can be instantaneously extracted from a point on the baseline or power line'. The term 'instantaneous' refers to the time it takes for typical power load equipment to complete the startup process and enter steady-state operation, specifically ranging from a few seconds to several tens of seconds. Electrical inertia in existing power systems also operates within this time frame. Therefore, the ability to supply power over several hours is different in both meaning and unit (J), as it is the electrical energy that can be supplied. Consequently, the unit for electrical inertia in DC grids should be watt (W).

If we consider the electrical equivalent circuit of a battery, as shown in Figure 4a, the maximum power P_{max} that can be extracted from the battery terminals is given by the following Equation (4).

$$P_{max} = \frac{E^2}{4r_i}, \quad (4)$$

When directly connecting batteries to the baseline, there is a certain amount of conductor resistance in the baseline, so the electrical inertia depends on the thickness of the power line in the baseline, and the maximum power that can be extracted varies depending on the point on the baseline. The maximum power that can be extracted decreases as one moves further away from the directly connected battery, and the electrical inertia weakens. At a point sandwiched between two directly connected batteries, contributions to electrical inertia from both batteries can be obtained.

Figure 5 shows the calculation results of electrical inertia (the maximum power that can be extracted from that point) when a 51.2 V, 60 Ah lithium iron phosphate ($LiFePO_4$) battery with an internal resistance measurement value of 50 mΩ is directly connected to a baseline with a cross-linked polyethylene insulated vinyl sheath (CV) cable of 14 SQ thickness (1.34 Ω/km), where N represents the number of battery divisions. However, it is assumed that when dividing the battery, the value of internal resistance increases in proportion to the number of divisions. When directly connecting distributed batteries with the same total capacity, it is found that it is better to divide them and distribute them evenly along the baseline rather than connecting a large-capacity battery to one point to ensure uniformly large electrical inertia throughout the entire baseline. If divided into 40 parts (N = 40), small batteries with 1.5 Ah each are placed on the baseline at 25 m intervals, enabling the provision of electrical inertia of more than 10 kW throughout a 1 km baseline. The duration of the electrical inertia also lasts for more than 10 min in the discharge range suitable for droop control of 3C or less, which is sufficient for the startup of typical equipment.

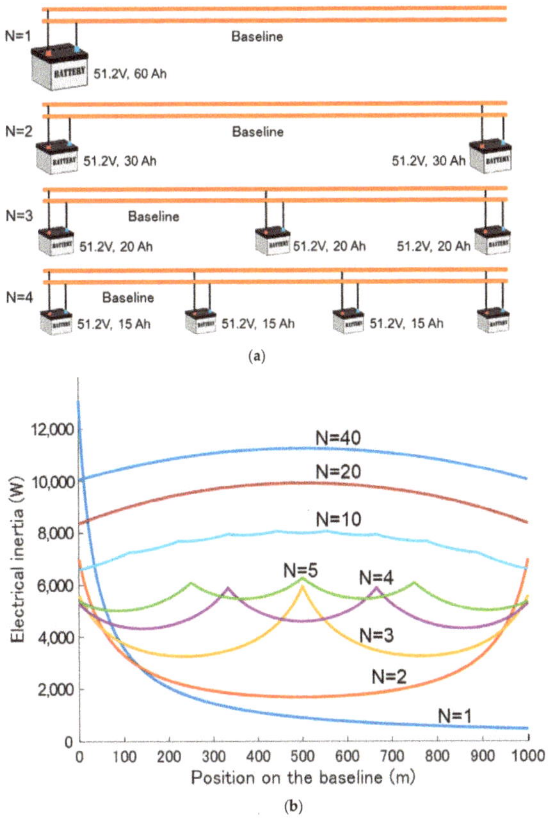

Figure 5. Calculation of electrical inertia for number of battery divisions N: (**a**) divided battery placement on the baseline; (**b**) calculated electrical inertia on the baseline.

The magnitude of electrical inertia in DC microgrids directly connected to the baseline is determined by the internal resistance of the batteries used and is not directly related to battery capacity. However, the value of internal resistance of the batteries is inversely proportional to battery capacity. The discharge response speed of batteries varies depending on the type of battery, but in the case of lithium-ion batteries, it is less than a few microseconds. Therefore, since power transmission through power lines occurs at nearly the speed of light, electrical inertia reaches the power load within a few microseconds after switching on the equipment. However, when a DC/DC converter for voltage conversion is used between the batteries and the baseline, the response speed slows to several milliseconds or more due to the electrical control of the DC/DC converter, and the maximum power is further limited by output devices, which may prevent the inherent electrical inertia of the batteries from being fully transmitted to the baseline. This is the primary reason for directly connecting batteries to the baseline.

6. Spontaneous Equalization of Each Battery SoC

In the proposed battery direct-connected DC grid, distributed batteries with the same terminal voltage are connected by a single baseline, so even if there is a temporary difference in terminal voltage, it will equalize over time. This has been demonstrated by my testbed and will be discussed in Section 9. In this case, the terminal voltage of the battery reflects the battery's SoC, assuming that the terminal voltage of a battery with a higher SoC is high than that of a battery with a lower SoC. This is generally the relationship in typical batteries. As a simple calculation, if a potential difference of 1 V occurs between adjacent batteries, and the electrical resistance of the baseline connecting the batteries is 0.1 Ω (corresponding to the conductor resistance of a 100 m length of 14 SG CV cable), if the baseline voltage is 400 V, 2 kW of power will flow from the battery with a higher terminal voltage to the one with a lower terminal voltage. This results in the spontaneous equalization of the SoC.

In typical microgrids, the SoC of each distributed battery is monitored by a controller installed somewhere, and if there is an imbalance in the SoC between batteries, the controller commands the DC/DC converter connected to the battery to artificially equalize the SoC. When directly connecting batteries to the baseline, such operations are unnecessary for SoC equalization.

This spontaneous SoC equalization by direct connection to the baseline is independent of the capacity of the distributed batteries. Therefore, even if batteries with different capacities are connected to the baseline, spontaneous equalization of SoC occurs for all batteries. Furthermore, even if the batteries are degraded, the typical battery degradation mode only increases internal resistance and does not significantly change the electromotive force value, so there is no significant problem even with slightly degraded batteries.

7. Locality of Autonomous Decentralized Control and a Weakly Coupled Grid

Autonomous distributed control is a method in which each device using the grid controls autonomously, so the control must be localized. In other words, the control of one device should not extend to neighboring devices, and there should be no interference between their controls. For example, a charge controller is used to charge the battery with power generated by solar panels. The charge controller monitors the terminal voltage of the battery being controlled and decides whether to charge or stop charging based on that terminal voltage. Therefore, as shown in Figure 6a, when a single charge controller controls the charging of each battery, this allows for the proper use of the charge controller and poses no problem. However, in this case, the batteries are not connected by power lines, so they are in 'island mode' and no power exchange occurs between them.

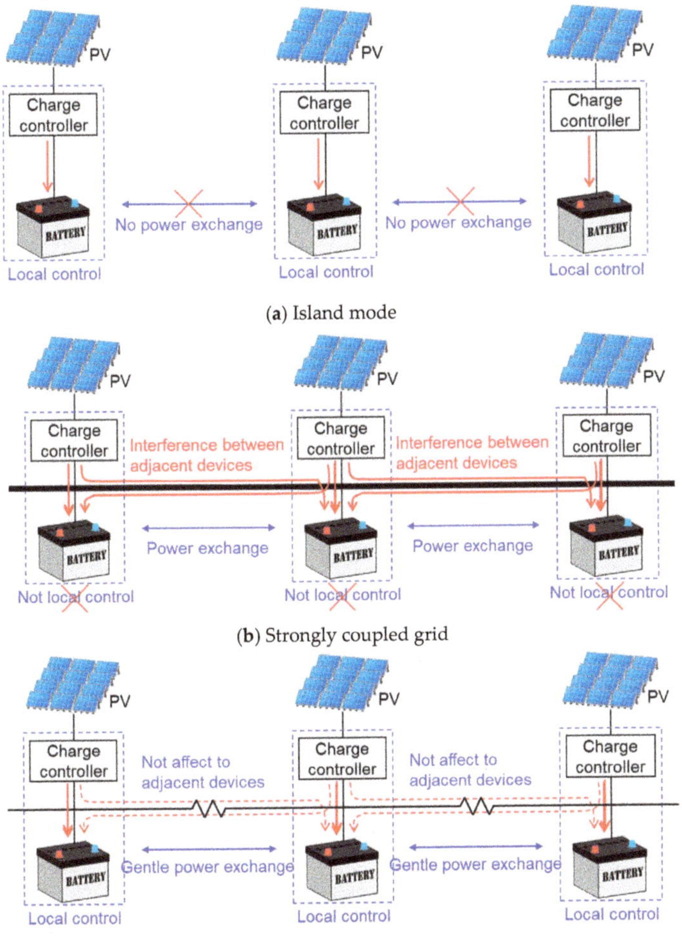

Figure 6. Differences in charging control methods depending on the strength of electrical coupling between batteries: (**a**) island mode; (**b**) strongly coupled grid; (**c**) weakly coupled DC grid.

However, as shown in Figure 6b, when batteries are directly connected to the baseline, they may be affected by the terminal voltage of adjacent directly connected batteries, leading to incorrect charging control as the charge controller cannot accurately grasp the terminal voltage of the intended battery. Similarly, the charge control might extend to adjacent batteries, causing them to be charged as well. Thus, to achieve ADCC, the control must be localized and should not affect neighboring devices. To prevent this, I propose a 'weakly coupled grid' (Figure 6c) with weak electrical coupling between devices connected to the grid.

In existing power grids, generators at power plants, transformers at substations, and power loads at consumers are electrically strongly coupled, forming a tightly coupled grid. Consequently, the baseline voltage is considered constant everywhere, and centralized control is performed. In contrast, my proposed battery directly connected to the DC grid requires a loosely coupled condition where each device is electrically weakly coupled. The strength of the electrical coupling between each device is an indicator of the capacity of the power lines connecting those devices, relative to the rated power of each device (generation power for generators, charge/discharge power for batteries, and consumption power for power loads). Therefore, the lower the electrical resistance of the power lines connecting the devices, the higher the voltage, and the larger the allowable current, the stronger the coupling. As a result, the weakly coupled grid I propose, allowing for a small deviation of the baseline voltage to which each device is connected, even if they are connected by a single baseline. Having devices weakly coupled electrically means each device can maintain the independence of ADCC in the short term while allowing for relatively slow power exchange between devices in the long term.

8. Grid Configuration and Safety Measure

The method of configuring DC baselines with distributed and directly connected batteries is shown in Figure 7. In this configuration, small batteries are distributed and directly connected to the positive and negative baselines, with the midpoint of those batteries grounded. For example, in the case of a baseline voltage of about 50 V, two batteries with an equal terminal voltage of 25 V are connected in series, and the midpoint is grounded. This enhances resistance to lightning strikes and electromagnetic pulse (EMP) attacks and ensures safety against leakage currents. Power generation equipment such as solar panels and power load devices is connected to the positive and negative lines of the baseline to send or receive power. Therefore, if there is no ground fault causing leakage on one of the baseline lines, the same magnitude of current flows in the opposite direction in the positive and negative lines, preventing any external leakage magnetic field. Additionally, if a ground fault occurs, the imbalance in current flow between the positive and negative lines can be easily detected by a clamp-type leakage detector or similar device.

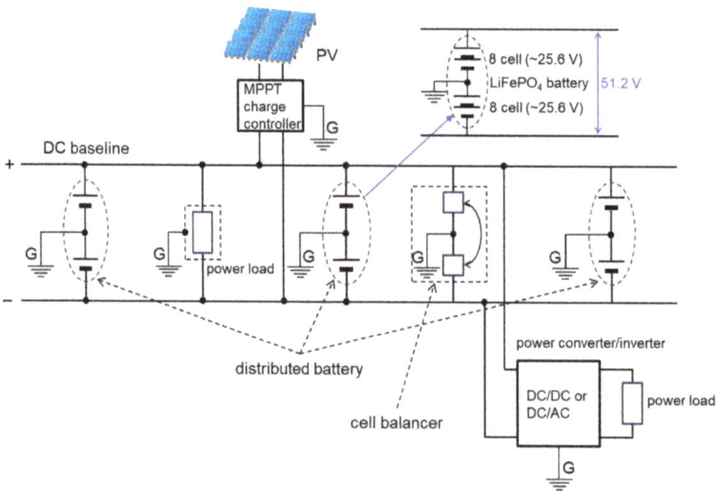

Figure 7. Configuration of the DC baseline.

In this setup, the two lines of the baseline are fixed at +25 V and −25 V with respect to the ground potential, so in the unlikely event of electric shock due to leakage, the voltage shock will not exceed this value. Of course, if both poles of the baseline are touched, one will experience the full 50 V baseline voltage.

In the event of a direct lightning strike on the baseline, a momentary high voltage will occur, possibly damaging storage batteries and equipment near the strike point. However, the damaged storage batteries could serve as a fuse, potentially protecting other equipment. In the case of induced lightning, the directly connected storage batteries function as large capacitors, absorbing the surge current caused by the lightning. Similarly, the effects of magnetic storms or EMP attacks are mitigated due to this function.

Since the same magnitude of current flows in both the positive and negative lines of the baseline, the positive and negative sides of the directly connected batteries always maintain the same SoC and an equal voltage magnitude with opposite polarity relative to the midpoint. To correct any imbalance, a cell balancer with correction capabilities is desirable [38,39].

When directly connecting batteries to the baseline, it is necessary to match the terminal voltage to the baseline voltage. This is not particularly difficult since various types of batteries with different terminal voltages are widely available in the market. The LiFePO4 batteries used in my constructed microgrid have a cell voltage of 3.2 V, with many commercially available options of 25.6 V (8 cells in series) or 51.2 V (16 cells in series). Therefore, as shown in the figure, by connecting these in series, storage batteries that correspond to baseline voltages of 50 V or 400 V can be configured.

In the proposed DC grid, which is characterized by the distribution and direct connection of batteries to the baseline to provide electrical inertia, safety measures are indispensable because large currents flow during ground faults or short circuits in the baseline, potentially leading to major accidents [40,41]. Of course, existing AC power grids are also equipped with circuit breakers (CBs) and fuses in various locations to prepare for such accidents. While existing ultra-high voltage circuits in power grids require special CBs to cut the arc discharge that occurs during circuit disconnection, the DC microgrid envisioned here has a baseline voltage of around several hundred volts, so such special CBs are not considered necessary. However, since many commercially available CBs are for AC use, CBs that can be used with DC are required. When a short circuit or ground fault occurs in the baseline, an abnormal large current flows momentarily. The mechanism to detect this and cut the circuit is the same as existing ones, but a considerably large current capacity (hundreds of amperes) and rapid response are required.

As an example, various safety measures for the DC grid with a baseline voltage of 400 V are shown in Figure 8. By installing overcurrent CBs that detect and cut overcurrent at various locations in the baseline, the shorted section can be isolated on both sides from the baseline, preventing accident escalation. In this way, the installation of overcurrent CBs along the baseline is feasible only in weakly coupled grids. In contrast, in strongly coupled existing grids, it is uncommon to install overcurrent CBs in the middle of transmission or distribution lines.

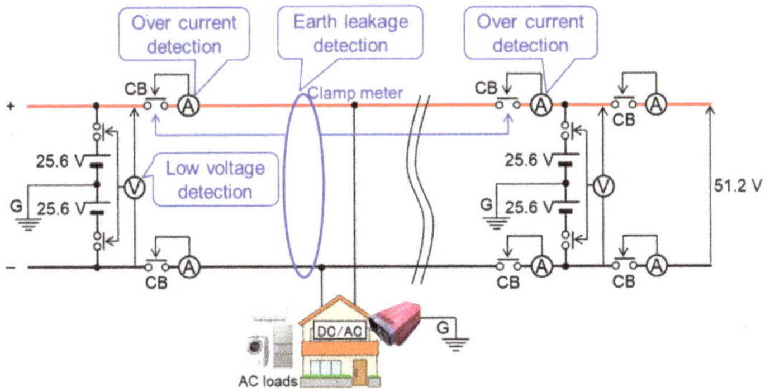

Figure 8. Safety measures for microgrids.

At the very least, as shown in the figure, overcurrent CBs should be installed at all connection points between the distributed batteries and the baseline. In my constructed microgrid, 63 A CBs are installed between each battery and the baseline. Therefore, it is possible to replace each battery (hot swap) without stopping the grid, and CBs are also installed between each power box and the baseline, allowing each power box to be freely connected or disconnected from the baseline.

Furthermore, when a short circuit or ground fault occurs in the baseline, a large discharge current occurs from the batteries directly connected to the nearby baseline, leading to an abnormal drop in terminal voltage. Therefore, it is also effective to install abnormal voltage CBs that detect and cut the circuit upon detecting an abnormal battery terminal voltage. This is also effective for protecting batteries from overcharging or over-discharging. Some lithium-ion batteries are equipped with a BMS function that disconnects from the terminals to protect the battery when the terminal voltage drops below a specified value.

On the other hand, ground fault accidents can be addressed with a breaker based on the same principle as commercially available AC residual current devices (RCDs). In a neutral point grounded DC baseline with the baseline voltage's neutral point grounded as shown in the figure, the magnitude of current flowing in the positive and negative sides of the baseline should normally be equal, but if a ground fault occurs somewhere, this becomes unbalanced and can be detected by a conventional clamp meter or RCD. Therefore, countermeasures can be taken by installing such devices at various locations along the baseline.

9. Demonstration Experiment

To verify the feasibility of the battery directly connected DC microgrid, I constructed a testbed of the microgrid on my university campus, as shown in Figure 9. Power boxes, which house $LiFePO_4$ batteries with a rated terminal voltage of 51.2 V, are installed on the rooftops of four school buildings, namely Buildings 1–4 (B1–B4). These are connected by a power baseline made up of 14 SQ CV cables. Additionally, 800 W solar panels are installed on the rooftops of Buildings 1–3 (B1–B3) and connected to the batteries within each power box through MPPT charge controllers. The baseline is laid in a T-shape with Building 4 (B4) serving as a hub and Buildings 1–3 (B1–B3) being the endpoints of the baseline. The B4 hub is equipped with a battery with double the capacity (240 Ah) compared to the batteries in B1–B3, although it does not have solar panels installed.

The reason for installing the power boxes outdoors on the rooftops of school buildings is to mitigate the risk of fire from lithium-ion batteries, to comply with fire safety regulations, and to prepare for future replacements of existing power distribution grids with DC systems, wherein distributed storage batteries are envisioned to be installed at pole transformer locations. Consequently, the power boxes are waterproofed, lined with insulation on the inside, and further covered with a silver sheet. The batteries have shown no significant damage or degradation even though the outdoor temperatures can drop below freezing in winter and exceed 40 °C during summer days. The power boxes also contain MPPT charge controllers, DC/AC inverters, mini-PCs for monitoring, and switching hubs for wired network links, all of which have been installed for over three years and continue to function normally.

This microgrid operates off-grid (island mode), entirely powered by renewable energy (solar power). The power loads consist of distributed micro-datacenters installed on the rooftops of buildings B1–B3, which consume approximately 100 W of power on a regular basis, and electric vehicle (EV) charging spots. When the baseline voltage is sufficiently high, the EVs are occasionally charged. Table 1 shows the specifications of the microgrid.

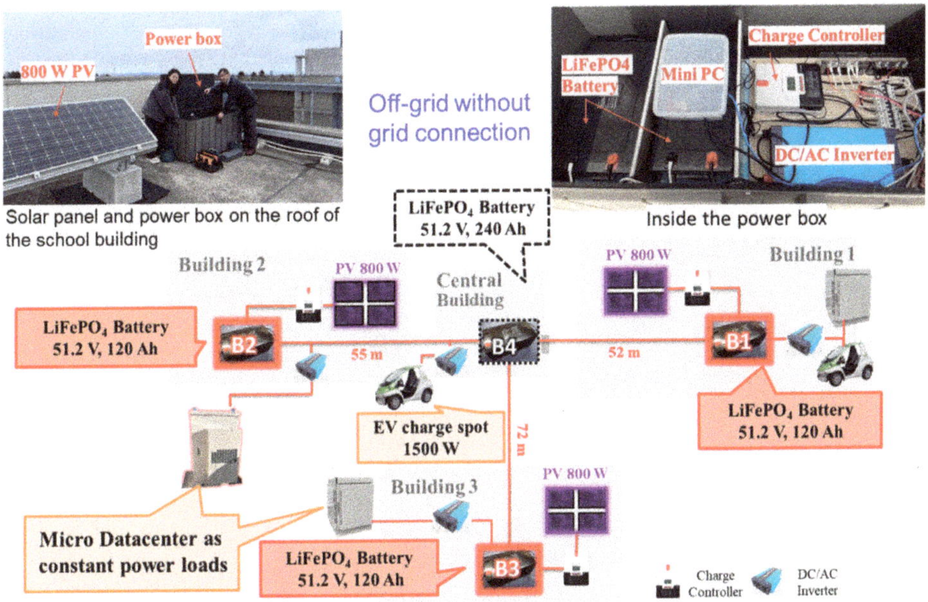

Figure 9. Testbed of the battery directly connected DC microgrids.

Table 1. The specifications of the constructed DC microgrid.

	Battery	Solar Panel	DC/AC Inverter	Monitoring
Building 1	$LiFePO_4$ 51.2 V, 60 Ah × 2	800 W	100 V, 3 kW for EV charging	Battery voltage Battery temperature Charging current PV power generation
Building 2	$LiFePO_4$ 51.2 V, 60 Ah × 2	800 W	100 V, 1.5 kW	Battery voltage Battery temperature Charging current PV power generation
Building 3	$LiFePO_4$ 51.2 V, 60 Ah × 2	800 W	100 V, 1.5 kW	Battery voltage Battery temperature Charging current PV power generation
Building 4	$LiFePO_4$ 51.2 V, 60 Ah × 4	–	100 V, 3 kW for EV charging	Battery voltage Baseline current
Entire grid	$LiFePO_4$ 51.2 V, 600 Ah (~30 kWh)	2.4 kW	–	Grid Code: Baseline voltage: 51~56 V

The devices connected to this grid operate autonomously in a distributed cooperative manner without exchanging information via communication lines. For example, the charge controller of the solar panel performs MPPT control when the terminal voltage of the connected battery is 56 V or lower. However, when it seems likely to exceed 56 V, it switches to float mode to maintain 55.6 V, to prevent overcharging or battery damage.

Additionally, when the battery terminal voltage drops below 51 V, a low-voltage alarm is sent to the administrator via email, leading to manual suppression of power load usage. If the battery voltage continues to drop further, the battery management system (BMS) within the battery stops output to the terminals to prevent over-discharge damage. As an off-grid power source, this microgrid handles such situations by disconnecting power loads before reaching this final measure. Therefore, the baseline voltage fluctuates within the range of 51 V to 56 V even during normal operation, requiring power loads to be designed to operate within this voltage range. In practice, the DC/AC converters used in the system can operate within this voltage range. (For reference, the voltage range of Japan's household 100 V indoor wiring is regulated by Article 38 of the Electricity Business Act Enforcement Regulations to be within the range of 95 to 107 V, not exceeding $\pm$6 V of 101 V).

In this grid, each device connected to the grid is required to comply with the grid usage rule (Grid Code) regarding the baseline voltage, which is to operate within the range of 51 V to 56 V. Therefore, each device autonomously controls its operation to avoid deviating from that range. For example, if the baseline voltage is close to the upper limit of 56 V, and further power supply might exceed the appropriate range, solar power generators will refrain from supplying power, even if they have the potential to generate it. Similarly, if the baseline voltage is close to the lower limit of 51 V, and further power consumption might fall below the appropriate range, the load will not start operating.

Such an ADCC grid might appear imperfect due to its limitations in overall optimization. However, output suppression of renewable energy power generation in existing grids is already common. The critical point is to minimize the frequency of such occurrences. The grid should be designed to ensure that these events are exceedingly rare. In addition, since the batteries of each power box are directly connected to the baseline, any voltage difference at the terminal voltage will naturally be equalized through power exchange via the baseline.

The charge controllers for solar panels are USB-connected to the mini-PCs within the power boxes, and through a wired LAN laid along the power baseline, the generated power, charging power, temperature inside the box, etc., can be monitored in real-time. This setup is solely for monitoring purposes and does not involve any specific control. Figure 10 shows the monitoring screen of the microgrid, allowing monitoring at 1 s intervals for the battery terminal voltage, and at 1 min intervals for the battery temperature, solar panel generation power, power consumption of the connected loads, etc.

The $LiFePO_4$ batteries used in my microgrid are configured with 16 series-connected and 12 parallel-connected 26,650 cells (3.2 V, 3.3 Ah). The relationship between the terminal voltage and SoC (approximate) at no load is estimated as shown in Figure 11. Therefore, the SoC of the batteries near the measurement point of the baseline can be roughly estimated from the baseline voltage. This means that each device connected to the baseline can understand the power situation (the SoC of the distributed battery) around it by sensing the baseline voltage.

Figure 12 shows the time variations in solar power generation and battery terminal voltage (equal to the baseline voltage at that point) at B1, B2, and B3 measured over the course of one month, January 2023 [42]. Multiple solar panels are connected to each battery via charge controllers; however, due to the moderate conductor resistance of the CV cables between each battery, they are loosely coupled, ensuring the independence of each charge controller's regulation. Each device autonomously adheres to the grid regulations, resulting in the battery terminal voltage of each power box ranging from 51 to 56 V. Consequently, even if temporary discrepancies occur due to activities such as EV charging, the terminal voltage of each battery naturally equalizes over time.

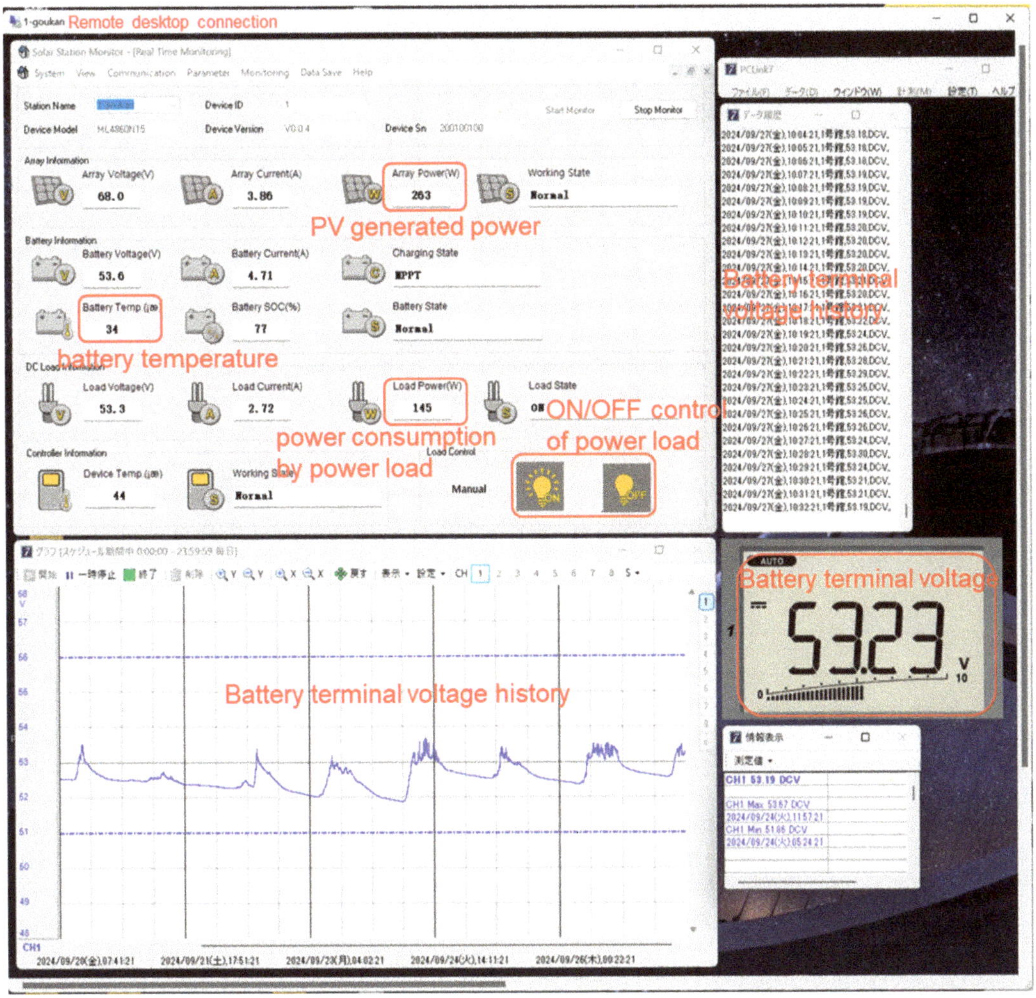

Figure 10. Monitor screen of the microgrid operation.

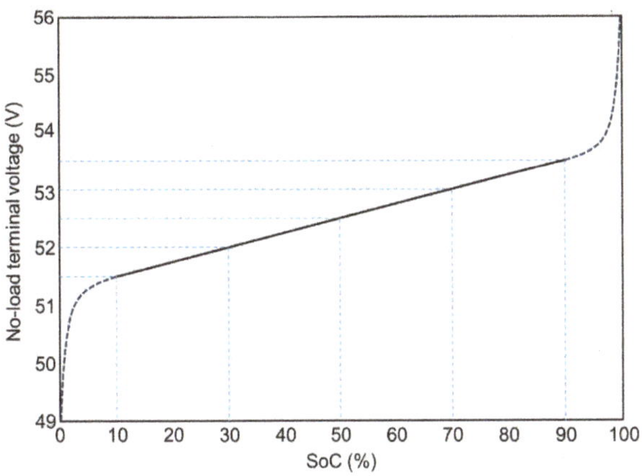

Figure 11. Relationship between battery terminal voltage and SoC.

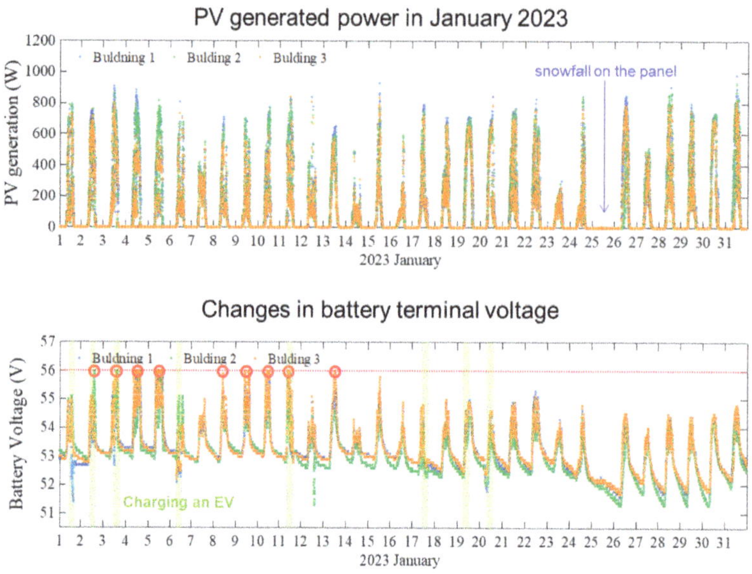

Figure 12. System operation record for one month, January 2023.

In this microgrid, multiple small batteries are directly connected to the baseline, and voltage is always applied to the baseline, allowing for instantaneous power extraction from any point on the baseline. Furthermore, the power supply to the load is coordinated by multiple batteries in the vicinity, distributing the power load across multiple batteries (coordinated power supply). To demonstrate this, an experiment was conducted on a sunny day, 14 April 2023, from 9:40 A.M. to 11:00 A.M., where power was artificially extracted from the B4 power box using an electronic load device.

Figure 13 shows the graphs of (a) solar power generation, (b) the terminal voltage of each battery, and (c) the inflow and outflow current of the B4 power box over time, as recorded during that period. The three buildings, B1 to B3, where the solar panels are installed are adjacent to each other and all have the same rated (800 W) solar panels, resulting in similar solar power generation, as shown in Figure 13a. At the start of the measurement period, the batteries in B1 to B3 were receiving 550–600 W of generated power from each solar panel. The terminal voltage of the batteries in B1 and B3 was high, at around 54 V. On the other hand, the batteries in B4, which did not have solar panels, and B2, which consumed power due to the micro-datacenter, had slightly lower voltages (Figure 13b).

In the experiment, the power consumption of the electronic load device connected to B4 was switched every 10 min in the following order: 0 W, 1200 W, 200 W, 800 W, 600 W, 2000 W, 1500 W, and 0 W. Figure 13c shows that a large current was discharged from the B4 battery, directly connected to the electronic load. In the graph, the positive values on the vertical axis for B1 to B4 represent discharging from each battery, and the negative values represent charging to each battery. At the beginning of the experiment, B4's battery was being charged by the batteries in B1 and B3. The graph indicates that when a large load, such as 2 kW, was connected, the nearby batteries coordinated to supply power (coordinated power supply), supporting the validity of droop control by the battery itself.

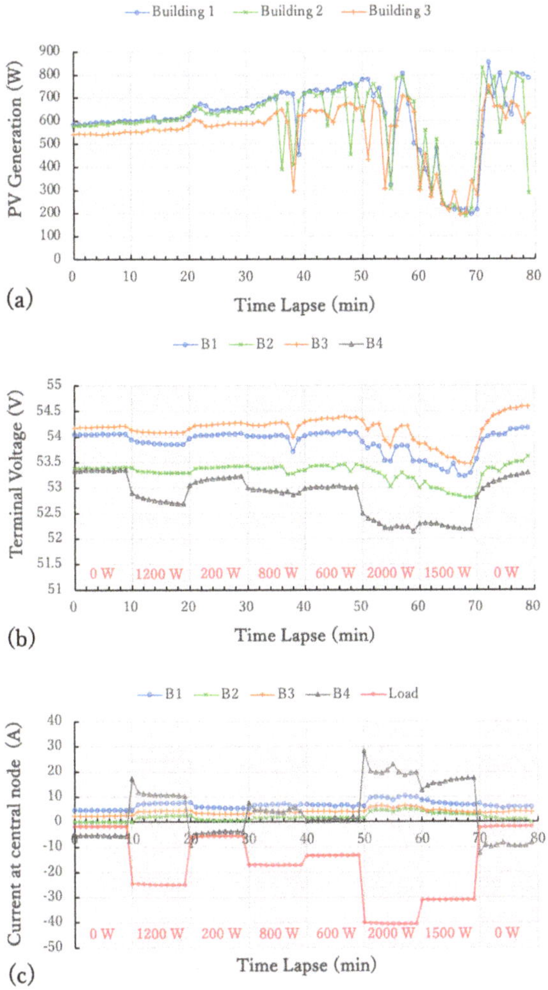

Figure 13. Graph of time change in power load experiment: (**a**) solar power generation, (**b**) terminal voltage of each battery, and (**c**) inflow and outflow current of the B4 power box over time.

In coordinated power supply, the distribution ratio of power supply from each battery is determined solely by the voltage difference at the terminals of each battery and the electrical resistance of the power lines connecting them. Figure 14a shows the simplified equivalent circuit of the baseline and batteries in my DC microgrid. Each battery is represented by a DC voltage source corresponding to its electromotive force and internal resistance, while the baseline cables are modeled as electrical resistances with lumped constants representing their conductor resistance. V_1 to V_4 and r_1 to r_4 represent the electromotive force of each battery (with variations depending on their SoC) and their internal resistance, while R_1 to R_3 mainly represent the conductor resistance of the baseline cables connecting the batteries, and R_4 represents contact resistances such as terminal connections. Table 2 shows measured parameter values of the equivalent circuit for my microgrid. Based on this equivalent circuit, the voltage at each node and the current flowing through each branch can be easily calculated using DC circuit theory. Figure 14b compares the measured baseline currents I_1 to I_4 at B4 with their calculations with the simplified equivalent circuit model, and the two align closely. This demonstrates that the state of the grid can be accurately described using such a simple equivalent circuit model and straightforward calculations, which is one of the features of a battery directly connected DC grid.

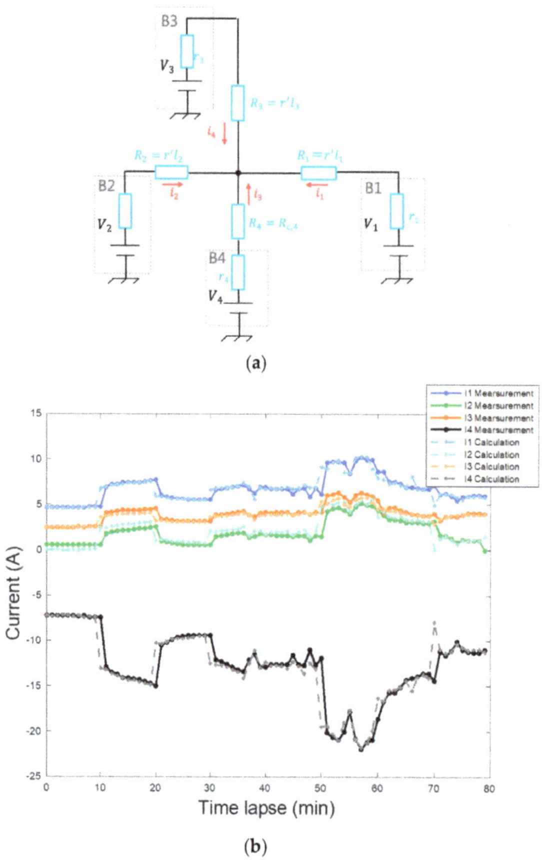

Figure 14. (**a**) Simplified equivalent circuit of the baseline and batteries in my DC microgrid, (**b**) measured baseline currents I_1 to I_4 at B4 and calculations with the simplified equivalent circuit model.

Table 2. Measured parameter values of the equivalent circuit for my microgrid.

r_1~r_3	r_4	R_1	R_2	R_3	R_4
75 mΩ	32 mΩ	142.5 mΩ	170 mΩ	320 mΩ	5.5 mΩ

In this grid, the baseline has a large electrical inertia, so that a large amount of power can be always extracted from any point on the baseline. Next, experiments applying power loads to the batteries on B1 to B3 were conducted to investigate the power supply distribution ratio from each battery in response to the applied loads.

Figure 15 illustrates the power supplied by each battery, including the newly measured results for B4, with different colors used to indicate the contributions. Positive values on the vertical axis represent the discharge power of the batteries, while negative values indicate charging power. During the experiments, solar generated power also flowed into the batteries at the nodes B1 to B3, so the power generation needed to be subtracted to calculate the applied load values. Consequently, the red line labeled "Net Demand" represents the net load power after subtracting the solar power generation. A negative value for this line indicates that the solar power generation exceeded the applied load power. From Figure 15, it can be observed that the battery connected to the applied load naturally had the highest contribution ratio. However, nearby batteries also shared a substantial portion of the load, demonstrating the effectiveness of coordinated power supply.

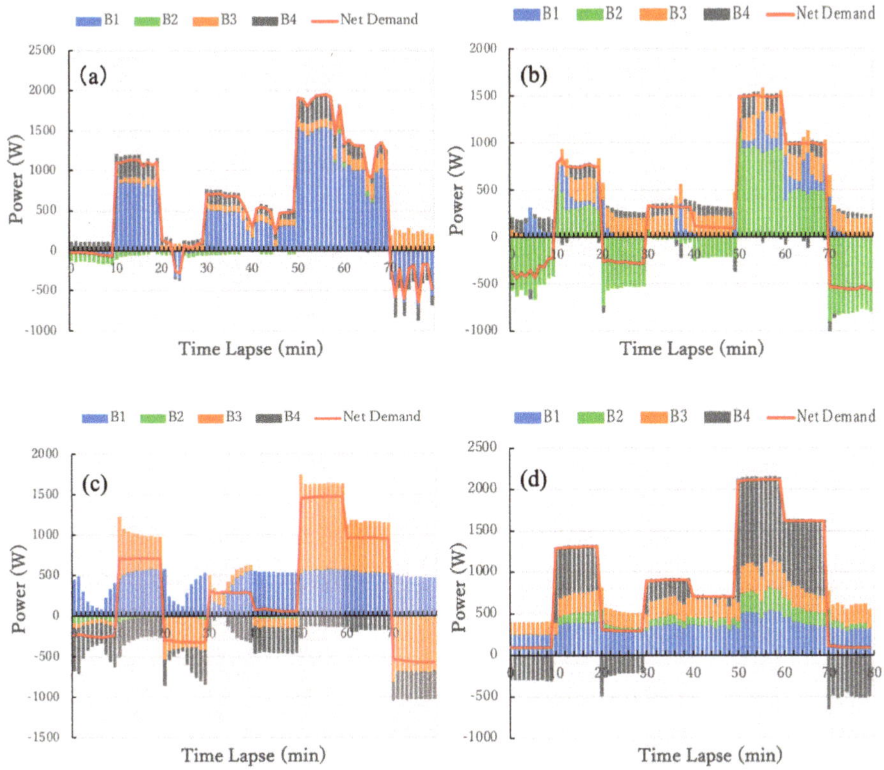

Figure 15. Power supply distribution of each battery when the power load is connected to (**a**) B1, (**b**) B2, (**c**) B3, and (**d**) B4.

Finally, I would like to mention the disaster resilience of this microgrid. The greatest feature of the weakly coupled ADCC DC grid is that even if a part of the baseline is cut and the grid is split into two, the two separated grids can continue to function normally as independent ADCC grids. In my testbed, CBs are installed between each power box and the baseline, and it has been confirmed that there is no change in the grid when they are disconnected at any moment. The two grids that are separated can both continue to operate normally. Additionally, if the baseline voltages of the two grids are almost equal, they can be reconnected to form a single grid again. In other words, the grid can be disconnected and reconnected at any time with a single touch, which also underscores its strong disaster resilience.

10. Interconnection with Existing Power Grid

In microgrids that rely primarily on unstable renewable energy power generation, stable power supply remains a significant challenge. Conversely, existing power grid systems are vulnerable to natural disasters and face destabilization issues as the integration of renewable energy progresses. By interconnecting with existing power grid systems, I aim to cover the shortcomings of both systems, achieving a green, resilient, and stable power supply.

Figure 16 illustrates the method of interconnection between both grids. Within distributed regional microgrids, the goal is to utilize locally generated renewable energy, such as small-scale solar, wind, and hydroelectric power, to the fullest extent, while maintaining power supply through the provision of engine-driven emergency power sources in case of power shortages. However, equipping each microgrid with its own emergency power source is not efficient. By interconnecting with existing power grid systems, power can be received from the grid during shortages. Therefore, while each microgrid aims for partial optimization of power distribution through ADCC, the existing power grid system aims for overall optimization by equalizing power distribution among microgrids. This division of roles leverages the strengths of both microgrids and existing power grid systems. In this case, short-term and small-scale power fluctuations are addressed by the electrical inertia within each microgrid (momentary response), while long-term and large-scale fluctuations are addressed by power supply from the grid (sustained response).

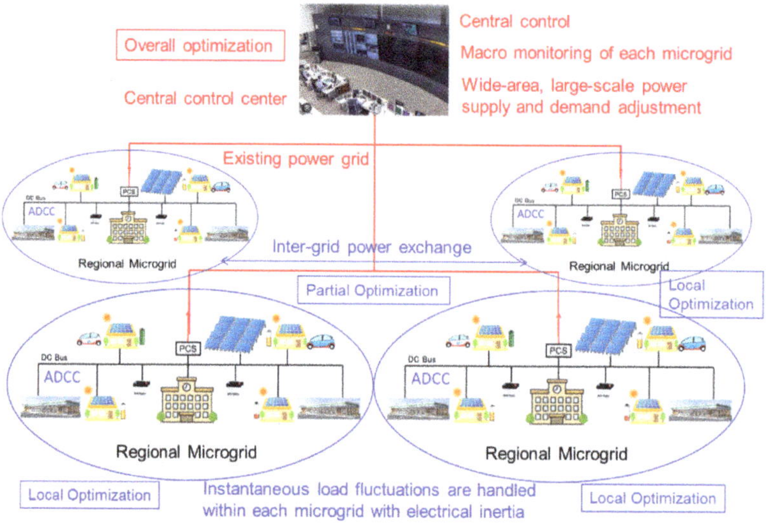

Figure 16. Interconnection with existing power grid.

Interconnection with the existing power grid also enables power exchange between microgrids. If renewable energy generation within a microgrid exceeds consumption and storage batteries become fully charged, off-grid systems will have to halt power generation. However, with interconnection, surplus power can be fed back into the grid and utilized by microgrids experiencing power shortages, thereby effectively utilizing the renewable energy potential without waste. Utilizing the existing power grid allows for wide-area power exchange and realizes overall optimization of power distribution.

11. Conclusions

This paper proposes a novel approach for constructing a DC grid by directly connecting small batteries along the DC bus in a distributed manner and introduces a method for achieving autonomous decentralized cooperative control (ADCC).

Conventional DC grid systems have treated the DC baseline, which is situated at the lowest physical layer of the grid's system hierarchy, as mere wiring for power exchange among devices. Consequently, essential grid-stabilizing functions such as electrical inertia and droop control for coordinated operation between distributed power sources were implemented within the upper control layer via DC/DC converters with pseudo-inertia or configured droop parameters. In contrast, the battery-integrated DC baseline proposed herein is not just a passive conductor but serves as a "functional powerline" by endowing it with electrical inertia and enabling functions such as coordinated power supply and distribution balancing among connected batteries. This was realized by leveraging the intrinsic droop characteristics of batteries. By directly connecting batteries to the baseline, their inherent droop properties were utilized to achieve coordinated operation among distributed batteries without relying on DC/DC converters. Additionally, this architecture allows for significant electrical inertia to permeate throughout the grid, contributing to grid stability. Furthermore, spontaneous power exchange among batteries directly connected to the baseline naturally resolves temporary imbalances in state of charge (SoC) over time.

We also explored the ADCC method tailored for DC grids employing this battery-integrated baseline. For ADCC to be realized, it is crucial for each device connected to the grid to grasp the state of the grid. Unlike centralized control, it is difficult for individual devices to have a comprehensive understanding of the state of the entire grid, but they can have some understanding of the state of the grid in their own neighborhood. In the proposed architecture, the SoCs of nearby batteries are reflected in the baseline voltage. This enables individual devices to infer the local grid state (e.g., power conditions) from the baseline voltage. Moreover, maintaining the independence of each device's control is imperative, for which methods to ensure the locality of device control were examined. I identified that weak electrical coupling among devices connected to the baseline is essential to achieving ADCC. In other words, electrical control should not extend beyond adjacent devices, while still allowing for gradual power exchange over the long term, necessitating an appropriate coupling strength.

To validate the feasibility of this approach, a DC microgrid testbed with battery-integrated baselines was constructed on a university campus, and the fluctuations of photovoltaic (PV) generation and baseline voltage were monitored over a one-month period. The results confirmed the locality of PV charging control for individual distributed batteries, demonstrating that PV charge controllers accurately regulated only their respective batteries. Consequently, the baseline voltage was maintained within the optimal range of 51–56 V, and temporary SoC imbalances caused by electric vehicle (EV) charging naturally equalized over time. Furthermore, experiments with electronic loads at various baseline points confirmed the coordinated operation of distributed batteries near the loads, effectively supplying power collaboratively.

While numerous advantages of the proposed architecture have been demonstrated, challenges remain. Firstly, the terminal voltage of batteries directly connected to the baseline must match the baseline voltage within a certain range. Preparing enough of such batteries is a potential issue. My testbed was small, and I built a baseline to fit 10 available batteries with terminal voltages of 51.2 V. Whether suitable batteries are widely available for the commonly utilized baseline voltage range of 350–400 V remains uncertain. Additionally, my testbed utilized new batteries in experiments spanning less than three years. Long-term issues related to battery degradation over 10 or 20 years warrant further investigation.

Safety concerns also present a challenge. While direct connecting batteries provide significant electrical inertia, they also pose risks, such as substantial current flow from nearby batteries in the event of a short circuit or ground fault. Although safety measures were incorporated into my testbed, and no accidents occurred during testing, the effectiveness of these measures under actual fault conditions remains to be confirmed.

Despite these challenges, the biggest feature of this method is that it delegates basic grid-operating functions such as electrical inertia and power distribution balancing, which were conventionally handled by the higher-level control layer, to the physical layer of the baseline. This approach alleviates the burden on the control layer, allowing it to focus on more advanced and efficient operations. By integrating the proposed "functional powerline" into existing DC grids to form a composite system, basic control tasks such as grid stabilization and power distribution balancing can be managed at the baseline level. Consequently, the higher control layers and application layers of the grid system can be dedicated to more sophisticated applications, including energy management systems (EMSs) or power trading, facilitating the development of application-centric systems.

The proposed ADCC-based DC microgrid demonstrates resilience, as even in cases of partial system failure, the remaining components can maintain functionality to some extent. This characteristic makes it particularly suitable for residential areas in disaster-prone regions. Equipping individual households with small batteries interconnected via weak DC baselines will enable sharing of locally generated power, such as solar energy, among neighboring households—potentially paving the way toward a resilient and community-based power infrastructure.

Funding: This research was funded by the JST OPERA Prog. (Grant Number JPMJOP1852).

Data Availability Statement: The original contributions presented in this study are included in the article. Further inquiries can be directed to the corresponding author.

Acknowledgments: The author would like to extend a special thanks to Liu Ke, who was a student in the author's laboratory. Part of the content of this paper is based on the research conducted as part of her thesis under the guidance of the author. Additionally, through the JST OPERA project, the author had many valuable discussions with K. Iwatsuki and T. Otsuji. The author would like to express deep gratitude to them once again.

Conflicts of Interest: The author declares no conflicts of interest. The funders had no role in the design of this study; in the collection, analyses, or interpretation of data; in the writing of the manuscript; or in the decision to publish the results.

Abbreviations

The following abbreviations are used in this manuscript:

AC	alternating current
ADCC	autonomous decentralized cooperative control
BMS	battery management system
CB	circuit breaker
CO_2	carbon dioxide
CV	cross-linked polyethylene insulation vinyl sheath
DC	direct current
EMP	electromagnetic pulse
EV	electric vehicle
LAN	local area network
MPPT	maximum power-point tracking
PC	personal computer
PV	photovoltaic (solar cell)
RCD	residual current device

SoC	state of charge for batteries
SQ	square meters
USB	universal serial bus

References

1. Marnay, C.; Xu, T.; Hatziargyriou, N.D.; Hirase, Y.; Mendoza-Araya, P. Microgrids 2023 editorial. *Appl. Energy* **2023**, *352*, 121981. [CrossRef]
2. Uddin, M.; Mo, H.; Dong, D.; Elsawah, S.; Zhu, J.; Guerrero, J.M. Microgrids: A review, outstanding issue and future trends. *Energy Storage Rev.* **2023**, *49*, 101127. [CrossRef]
3. Punitha, S.; Subramaniam, N.P.; Vimal Raj, P.; Ajay, D. A comprehensive review of microgrid challenges in architectures, mitigation approaches, and future directions. *J. Electr. Syst. Inf. Technol.* **2024**, *11*, 60. [CrossRef]
4. Miyamoto, Y.; Hasegawa, M.; Kyoto, T. Isumi-City regional micro-grid current situation. *Jpn. Sol. Energy Soc. JSES Conf.* **2024**, 31–32. (In Japanese) [CrossRef]
5. Regional Microgrids Program in Australia, Australian Government Australian Renewable Energy Agency. Available online: https://arena.gov.au/funding/rmp/ (accessed on 2 March 2025).
6. Elsayed, A.T.; Mohamed, A.A.; Mohammed, O.A. DC microgrids and distribution systems: An overview. *Electr. Power Syst. Res.* **2015**, *119*, 407–417. [CrossRef]
7. Dragičević, T.; Lu, X.; Vasquez, J.C.; Guerrero, J.M. DC Microgrids—Part I: A Review of Control Strategies and Stabilization Techniques. *IEEE Trans. Power Electron.* **2015**, *31*, 4876–4891. [CrossRef]
8. Dragičević, T.; Lu, X.; Vasquez, J.C.; Guerrero, J.M. DC Microgrids—Part II: A Review of Power Architectures, Applications, and Standardization Issues. *IEEE Trans. Power Electron.* **2016**, *31*, 3528–3549. [CrossRef]
9. Kumara, J.; Agarwal, A.; Singh, N. Design, operation and control of a vast DC microgrid for integration of renewable energy sources. *Renew. Energy Focus* **2020**, *34*, 17–36. [CrossRef]
10. Al-Ismail, F.S. DC Microgrid Planning, Operation, and Control: A Comprehensive Review. *IEEE Access* **2021**, *9*, 36154–36172. [CrossRef]
11. Rangarajan, S.S.; Raman, R.; Singh, A.; Shiva, C.K.; Kumar, R.; Sadhu, P.K.; Collins, E.R.; Senjyu, T. DC Microgrids: A Propitious Smart Grid Paradigm for Smart Cities. *Smart Cities* **2023**, *6*, 1690–1718. [CrossRef]
12. Lidula, N.W.A.; Rajapakse, A.D. Microgrids research: A review of experimental microgrids and test systems. *Renew. Sustain. Energy Rev.* **2011**, *15*, 186–202. [CrossRef]
13. Hossain, E.; Kabalci, E.; Bayindir, R.; Perez, R. Microgrid testbeds around the world: State of art. *Energy Convers. Manag.* **2014**, *86*, 132–153. [CrossRef]
14. Cagnano, A.; Tuglie, D.E.; Mancarella, P. Microgrids: Overview and guidelines for practical implementations and operation. *Appl. Energy* **2020**, *258*, 114039. [CrossRef]
15. Cañizares, C.A.; Palma-Behnke, R.; Olivares, D.E.; Mehrizi-Sani, A.; Etemadi, A.H.; Iravani, R.; Kazerani, M.; Hajimiragha, A.H.; Gomis-Bellmunt, O.; Saeedifard, M.; et al. Trends in Microgrid Control. *IEEE Trans. Smart Grid* **2014**, *5*, 1905–1919. [CrossRef]
16. Meng, L.; Shafiee, Q.; Trecate, G.F.; Karimi, H.; Fulwani, D.; Lu, X.; Guerrero, J.M. Review on Control of DC Microgrids and Multiple Microgrid Clusters. *IEEE J. Emerg. Sel. Top. Power Electron.* **2017**, *5*, 928–948. [CrossRef]
17. Kuma, J.; Agarwal, A.; Agarwal, V. A review on overall control of DC microgrids. *J. Energy Storage* **2019**, *21*, 113–138. [CrossRef]
18. Papadimitriou, C.N.; Zountouridou, E.I.; Hatziargyriou, N.D. Review of hierarchical control in DC microgrids. *Electr. Power Syst. Res.* **2015**, *122*, 159–167. [CrossRef]
19. Guerrero, J.M.; Vasquez, J.C.; Matas, J.; García de Vicuña, L.; Castilla, M. Hierarchical Control of Droop-Controlled AC and DC Microgrids—A General Approach Toward Standardization. *IEEE Trans. Ind. Electron.* **2011**, *58*, 158–172. [CrossRef]
20. Wang, C.J.P.; Xiao, J.; Tang, Y.; Choo, F.H. Implementation of Hierarchical Control in DC Microgrids. *IEEE Trans. Ind. Electron.* **2014**, *61*, 4032–4042. [CrossRef]
21. Shuai, Z.; Fang, J.; Ning, F.; Shen, Z.J. Hierarchical structure and bus voltage control of DC microgrid. *Renew. Sustain. Energy Rev.* **2018**, *82*, 3670–3682. [CrossRef]
22. Kaur, A.; Kaushal, J.; Basak, P. A review on microgrid central controller. *Renew. Sustain. Energy Rev.* **2016**, *55*, 338–345. [CrossRef]
23. Wang, B.; Sechilariu, M.; Locment, F. Intelligent DC Microgrid with Smart Grid Communications: Control Strategy Consideration and Design. *IEEE Trans. Smart Grid* **2012**, *3*, 1248–2156. [CrossRef]
24. Saleh, M.; Esa, Y.; Mohamed, A.A. Communication-Based Control for DC Microgrids. *IEEE Trans. Smart Grid* **2019**, *10*, 2180–2195. [CrossRef]
25. Mehdi, M.; Kim, C.-H.; Saad, M. Robust Centralized Control for DC Islanded Microgrid Considering Communication Network Delay. *IEEE Access* **2020**, *8*, 77765–77778. [CrossRef]
26. Azizi, A.; Peyghami, S.; Mokhtari, H.; Blaabjerg, F. Autonomous and decentralized load sharing and energy management approach for DC microgrids. *Electr. Power Syst. Res.* **2019**, *177*, 106009. [CrossRef]

27. Farshad, A. Autonomous DC Microgrid with Self-Configurable Feasibility. Master's Thesis, Michigan Technological University, Houghton, MI, USA, 2016. [CrossRef]
28. Yoshida, N.; Kikuchi, A.; Shimada, T.; Ide, K. Autonomous Decentralized DC Bus Voltage Control using DC Multi Power Units in DC Microgrid Applications. *IEEJ Trans. Ind. Appl.* **2023**, *143*, 564–569. [CrossRef]
29. Chen, D.; Xu, L. Autonomous DC Voltage Control of a DC Microgrid with Multiple Slack Terminals. *IEEE Trans. Power Syst.* **2012**, *27*, 1897–1905. [CrossRef]
30. Chen, D.; Xu, L.; Yao, L. DC Voltage Variation Based Autonomous Control of DC Microgrids. *IEEE Trans. Power Deliv.* **2013**, *28*, 637–648. [CrossRef]
31. Vijayaragavan, R.; Umamaheswari, B. Decentralized control of autonomous DC microgrids with composite loads: An approach using optimal control. *Electr. Eng.* **2021**, *103*, 2871–2885. [CrossRef]
32. Richard, L.; Boudinet, C.; Ranaivoson, S.A.; Rabarivao, J.O.; Befeno, A.E.; Frey, D.; Alvarez-Hérault, M.-C.; Raison, B.; Saincy, N. Development of a DC Microgrid with Decentralized Production and Storage: From the Lab to Field Deployment in Rural Africa. *Energies* **2022**, *15*, 6727. [CrossRef]
33. Lu, X.; Guerrero, J.M.; Sun, K.; Vasquez, J.C. An Improved Droop Control Method for DC Microgrids Based on Low Bandwidth Communication with DC Bus Voltage Restoration and Enhanced Current Sharing Accuracy. *IEEE Trans. Power Electron.* **2014**, *29*, 1800–1812. [CrossRef]
34. Peyghami, S.; Mokhtari, H.; Loh, P.C.; Davari, P.; Blaabjerg, F. Distributed Primary and Secondary Power Sharing in a Droop-Controlled LVDC Microgrid with Merged AC and DC Characteristics. *IEEE Trans. Smart Grid* **2018**, *9*, 2284–2294. [CrossRef]
35. Dragicevi, T.; Guerrero, J.M.; Vasquez, J.C.; Skrlec, D. Supervisory Control of an Adaptive-Droop Regulated DC Microgrid With Battery Management Capability. *IEEE Trans. Power Electron.* **2014**, *29*, 695–706. [CrossRef]
36. Zhu, X.; Meng, F.; Xie, Z.; Yue, Y. An Inertia and Damping Control Method of DC–DC Converter in DC Microgrids. *IEEE Trans. Energy Convers.* **2020**, *35*, 799–807. [CrossRef]
37. Wu, W.; Chen, Y.; Luo, A.; Zhou, L.; Zhou, X.; Yang, L.; Dong, Y.; Guerrero, J.M. A Virtual Inertia Control Strategy for DC Microgrids Analogized with Virtual Synchronous Machines. *IEEE Trans. Ind. Electron.* **2017**, *64*, 6005–6016. [CrossRef]
38. Chowdhury, S.; Shaheed, M.N.B.; Sozer, Y. State-of-Charge Balancing Control for Modular Battery System with Output DC Bus Regulation. *IEEE Trans. Transp. Electrif.* **2021**, *7*, 2181–2193. [CrossRef]
39. Lu, X.; Sun, K.; Guerrero, J.M.; Vasquez, J.C.; Huang, L. State-of-Charge Balance Using Adaptive Droop Control for Distributed Energy Storage Systems in DC Microgrid Applications. *IEEE Trans. Ind. Electron.* **2014**, *61*, 2804–2815. [CrossRef]
40. Mohammadi, J.; Ajaei, F.B.; Stevens, G. Grounding the DC Microgrid. *IEEE Trans. Ind. Appl.* **2019**, *55*, 4490–4499. [CrossRef]
41. de Oliveira, T.R.; Bolzon, A.S.; Donoso-Garcia, P.F. Grounding and safety considerations for residential DC microgrids. In Proceedings of the IECON 2014—40th Annual Conference of the IEEE Industrial Electronics Society, Dallas, TX, USA, 29 October 2014. [CrossRef]
42. Liu, K.; Yamada, H.; Iwatsuki, K.; Otsuji, T. Experimental Verification and Simulation Analysis of a Battery Directly Connected DC-Microgrid System. *Int. J. Electr. Electron. Eng. Telecommun.* **2023**, *12*, 326–333. [CrossRef]

electronics

MDPI

Article

Human-Centric Microgrid Optimization: A Two-Time-Scale Framework Integrating Consumer Behavior

Ke Zeng [1], Hanqing Yang [1,*], Tieshan Li [1,2] and Yue Long [1]

1 School of Automation Engineering, University of Electronic Science and Technology of China, Chengdu 611731, China; 202221060738@uestc.com (K.Z.); litieshan073@uestc.edu.cn (T.L.); longyue@uestc.edu.cn (Y.L.)

2 Laboratory of Electromagnetic Space Cognition and Intelligent Control, Beijing 100089, China

* Correspondence: hqyang5517@uestc.edu.cn

Abstract: This paper presents a two-time-scale human-centric microgrid optimization framework, developed based on singular perturbation theory. First, a comprehensive model is constructed, integrating the electrical characteristics of microgrid components with the evolutionary dynamics of consumer behavior. Subsequently, the system is decomposed into distinct fast and slow time scales using singular perturbation theory, enabling the effective separation of rapid electrical responses from the slower motivational dynamics of consumer behavior. Tailored optimal control strategies are then formulated for each time scale to ensure the rapid stabilization of fast system dynamics in response to transient disturbances and the gradual optimization of slow system dynamics under steady-state conditions. Finally, the proposed approach is validated through preliminary numerical simulations, which demonstrate its potential effectiveness in maintaining microgrid stability under transient conditions, facilitating behavioral adaptation, and improving operational efficiency of the microgrid.

Keywords: consumer behavior dynamics; human-centric microgrid systems; optimal control strategies; singular perturbation theory; two-time-scale optimization

1. Introduction

The increasing demand for clean energy and the shift toward decentralized power systems have made microgrids a critical component of modern energy infrastructure. Microgrids offer enhanced reliability, flexibility, and efficiency by integrating distributed energy resources (DERs) such as solar and wind power [1–5]. The development of operational architectures and energy management systems for multiple microgrid clusters further enhances their potential in complex energy systems. Additionally, recent advancements in microgrid planning, operation, and control have laid the foundation for the application and promotion of various control methods [6–9]. However, beyond these technological developments, consumer behavior plays an equally crucial role in shaping the future of microgrid systems.

In response to evolving energy demands, traditional energy management models have transformed into intelligent, human-centric, and distributed frameworks, where consumer behavior plays a pivotal role in optimizing energy consumption and system efficiency [10–15]. Guttromson et al. [16] first introduced a personalized residential energy model in 2003, emphasizing the influence of appliance usage and consumer behavior on energy efficiency, which was later advanced by Aksanli et al. [17] through user-driven smart scheduling to optimize consumption. Moreover, the rise of distributed power trading and cooperative mechanisms has introduced new challenges to self-regulation in energy internet systems. Teng et al. [18] demonstrated how consumer behavior mitigates price fluctuations through distributed trading, an area further explored by Tushar et al. [19] using game theory in P2P energy trading. Moreover, Cintuglu et al. [20] analyzed its effects on microgrid frequency control and stability. Furthermore, active consumer participation is crucial for the success of community energy projects. O'Neill-Carrillo et al. [21] demonstrated the shift from passive consumers to active participants in Caribbean energy projects, which significantly boosted local socio-economic development, while Fragniere et al. [22] examined user acceptance through ethnographic methods and the SECI model, proposing strategies for microgrid adoption. Indeed, in microgrid control, consumer behavior plays a critical role in system dynamics. Cucuzzella et al. [23] were the first to integrate psychological and social motivations into the distributed control of DC microgrids, revealing the profound impact of consumer behavior on system stability. Building on this, Feng et al. [24] introduced a human-cyber-physical framework to further explore the interaction between consumer behavior and microgrid dynamics. Doumen et al. [25] quantified the effects of different decision-making motives on distribution grid operations through simulations, offering new insights into microgrid control strategies.

Although consumer behavior is critical in microgrid management, the dynamic complexity of these systems equally demands sophisticated control strategies. Previous research has predominantly focused on single time-scale control [26–31]. However, in recent years, multi-time-scale control has garnered increasing attention and become a topic of significant interest. In theoretical research, the foundation of two-time-scale control was laid by Kokotović et al. [32], who utilized singular perturbation theory to decouple rapid and slow system dynamics. Litkouhi and Khalil [33] later refined this approach by extending it to multirate control specifically in discrete-time systems. Building on these foundations, Zhang et al. [34] and Yu et al. [35] introduced event-triggered mechanisms to enhance the robustness of cyber-physical systems, asynchronously regulating fast and slow dynamics. The framework was further expanded in [36] to account for uncertainties in nonlinear stochastic systems, offering a more comprehensive control strategy. In the realm of microgrid applications, Zhang et al. [37] proposed a two-time-scale energy management model that integrates day-ahead and intraday scheduling, optimizing both economic efficiency and system reliability under the fluctuations of renewable energy. This concept was advanced in [38] with a hierarchical frequency stability control strategy, specifically designed for islanded microgrids, utilizing multi-time scale to ensure rapid dynamic responses and long-term stability. Furthermore, Bao et al. [39] and Zhang et al. [40] demonstrated the effectiveness of multi-time-scale coordination in optimizing the management of diverse energy resources, thereby improving overall system efficiency and stability. Further advancements have tackled the increasingly complex interactions within microgrids. Hua et al. [41] introduced a three-time-scale hybrid control algorithm to manage the interaction between inverters and batteries in islanded microgrids, ensuring large-signal stability and operational reliability.

Despite the fact that prior studies have offered valuable insights into energy management, they have largely overlooked the intricate interplay between consumer behavior and microgrid dynamics, particularly across two-time scales. This gap in the literature highlights the need for a more integrated control approach. Consequently, to bridge this gap, we propose a novel two-time-scale control framework that explicitly incorporates consumer behavior into microgrid dynamics. By leveraging singular perturbation theory, we decouple the microgrid system into fast and slow dynamics, thereby facilitating the development of optimized control strategies that specifically address the distinct challenges posed by each time scale. Fast dynamics address immediate physical system responses, while slow dynamics capture the gradual evolution of consumer behavior influenced by motivations and interventions. The main contributions of this paper are as follows:

1. A novel hierarchical optimization control framework that independently manages system dynamics across different time scales. It effectively handles the interaction between electrical dynamics and consumer behavior in microgrids and can be extended to other complex multi-time-scale systems. This framework ensures system stability and operational efficiency in dynamic environments.
2. A two-time-scale control method is introduced for microgrids integrating consumer behavior. Using singular perturbation theory, it decouples fast electrical responses from slower consumer behavior evolution, allowing optimized control for both transient and steady-state dynamics, improving response speed and long-term stability.

The remainder of this paper is organized as follows: Section 2 introduces the preliminaries and problem formulation, including integrated system modeling and the decoupling approach. Section 3 elaborates on the primary findings, including the optimization challenges for both fast and slow dynamics, along with the design of optimal controllers, and it offers a comprehensive stability analysis of the proposed control strategy. Section 4 presents numerical simulations to validate our method, followed by conclusions in Section 5.

2. Preliminaries and Problem Formulation

This chapter introduces a framework that integrates physical systems with consumer behavior. By first presenting both physical and behavioral models, the chapter elucidates how consumer behavior is incorporated into system dynamics. Subsequently, singular perturbation theory is employed to decouple fast and slow dynamics, laying a solid foundation for optimal control strategies.

2.1. System Modeling

The microgrid system analyzed in this study, as shown in Figure 1, consists of N interconnected prosumers, each comprising a distributed generation unit (DGU) and associated loads. These prosumers are connected through communication networks and physical links to enable the exchange of energy and information, closely simulating real-world operational scenarios. Within this framework, the load characteristics of each prosumer are dynamically influenced by consumer behavior. Consumer behavior is shaped by both external social interventions (such as economic incentives, price regulations, and policy initiatives) and intrinsic motivations. These intrinsic motivations arise from the interaction of two dominant value systems: hedonic values, which prioritize personal satisfaction and immediate gratification, and biospheric values, which emphasize environmental responsibility and sustainable development. Together, these factors determine consumers' energy usage patterns and consistently influence the load characteristics of the microgrid.

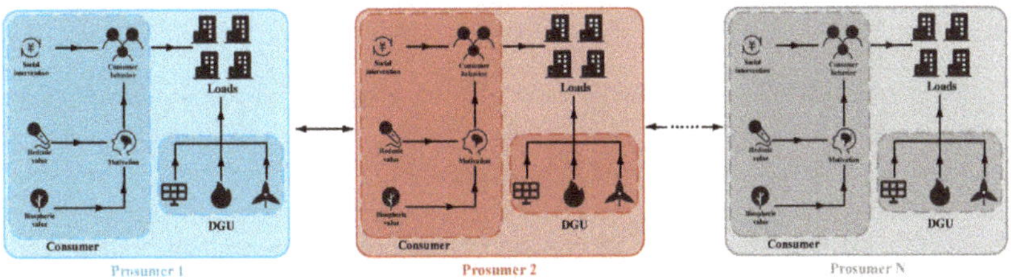

Figure 1. Two-time-scale microgrid framework: integrating consumer behavior with load and generation dynamics.

The internal electrical configuration of each prosumer is shown in Figure 2. The distributed generation unit DGU consists of a power source and a voltage source converter (VSC), which together form the core components for energy conversion and regulation. The load connected to the DGU represents consumer demands, and prosumers are interconnected via transmission lines, enabling bidirectional energy and information exchange within the microgrid. As depicted in Figure 2, the dynamic processes at the physical level of the microgrid proceed as follows:

$$C_f\dot{V}_d = -R_s^{-1}V_d + \omega_r C_f V_q + I_{td} + \mathcal{B}I_d - I_{Ld}g, \quad (1a)$$
$$C_f\dot{V}_q = -\omega_r C_f V_d - R_s^{-1}V_q + I_{tq} + \mathcal{B}I_q - I_{Lq}g, \quad (1b)$$
$$L_f\dot{I}_{td} = -V_d - R_f I_{td} + \omega_r L_f I_{tq} + u_d, \quad (1c)$$
$$L_f\dot{I}_{tq} = -V_q - R_f I_{tq} - \omega_r L_f I_{td} + u_q, \quad (1d)$$
$$L\dot{I}_d = -\mathcal{B}^T V_d - RI_d + \omega_r LI_q, \quad (1e)$$
$$L\dot{I}_q = -\mathcal{B}^T V_q - \omega_r LI_d - RI_q, \quad (1f)$$

where the subscripts d and q denote the direct and quadrature components, respectively. C_f and L_f represent the filter capacitance and inductance, which define the system's filtering characteristics. g represents the consumer load, while R_f and R_s are the system and load resistances, respectively, which influence power dissipation. V_d and V_q represent the load voltages, while I_{td} and I_{tq} are the output currents of the DGU. I_d and I_q represent the currents along the transmission lines, with R and L corresponding to the resistance and inductance of the transmission lines, respectively. The system dynamics are governed by the input voltages u, while $\mathcal{B}$, the edge-node incidence matrix of the prosumer's communication topology, defines the structure of information exchange between nodes. The angular frequency ω_r characterizes the system's operational frequency.

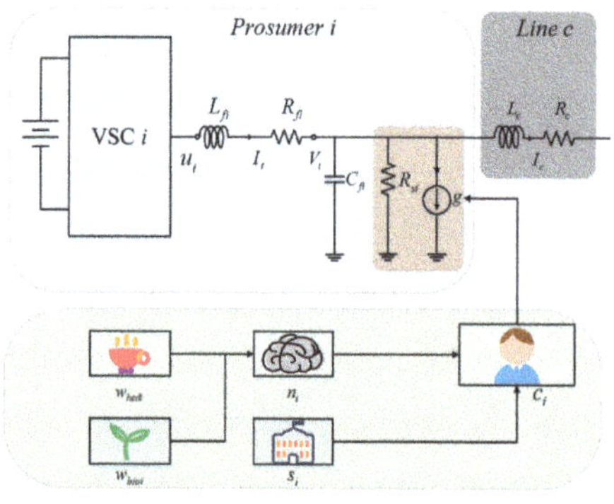

Figure 2. Schematic representation of the internal electrical configuration of a prosumer in a microgrid.

The internal consumer behavior dynamics of each prosumer are illustrated in Figure 2. The operation of the microgrid is influenced by both the dynamics of the physical layer and consumer behavior. As described in the literature [24], consumer behavior is shaped by motivations, social interventions, and personal values, all of which impact energy consumption patterns. Decisions are driven by a combination of external interventions and personal values. To model this, a consumer behavior framework is proposed to capture the dynamic interaction between motivations and interventions, as follows:

$$\dot{c} = B(n - c - Fs), \tag{2a}$$
$$\dot{n} = D(w_{\text{hed}} - n) + E(w_{\text{bio}} - n), \tag{2b}$$

where the variable c describes the dynamic evolution of behavior, representing the response of consumers to external influences or interventions. n represents the motivational variable, shaped by diverse values, including the hedonic value w_{hed} and biospheric value w_{bio}, both of which jointly affect the decision-making process of consumers. The variable s quantifies the strength of social interventions, while the matrix F governs the intensity of these interventions. The parameter B dictates the rate at which behavioral changes occur. Parameters D and E serve as weights and time constants, regulating the combined effects of interventions and values on behavior. Together, these parameters capture the complex interplay between personal values, external interventions, and the dynamics of consumer behavior.

Building on the consumer behavior model, the integration of behavioral and physical layers is achieved by substituting the behavioral variable c with the load control input g, thereby converting abstract behavioral dynamics into concrete physical signals. This transformation seamlessly integrates consumer energy decisions into the microgrid's load control mechanism, enabling the system to adjust loads adaptively based on real-time demand. The integrated system equations are as follows:

$$C_f \dot{V}_d = -R_s^{-1} V_d + \omega_r C_f V_q + I_{td} + \mathcal{B} I_d - I_{Ld} c, \tag{3a}$$
$$C_f \dot{V}_q = -\omega_r C_f V_d - R_s^{-1} V_q + I_{tq} + \mathcal{B} I_q - I_{Lq} c, \tag{3b}$$
$$L_f \dot{I}_{td} = -V_d - R_f I_{td} + \omega_r L_f I_{tq} + u_d, \tag{3c}$$
$$L_f \dot{I}_{tq} = -V_q - R_f I_{tq} - \omega_r L_f I_{td} + u_q, \tag{3d}$$
$$L \dot{I}_d = -\mathcal{B}^T V_d - R I_d + \omega_r L I_q, \tag{3e}$$
$$L \dot{I}_q = -\mathcal{B}^T V_q - \omega_r L I_d - R I_q, \tag{3f}$$
$$\dot{c} = B(n - c - Fs), \tag{3g}$$
$$\dot{n} = D(w_{\text{hed}} - n) + E(w_{\text{bio}} - n). \tag{3h}$$

2.2. System Decoupling

The integrated system equations can be expressed as

$$\begin{cases} \dot{\mathbf{y}} = A_{yy}\mathbf{y} + A_{yz}\mathbf{z} + G_y \mathbf{u} + B_y' \mathbf{x}, \\ \dot{\mathbf{z}} = A_{zy}\mathbf{y} + A_{zz}\mathbf{z}, \\ \dot{\mathbf{x}} = H\mathbf{w} - P\mathbf{x}, \end{cases} \tag{4}$$

where $\mathbf{y} = \begin{bmatrix} V_d^T & V_q^T & I_{td}^T & I_{tq}^T & c^T \end{bmatrix}^T$ and $\mathbf{z} = \begin{bmatrix} I_d^T & I_q^T \end{bmatrix}^T$ represent the electrical responses and behavioral dynamics, respectively, while the control input vector is given by $\mathbf{u} = \begin{bmatrix} u_d^T & u_q^T & s^T \end{bmatrix}^T$. The system is influenced by intrinsic consumer values, represented by the vector $\mathbf{w} = \begin{bmatrix} w_{\text{hed}}^T & w_{\text{bio}}^T \end{bmatrix}^T$, and $\mathbf{x} = \begin{bmatrix} x \end{bmatrix}$ denotes the motivational variables. Additionally, the matrices $H = \begin{bmatrix} D & E \end{bmatrix}$ and $P = \begin{bmatrix} D + E \end{bmatrix}$ describe the relationships between the system's behavioral dynamics and external factors, $\mathbf{B}' = \begin{bmatrix} 0 & 0 & 0 & 0 & B \end{bmatrix}^T$, where A_{yy} is defined as follows:

$$A_{yy} = \begin{bmatrix} -\frac{1}{R_s C_f} & \omega_r & \frac{1}{C_f} & 0 & -\frac{I_{Lx}}{C_f} \\ -\omega_r & -\frac{1}{R_s C_f} & 0 & \frac{1}{C_f} & -\frac{I_{Ly}}{C_f} \\ -\frac{1}{L_f} & 0 & -\frac{R_f}{L_f} & \omega_r & 0 \\ 0 & -\frac{1}{L_f} & -\omega_r & -\frac{R_f}{L_f} & 0 \\ 0 & 0 & 0 & 0 & -B \end{bmatrix},$$

where A_{yz} and A_{zy} are defined as follows:

$$A_{yz} = \begin{bmatrix} \frac{B}{C_f} & 0 & 0 & 0 \\ 0 & \frac{B}{C_f} & 0 & 0 \end{bmatrix}^T, \quad A_{zy} = \begin{bmatrix} -\frac{B^T}{L} & 0 & 0 & 0 & 0 \\ 0 & -\frac{B^T}{L} & 0 & 0 & 0 \end{bmatrix},$$

where A_{zz} and G_y are defined as follows:

$$A_{zz} = \begin{bmatrix} -\frac{R}{L} & \omega_r \\ -\omega_r & -\frac{R}{L} \end{bmatrix}, \quad G_y = \begin{bmatrix} 0 & 0 & 0 \\ 0 & 0 & 0 \\ \frac{1}{L_f} & 0 & 0 \\ 0 & \frac{1}{L_f} & 0 \\ 0 & 0 & -BF \end{bmatrix}.$$

In the proposed framework, the variable n evolves on a slower time scale compared to other variables, resulting in a two-time-scale system. Traditional control methods face challenges in addressing this, necessitating a more tailored approach. By applying singular perturbation theory, we assume that fast variables reach a steady state in the slow system, while slow variables remain constant in the fast system. This enables the decomposition of the system into distinct fast and slow dynamic subsystems. The decoupled slow system is represented by the following equations:

$$\begin{cases} \dot{\mathbf{x}}_s = H\mathbf{w} - P\mathbf{x}_s, \\ 0 = A_{yy}\mathbf{y}_s + A_{yz}\mathbf{z}_s + G_y\mathbf{u}_s + B'_y\mathbf{x}_s, \\ 0 = A_{zy}\mathbf{y}_s + A_{zz}\mathbf{z}_s, \end{cases} \tag{5}$$

where the variables $\mathbf{x}_s$, $\mathbf{y}_s$, $\mathbf{z}_s$ and $\mathbf{u}_s$ represent the slow dynamic components of $\mathbf{x}$, $\mathbf{y}$, $\mathbf{z}$ and $\mathbf{u}$. In the steady state of the slow system, the following holds:

$$0 = H\mathbf{w} - P\mathbf{x}_s, \tag{6a}$$

$$0 = \left(A_{yy} - A_{yz}A_{zz}^{-1}A_{zy}\right)\mathbf{y}_s + G_y\mathbf{u}_s + B'_y\mathbf{x}_s, \tag{6b}$$

where $\mathbf{z}_s = -A_{zz}^{-1}A_{zy}\mathbf{y}_s$.

In the fast dynamic subsystem, it can be expressed as

$$\begin{aligned} \dot{\mathbf{y}}_f &= \dot{\mathbf{y}} - \dot{\mathbf{y}}_s \\ &= A_{yy}\mathbf{y} + A_{yz}\mathbf{z} + G_y\mathbf{u} + B'_y\mathbf{x} \\ &= A_{yy}\mathbf{y}_s + G_y\mathbf{u}_s + A_{yz}\mathbf{z}_s + B'_y\mathbf{x}_s + A_{yy}\mathbf{y}_f + G_y\mathbf{u}_f + A_{yz}\mathbf{z}_f, \end{aligned} \tag{7}$$

where the variables $\mathbf{x}_f$, $\mathbf{y}_f$, $\mathbf{z}_f$ and $\mathbf{u}_f$ represent the fast dynamic components of $\mathbf{x}$, $\mathbf{y}$, $\mathbf{z}$ and $\mathbf{u}$.

By substituting Equation (6b) into Equation (7), the independent fast dynamic equation can be derived as follows:

$$\dot{\mathbf{y}}_f = A_{yy}\mathbf{y}_f + G_y\mathbf{u}_f + A_{yz}\mathbf{z}_f. \tag{8}$$

Similarly,

$$\dot{\mathbf{z}}_f = A_{zy}\mathbf{y}_f + A_{zz}\mathbf{z}_f. \tag{9}$$

In the steady state of the fast system, the following holds:

$$0 = \left(A_{yy} - A_{yz}A_{zz}^{-1}A_{zy}\right)\mathbf{y}_f + G_y\mathbf{u}_f, \tag{10}$$

where the fast dynamic state variable is expressed as $\mathbf{z}_f = -A_{zz}^{-1}A_{zy}\mathbf{y}_f$. The currents I_{df} and I_{qf} are defined in terms of the voltages V_{df} and V_{qf}, where $I_{df} = -\alpha_1 V_{df} - \alpha_2 V_{qf}$ and $I_{qf} = -\alpha_2 V_{df} + \alpha_1 V_{qf}$. The parameters α_1 and α_2 are given by $\alpha_1 = \mathcal{B}R/(L_c^2\omega_r^2 + R^2)$ and $\alpha_2 = \mathcal{B}L_c\omega_r/(L_c^2\omega_r^2 + R^2)$, respectively.

3. Main Result

This section analyzes the optimization challenges arising from the decoupled fast and slow dynamics of the system. For each time scale, distinct optimization objectives are established, followed by the design of tailored controllers to address the unique characteristics of both the fast and slow dynamics.

3.1. Optimization Control Process

The optimization control process, as shown in Figure 3, is designed to ensure continuous adaptation and optimization based on real-time data. The process begins with the continuous monitoring of the system status, ensuring that all relevant parameters are updated in real time. These real-time variables are then input into the optimization controller, which adjusts the control signals accordingly. The optimization controller is responsible for updating control signals across both fast and slow time scales, ensuring the system can effectively respond to both immediate and gradual changes. Once the control signals are updated, the system executes the control strategy, applying the necessary adjustments to maintain optimal performance. Following execution, the system status is updated to reflect the latest changes, after which the process returns to the monitoring step. Through this iterative process, the system remains in a state of continuous optimization, effectively managing both fast and slow dynamic conditions.

3.2. Optimization for Slow Dynamics

In the system, slow dynamics govern long-term behaviors, prioritizing stability, efficiency, and adaptability, while fast dynamics manage transient responses to address short-term disturbances and achieve rapid stabilization. The system's optimization aims to balance these two objectives, improving both operational efficiency and adaptability. Optimizing slow dynamics focuses on aligning system performance with consumer demands, ensuring a balanced supply and demand, minimizing external interventions, reducing energy waste, and promoting sustainability. This approach takes into account factors such as economic incentives and environmental considerations, with the goal of achieving long-term stability and adaptability while minimizing external influence. The following optimization design outlines the strategy for achieving these goals:

$$\begin{aligned}\mathcal{F}_s = &\frac{\zeta}{2}\sum\theta_{ul}(n_{is} - c_{is})^2 + \frac{\kappa}{2}\sum\theta_{ci}\left(I_{tds}^2 + I_{tqs}^2\right) + \frac{\mu}{2}\|u_{ds}\|^2 + \frac{\nu}{2}\|u_{qs}\|^2 \\ &+ \frac{\rho}{2}\|s_s - s_r\|^2 + \frac{\tau}{2}\|V_{ds} - V_{dr}\|^2 + \frac{\phi}{2}\|V_{qs} - V_{qr}\|^2,\end{aligned} \tag{11}$$

where $\theta_{ci} \in \mathbb{R} > 0$ denotes the unit cost for prosumer i, while $\theta_{ul} \in \mathbb{R} > 0$ reflects the degree of satisfaction prosumer i derives from meeting their load demand. A higher value of θ_{uli} indicates a greater expectation for comfort from prosumer i. By minimizing the error, the objective is to guide the voltage towards its nominal value, thereby achieving precise voltage regulation and minimizing $|n_i s - c_i s|$. The parameters ζ, κ, μ, ν, ρ, τ, and ϕ are tunable constants, each serving to prioritize different control objectives.

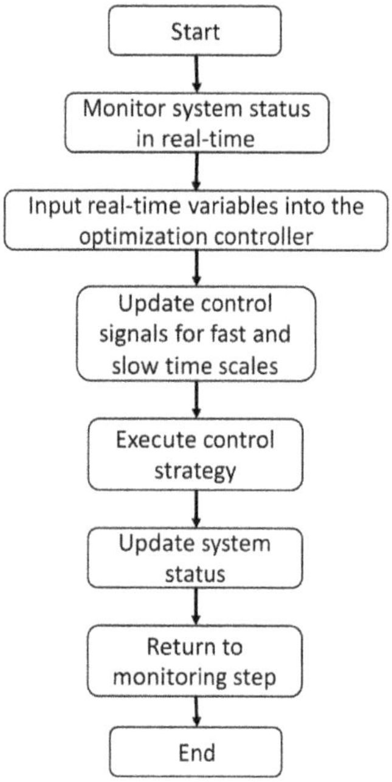

Figure 3. The optimization control process flowchart.

By incorporating steady-state conditions as constraints and aligning them with predefined optimization objectives, a comprehensive optimization problem is established. These equations enforce critical physical constraints, such as power balance, frequency stability, and voltage limits. Let $z_s^* = \left[x_s^{*T}\, y_s^{*T}\, u_s^{*T}\right]^T$ denote the optimization variable. The optimization problem for slow dynamics is then formulated as follows:

$$\min_{z_s^*} \quad \mathcal{F}_s(z_s^*) \tag{12a}$$

$$\text{s.t.} \quad 0 = H\mathbf{w} - P\mathbf{x}_s^* \tag{12b}$$

$$0 = \left(A_{yy} - A_{yz}A_{zz}^{-1}A_{zy}\right)\mathbf{y}_s^* + G_y\mathbf{u}_s^* + B_y'\mathbf{x}_s^*. \tag{12c}$$

Remark 1: In optimization theory, Karush–Kuhn–Tucker (KKT) conditions are essential tools for solving constrained optimization problems. Derived through the method of Lagrange multipliers, they describe the relationship between the objective function, the constraints, and the Lagrange multipliers [42]. The KKT conditions provide necessary optimality conditions for constrained optimization problems and are widely applied in solving various types of optimization tasks.

Using the KKT conditions, we derive the optimal solutions for the relevant variables, including the state variables $\bar{x}_s^*$ and $\bar{y}_s^*$, and the control variable $\bar{u}_s^*$. By setting the control variable to $u_s = \bar{u}_s^*$ during optimization, it reaches its optimal state, driving the state variable y_s towards the optimal value $\bar{y}_s^*$. Meanwhile, the state variable x_s naturally converges to its steady-state value $\bar{x}_s^*$. As a result, the system converges to the steady-state solutions $y_s = \bar{y}_s^*$ and $x_s = \bar{x}_s^*$, ensuring global optimal performance across time scales.

3.3. Optimization for Fast Dynamics

The optimization of fast dynamics aims to mitigate short-term disturbances, such as sudden load variations or external factors, ensuring a rapid system response and stability during transient conditions. The system must swiftly adapt to fluctuations to maintain both stability and performance. By optimizing fast dynamics, the system can quickly regain stability, thereby ensuring long-term operational efficiency. This approach enhances both dynamic responses and robustness in complex environments. Let $y_f = y - y_s$ represent the fast dynamic component, with the objective of driving these variables toward zero to guide the system toward its optimal steady-state values. Therefore, the following optimization objective function is proposed:

$$\begin{aligned}\mathcal{F}_f = &\frac{\zeta'}{2}\sum\theta'_{ul}\|c_{if}\|^2 + \frac{\kappa'}{2}\sum\theta'_{ci}\left(I^2_{tdf} + I^2_{tqf}\right) + \frac{\mu'}{2}\|u_{df}\|^2 + \frac{\nu'}{2}\|u_{qf}\|^2 \\ &+ \frac{\rho'}{2}\|s_f\|^2 + \frac{\tau'}{2}\|V_{df}\|^2 + \frac{\phi'}{2}\|V_{qf}\|^2,\end{aligned} \tag{13}$$

where the parameter θ'_{ul} reflects the influence of consumer load on behavior variable c_{if}, while θ'_{ci} captures the cost associated with adjusting current variables I_{tdf} and I_{tqf}. The parameters ζ', κ', ρ', τ', and ϕ' are weights that control the importance of behavior, current, and voltage deviations in the optimization process. Specifically, ζ' governs the behavior variable, ρ' controls social intervention, and κ', τ', and ϕ' ensure that current and voltage remain close to their desired values. Finally, μ' and ν' adjust the control inputs u_{df} and u_{qf}, ensuring optimal control effort. The overall goal of this objective function is to minimize deviations across behavior, current, and voltage, while optimizing control input to maintain system stability and performance in fast dynamics.

By incorporating the optimization objectives with the constraint conditions, the optimization problem for fast dynamics can be formulated as follows:

$$\min_{\hat{y}^*_f} \quad \mathcal{F}_f(\hat{y}^*_f), \tag{14a}$$

$$\text{s.t.} \quad 0 = \alpha'_1 V^*_{df} + \alpha'_2 V^*_{qf} + I^*_{tdf} - I^*_{Ldf} c^*_f, \tag{14b}$$

$$0 = \alpha'_3 V^*_{df} + \alpha'_1 V^*_{qf} + I^*_{tqf} - I^*_{Lqf} c^*_f, \tag{14c}$$

$$0 = -V^*_{df} - R_f I^*_{tdf} + \omega_r L_f I^*_{tqf} + u^*_{df}, \tag{14d}$$

$$0 = -V^*_{qf} - \omega_r L_f I^*_{tdf} - R_f I^*_{tqf} + u^*_{qf}, \tag{14e}$$

$$0 = Bc^*_f + Fs^*_f, \tag{14f}$$

where $\alpha'_1 = \left(-\frac{1}{R_s} - \alpha_1\right)$, $\alpha'_2 = \left(\omega_r C_f - \alpha_2\right)$, and $\alpha'_3 = \left(-\omega_r C_f - \alpha_2\right)$.

Let $\lambda := [\lambda_1^T, \lambda_2^T, \lambda_3^T, \lambda_4^T, \lambda_5^T]^T \in \mathbb{R}^{5N}$ denote the vector of the Lagrange multipliers corresponding to the constraints. Below is the explicit form of the Lagrangian for the given optimization problem:

$$\begin{aligned}\mathcal{L} = &\mathcal{F}_f(\hat{y}^*_f) + \lambda_1\left(\alpha'_1 V^*_{df} + \alpha'_2 V^*_{qf} + I^*_{tdf} - I^*_{Ldf} c^*_f\right) + \lambda_2\left(\alpha'_3 V^*_{df} + \alpha'_1 V^*_{qf} + I^*_{tqf} - I^*_{Lqf} c^*_f\right) \\ &+ \lambda_3\left(-V^*_{df} - R_f I^*_{tdf} + \omega_r L_f I^*_{tqf} + u^*_{df}\right) + \lambda_4\left(-V^*_{qf} - \omega_r L_f I^*_{tdf} - R_f I^*_{tqf} + u^*_{qf}\right) \\ &+ \lambda_5\left(Bc^*_f + Fs^*_f\right).\end{aligned} \tag{15}$$

The first-order optimality conditions for this optimization problem, as outlined by the KKT framework, are provided as follows:

$$0 = \tau'\overline{V}^{*}_{df} + \lambda_1\alpha'_1 - \lambda_2\alpha'_3 - \lambda_3, \tag{16a}$$
$$0 = \tau'\overline{V}^{*}_{qf} + \lambda_1\alpha'_2 + \lambda_2\alpha'_1 - \lambda_4, \tag{16b}$$
$$0 = \kappa'\overline{I}^{*}_{tdf} + \lambda_1 - \lambda_3 R_f - \lambda_4\omega_r L_f, \tag{16c}$$
$$0 = \kappa'\overline{I}^{*}_{tqf} + \lambda_2 - \lambda_3\omega_r L_f - \lambda_4 R_f, \tag{16d}$$
$$0 = \zeta'\overline{c}^{*}_{f} - \lambda_1\overline{I}^{*}_{Ldf} - \lambda_2\overline{I}^{*}_{Lqf} + \lambda_5 B, \tag{16e}$$
$$0 = \mu'\overline{u}^{*}_{df} + \lambda_3, \tag{16f}$$
$$0 = \nu'\overline{u}^{*}_{qf} + \lambda_4, \tag{16g}$$
$$0 = \rho'\overline{s}^{*}_{f} + F\lambda_5, \tag{16h}$$
$$0 = \alpha'_1\overline{V}^{*}_{df} + \alpha'_2\overline{V}^{*}_{qf} + \overline{I}^{*}_{tdf} - \overline{I}^{*}_{Ldf}\overline{c}^{*}_{f}, \tag{16i}$$
$$0 = \alpha'_3\overline{V}^{*}_{df} + \alpha'_1\overline{V}^{*}_{qf} + \overline{I}^{*}_{tdf} - \overline{I}^{*}_{Lqf}\overline{c}^{*}_{f}, \tag{16j}$$
$$0 = -\overline{V}^{*}_{df} - R_f\overline{I}^{*}_{tdf} + \omega_r L_f\overline{I}^{*}_{tqf} + \overline{u}^{*}_{df}, \tag{16k}$$
$$0 = -\overline{V}^{*}_{qf} - \omega_r L_f\overline{I}^{*}_{tdf} - R_f\overline{I}^{*}_{tqf} + \overline{u}^{*}_{qf}, \tag{16l}$$
$$0 = B\overline{c}^{*}_{f} + F\overline{s}^{*}_{f}. \tag{16m}$$

To obtain an optimal solution, we use the derived KKT conditions as the foundation for designing a distributed control mechanism that ensures both local and global objectives are achieved. In this framework, controllers exchange information through a communication network that mirrors the physical network topology, utilizing primal-dual dynamic mechanisms for coordination across multiple nodes. The specific formulation is shown as follows:

$$-\theta_1\dot{V}^{*}_{df} = \tau' V^{*}_{df} + \lambda_1\alpha'_1 - \lambda_2\alpha'_2 - \lambda_3, \tag{17a}$$
$$-\theta_2\dot{V}^{*}_{qf} = \phi' V^{*}_{qf} + \lambda_1\alpha'_3 + \lambda_2\alpha'_1 - \lambda_4, \tag{17b}$$
$$-\theta_3\dot{I}^{*}_{tdf} = \kappa' I^{*}_{tdf} + \lambda_1 - \lambda_3 R_f - \lambda_4\omega_r L_f, \tag{17c}$$
$$-\theta_4\dot{I}^{*}_{tqf} = \kappa' I^{*}_{tqf} + \lambda_2 - \lambda_3\omega_r L_f - \lambda_4 R_f \tag{17d}$$
$$-\theta_5\dot{c}^{*}_{f} = \zeta' c^{*}_{f} - \lambda_1 I^{*}_{Ldf} - \lambda_2 I^{*}_{Lqf} + \lambda_5 B, \tag{17e}$$
$$-\theta_6\dot{u}^{*}_{df} = \mu' u^{*}_{df} + \lambda_3 + \gamma_1, \tag{17f}$$
$$-\theta_7\dot{u}^{*}_{qf} = \nu' u^{*}_{qf} + \lambda_4 + \gamma_2, \tag{17g}$$
$$-\theta_8\dot{s}^{*}_{f} = \rho' s^{*}_{f} + F\lambda_5 + \gamma_3, \tag{17h}$$
$$\theta_9\dot{\lambda}_1 = \alpha'_1 V^{*}_{df} + \alpha'_2 V^{*}_{qf} + I^{*}_{tdf} - I^{*}_{Ldf}c^{*}_{f}, \tag{17i}$$
$$\theta_{10}\dot{\lambda}_2 = \alpha'_3 V^{*}_{df} + \alpha'_1 V^{*}_{qf} + I^{*}_{tdf} - I^{*}_{Lqf}c^{*}_{f}, \tag{17j}$$
$$\theta_{11}\dot{\lambda}_3 = -V^{*}_{df} - R_f I^{*}_{tdf} + \omega_r L_f I^{*}_{tqf} + u^{*}_{df}, \tag{17k}$$
$$\theta_{12}\dot{\lambda}_4 = -V^{*}_{qf} - \omega_r L_f I^{*}_{tdf} - R_f I^{*}_{tqf} + u^{*}_{qf}, \tag{17l}$$
$$\theta_{13}\dot{\lambda}_5 = Bc^{*}_{f} + Fs^{*}_{f}, \tag{17m}$$

where the matrices $\theta_1, \theta_2, \theta_3, \theta_4, \cdots \in \mathbb{R}^{N\times N}$ are positive diagonal matrices, whose adjustment can modulate the dynamic response of the controller. Moreover, the vectors γ_1, γ_2, and γ_3 serve as control input ports, facilitating the interconnection between the controller and the physical system. It is important to emphasize that, because of the strict non-convexity of the objective function with respect to c^{*}_{f}, I^{*}_{dxf}, I^{*}_{dyf}, u^{*}_{xf}, u^{*}_{yf}, and s^{*}_{f}, the linearization of the constraints ensures that, given the constants γ_1, γ_2, and γ_3, the optimal solution $\bar{y}_f^*$ remains unique. This unique solution provides a reliable basis for optimizing control, ensuring both robustness and stability in practical implementation.

3.4. Stability Analysis

To ensure that the controller can accurately respond to the dynamic fluctuations of the physical layer while maintaining system stability and performance, we have implemented specific configurations for the control variables, external interventions, and the coupling parameters γ_1, γ_2, and γ_3, which serve to link the physical and network layers, as follows:

$$u_{df} = u_{df}^*, \quad u_{qf} = u_{qf}^*, \quad s_f = s_f^*, \tag{18}$$

$$\gamma_1 = I_{tdf}, \quad \gamma_2 = I_{tqf}, \quad \gamma_3 = -2F^T B^T P^T c, \tag{19}$$

where the matrix P is determined by the Lyapunov equation $B^T P + PB + Q = 0$, where P is the symmetric positive definite matrix to be solved, and Q is a given symmetric positive definite matrix.

The storage function for the physical layer is defined as follows:

$$\begin{aligned} S_p &= \frac{1}{2}\left(\dot{V}_{df}^T C_f \dot{V}_{df} + \dot{V}_{qf}^T C_f \dot{V}_{qf} + \dot{I}_{tdf}^T L_f \dot{I}_{tdf} + \dot{I}_{tqf}^T L_f \dot{I}_{tqf} + \dot{I}_{df}^T L_c \dot{I}_{df} + \dot{I}_{qf}^T L_c \dot{I}_{qf} + \dot{c}_f^T P \dot{c}_f\right) \\ &= -\dot{V}_{df}^T R_s^{-1} \dot{V}_{df} - \dot{V}_{qf}^T R_s^{-1} \dot{V}_{qf} - \dot{I}_{tdf}^T R_f \dot{I}_{tdf} - \dot{I}_{tqf}^T R_f \dot{I}_{tqf} + \dot{I}_{tqf}^T \dot{u}_{qf} - \dot{V}_{df}^T I_{Ld} \dot{c}_f \\ &\quad - \dot{V}_{qf}^T I_{Lq} \dot{c}_f - \dot{I}_{df}^T R \dot{I}_{df} - \dot{I}_{qf}^T R \dot{I}_{qf} - \dot{c}^T Q \dot{c} - 2\dot{c}^T PAF\dot{s} + \dot{I}_{tqf}^T \dot{u}_{df}. \end{aligned} \tag{20}$$

According to Young's inequality, we can derive the following:

$$-\dot{V}_d^T I_{Ld}\dot{c} - \dot{V}_q^T I_{Lq}\dot{c} \leq \frac{\dot{V}_d^T I_{Ld} \dot{V}_d}{2\epsilon_1} + \frac{\dot{V}_q^T I_{Lq} \dot{V}_q}{2\epsilon_2} + \dot{c}^T \frac{\epsilon_1 I_{Ld}}{2}\dot{c} + \dot{V}_q^T \frac{I_{Lq}}{2\epsilon_2}\dot{V}_q + \dot{c}^T \frac{\epsilon_2 I_{Lq}}{2}\dot{c}. \tag{21}$$

Substituting into the physical storage function yields the following:

$$\begin{aligned} \dot{S}_p &\leqslant -\dot{V}_d^T\left(R_s^{-1} - \frac{I_{Ld}}{2\epsilon_1}\right)\dot{V}_d - \dot{V}_q^T\left(R_s^{-1} - \frac{I_{Lq}}{2\epsilon_2}\right)\dot{V}_q - \dot{c}^T\left(Q - \frac{\epsilon_1 I_{Ld}}{2} - \frac{\epsilon_2 I_{Lq}}{2}\right)\dot{c} \\ &\quad - 2\dot{c}^T PAF\dot{s} + \dot{I}_{td}^T \dot{u}_d + \dot{I}_{tq}^T \dot{u}_q \\ &\leqslant -2\dot{c}^T PAF\dot{s} + \dot{I}_{td}^T \dot{u}_d + \dot{I}_{tq}^T \dot{u}_q. \end{aligned} \tag{22}$$

Similarly, the storage function for the optimized controller is as follows:

$$\begin{aligned} S_c = \frac{1}{2}\Big(&\dot{V}_d^{*T}\theta_1\dot{V}_d^* + \dot{V}_q^{*T}\theta_2\dot{V}_q^* + \dot{c}^{*T}\theta_3\dot{c}^* + \dot{I}_{td}^{*T}\theta_4\dot{I}_{td}^* + \dot{\lambda}_1^T\theta_9\dot{\lambda}_1 + \dot{\lambda}_2^T\theta_{10}\dot{\lambda}_2 + \dot{\lambda}_3^T\theta_{11}\dot{\lambda}_3 \\ &+ \dot{I}_{tq}^{*T}\theta_5\dot{I}_{tq}^* + \dot{u}_d^{*T}\theta_6\dot{u}_d^* + \dot{u}_q^{*T}\theta_7\dot{u}_q^* + \dot{s}^{*T}\theta_8\dot{s}^* + \dot{\lambda}_4^T\theta_{12}\dot{\lambda}_4 + \dot{\lambda}_5^T\theta_{13}\dot{\lambda}_5\Big). \end{aligned} \tag{23}$$

Its derivative is given by

$$\begin{aligned} \dot{S}_c &\leqslant -\dot{V}_d^{*T}\tau'\dot{V}_d - \dot{V}_q^{*T}\phi'\dot{V}_q - \dot{I}_{td}^{*T}\kappa'\dot{I}_{td}^* - \dot{I}_{tq}^{*T}\kappa'\dot{I}_{tq}^* + 2\dot{c}^T PAF\dot{s} - \dot{I}_{td}^T\dot{u}_d - \dot{I}_{tq}^T\dot{u}_q \\ &\quad - \dot{c}^{*T}\zeta'\dot{c}^* - \dot{u}_d^{*T}\mu'\dot{u}_d^* - \dot{u}_q^{*T}\nu'\dot{u}_q^* - \dot{s}^{*T}\rho'\dot{s}^* \\ &\leqslant 2\dot{c}^T PAF\dot{s} - \dot{I}_{td}^T\dot{u}_d - \dot{I}_{tq}^T\dot{u}_q. \end{aligned} \tag{24}$$

The storage function for the entire closed-loop system is as follows:

$$S = S_p + S_c. \tag{25}$$

Moreover, it holds that

$$\dot{S} = \dot{S_p} + \dot{S_c} \leqslant 0. \tag{26}$$

There exists a forward-invariant Ω, and in accordance with LaSalle's invariance principle, any solution with initial conditions lying within Ω will asymptotically converge to the maximal invariant set defined by $\Omega \cap \{(x_f, x_c) \in \mathbb{R}^{16N+2E} \mid \dot{y}_f = 0,\ b_f^* = 0,\ \dot{I}_{tdf}^* = 0,\ \dot{I}_{tqf}^* = 0,\ \dot{u}_d f^* = 0,\ \dot{u}_q f^* = 0,\ s_f^* = 0\}$. Subsequently, based on the corresponding conditions, it can be deduced that within this maximal invariant set, $\dot{\lambda}_1, \dot{\lambda}_2, \dot{\lambda}_3, \dot{\lambda}_4$ and $\dot{\lambda}_5$ all become zero. Finally, by examining steady-state constraints, it is evident that $\bar{x}_s$ is uniquely determined by $\bar{u}_s = \bar{u}_s^*$. Thus, we can conclude that, at a steady state, the physical state variables coincide with the corresponding optimization variables.

4. Numerical Simulation

To validate the efficacy of the proposed controller, we conducted a series of simulation experiments. The experimental setup consists of 10 prosumers, with the topology illustrated in Figure 4. The black arrows indicate the direction of current flow, while the dashed lines represent the communication links between the microgrids. Each prosumer is connected to its neighboring nodes via transmission lines. Specific parameters of the microgrids, such as voltage, current, and load impedance, along with those related to the consumer behavior model, are sourced from [24], as shown in Table 1. These parameters, derived from long-term monitoring of multiple distributed energy resources, closely mirror real-world operational conditions.

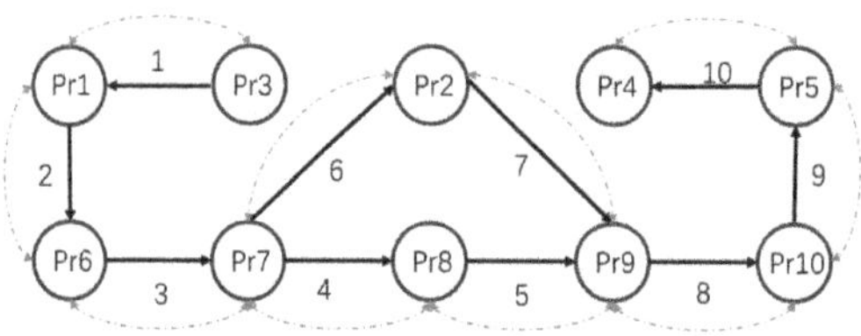

Figure 4. The topology of interconnected prosumers.

The parameters related to the consumer behavior model are as follows:

$$\begin{aligned}
D &= 10^{-5} \times \mathrm{diag}(1.4, 1.4, 1.6, 1.6, 1.44, 1.44, 1.6, 1.6, 1.48, 1.48), \\
E &= 10^{-5} \times \mathrm{diag}(1.6, 1.4, 0.6, 0.6, 1.62, 1.62, 0.6, 0.6, 1.64, 1.64), \\
w_{\text{hed}} &= [0.5\ 0.76\ 0.81\ 0.81\ 0.72\ 0.72\ 0.5\ 0.3\ 0.6\ 0.5]^T, \\
w_{\text{bio}} &= [0.4\ 0.6\ 0.6\ 0.6\ 0.7\ 0.7\ 0.5\ 0.3\ 0.6\ 0.5]^T.
\end{aligned}$$

Scenario 1: Optimization Effectiveness Under Normal Operating Conditions

To evaluate the effectiveness of the proposed control strategy under normal operating conditions, a baseline scenario was designed in which the system operates without external disturbances or sudden fluctuations. In this scenario, the microgrid follows typical load and generation patterns while consumer behavior remains stable. The primary objective is to validate the control strategy's performance optimization in a stable environment.

The simulation results, presented in Figures 5 and 6, illustrate the recalculated system parameters, including phase-a voltage V_a, current I_a, power output $P_v sc$, load power demand P_L, and consumer behavior variables. The controller demonstrates its effectiveness by maintaining the voltage V_a consistently at the target value, with fluctuations tightly constrained across both transient and steady-state phases. Likewise, the current I_a exhibits rapid dynamic stability, with oscillations diminishing swiftly to reach steady-state values.

Additionally, the power output P_vsc and load power P_L stabilize within a short time frame, highlighting the controller's exceptional dynamic adjustment capability and efficiency in steady-state operation.

Table 1. Physical parameters.

	Pros. 1	**Pros. 2**	**Pros. 3**	**Pros. 4**	**Pros. 5**
C_f (μF)	62.86	62.86	62.86	62.86	62.86
L_f (mH)	2.1	2.0	1.9	1.8	2.1
R_f (mΩ)	40.2	38.7	34.6	31.8	40.1
R_s (Ω)	11	13	11	13	11
I_{Ld} (A)	25	23	35	31	25
I_{Lq} (A)	−12	−15	−10	−18	−12
V_x^r (V)	$120\sqrt{2}$	$120\sqrt{2}$	$120\sqrt{2}$	$120\sqrt{2}$	$120\sqrt{2}$
V_y^r (V)	0	0	0	0	0
$\theta_{ci} = \theta_{ui}$	1	1	1	1	1
	Line 1	Line 2	Line 3	Line 4	Line 5
R_c (Ω)	0.25	0.27	0.24	0.26	0.25
L_c (μH)	1.2	1.3	1.8	2.1	1.2
	Pros. 6	Pros. 7	Pros. 8	Pros. 9	Pros. 10
C_f (μF)	62.86	62.86	62.86	62.86	62.86
L_f (mH)	2.02	1.9	1.82	2.1	2.02
R_f (mΩ)	38.7	34.6	31.8	40.2	38.7
R_s (Ω)	13	10.86	13	10.86	13
I_{Ld} (A)	23	35	31	25	23
I_{Lq} (A)	−15	−10	−18	−12	−15
V_x^r (V)	$120\sqrt{2}$	$120\sqrt{2}$	$120\sqrt{2}$	$120\sqrt{2}$	$120\sqrt{2}$
V_y^r (V)	0	0	0	0	0
$\theta_{ci} = \theta_{ui}$	1	1	1	1	1
	Line 6	Line 7	Line 8	Line 9	Line 10
R_c (Ω)	0.27	0.24	0.26	0.25	0.27
L_c (μH)	1.3	1.8	2.1	1.2	1.3

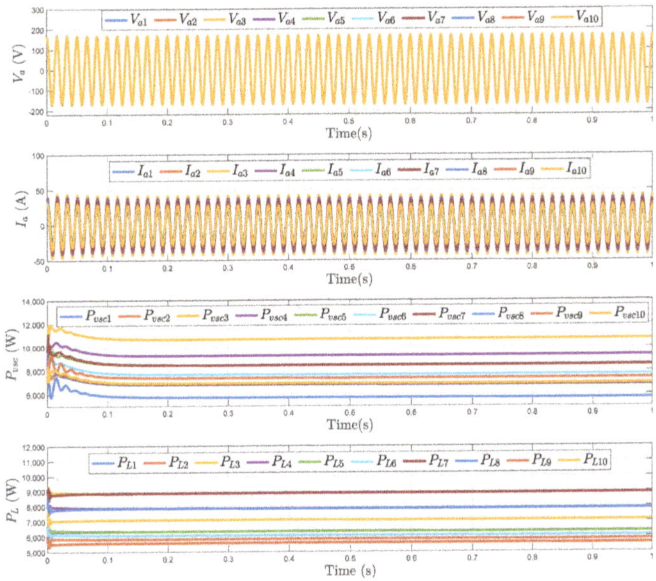

Figure 5. Scenario 1. Electrical Response of the System.

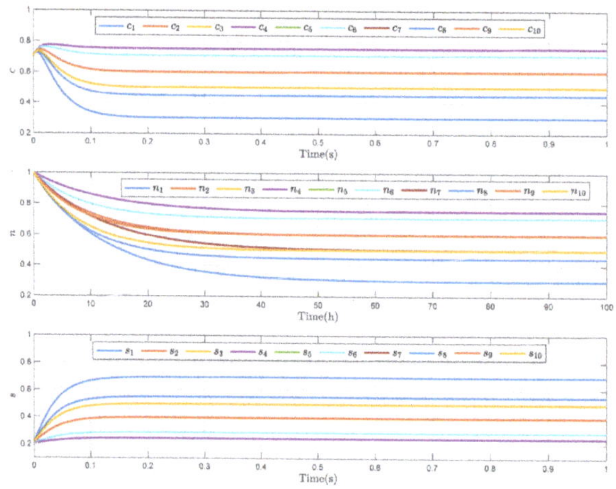

Figure 6. Scenario 1. Dynamics of Consumer Behavior, Motivation, and Social Intervention.

Additionally, the trends of consumer behavior c and motivational variables n reveal that, even with low levels of external intervention s, consumer behavior gradually adapts and converges to the desired steady-state values. This indicates the control strategy's effectiveness in optimizing long-term behavioral dynamics while minimizing intervention costs. By leveraging the controller's optimization capabilities, the system achieves improved operational efficiency while maintaining a harmonious balance between consumer behavior and system stability. These findings provide both theoretical and practical support for the sustainable operation of microgrids.

Scenario 2: Impact of Consumer Types on Simulation Outcomes

This scenario investigates the impact of different consumer value orientations on system performance through two simulation cases. In the first case, where consumer behavior is driven by hedonic values, simulation results in Figures 7 and 8 show a significant increase in energy consumption. This leads to a higher power output from the generation units P_vsc and a greater load power demand P_L, thereby increasing system stress. Despite this, the controller effectively maintains voltage V_a close to its target, and current I_a stabilizes quickly, demonstrating robustness under high-load conditions.

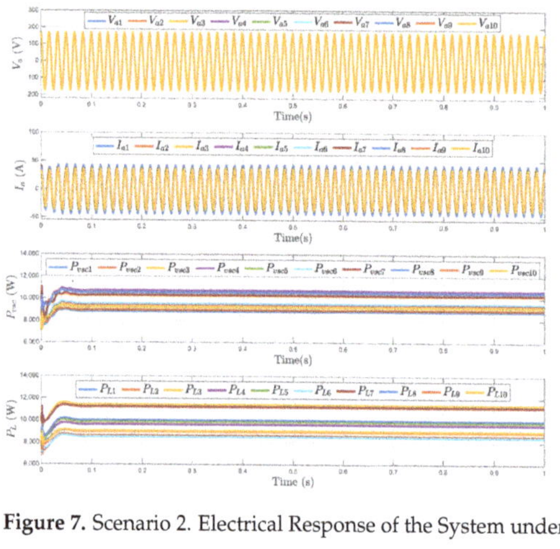

Figure 7. Scenario 2. Electrical Response of the System under Hedonic Value-Dominated Consumer.

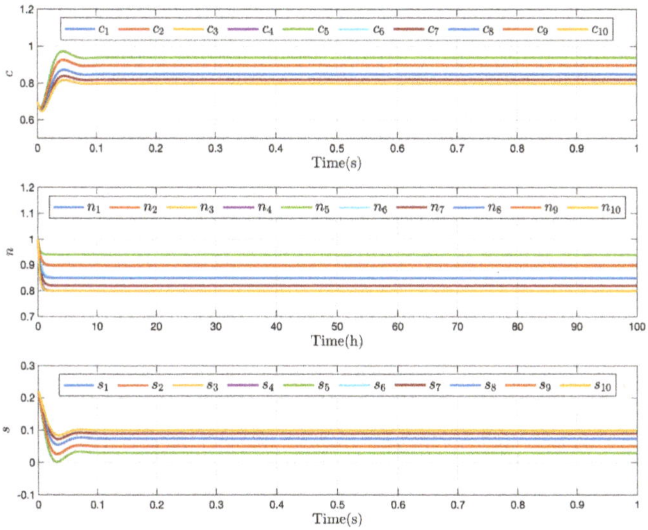

Figure 8. Scenario 2. Dynamics of Consumer Behavior, Motivation, and Social Intervention under Hedonic Value-Dominated Consumer.

In contrast, the second case assumes consumer behavior is driven by biospheric values, as illustrated in Figures 9 and 10. In this scenario, both power output and load demand decrease substantially. The controller keeps V_a at the target value, consistently optimizing performance across varying loads. Analysis of social intervention intensity s and consumer behavior c shows that consumers motivated by biospheric values converge to stable behavior with minimal intervention, while those driven by hedonic values require stronger interventions.

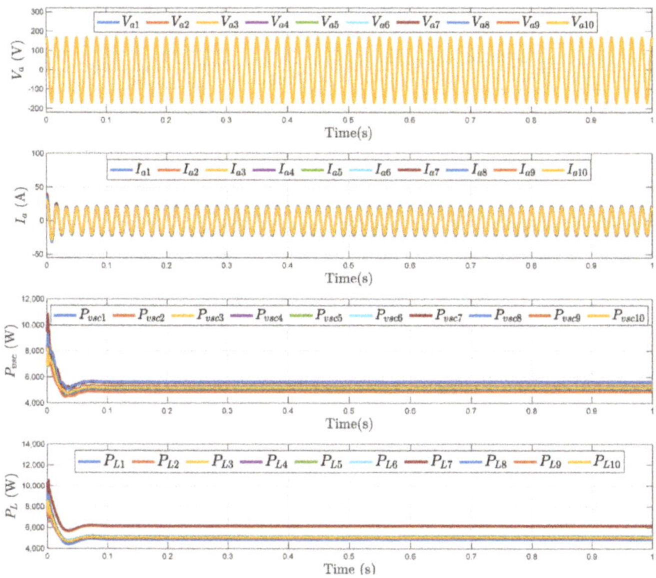

Figure 9. Scenario 2. Electrical Response of the System under Biospheric Value-Dominated Consumer.

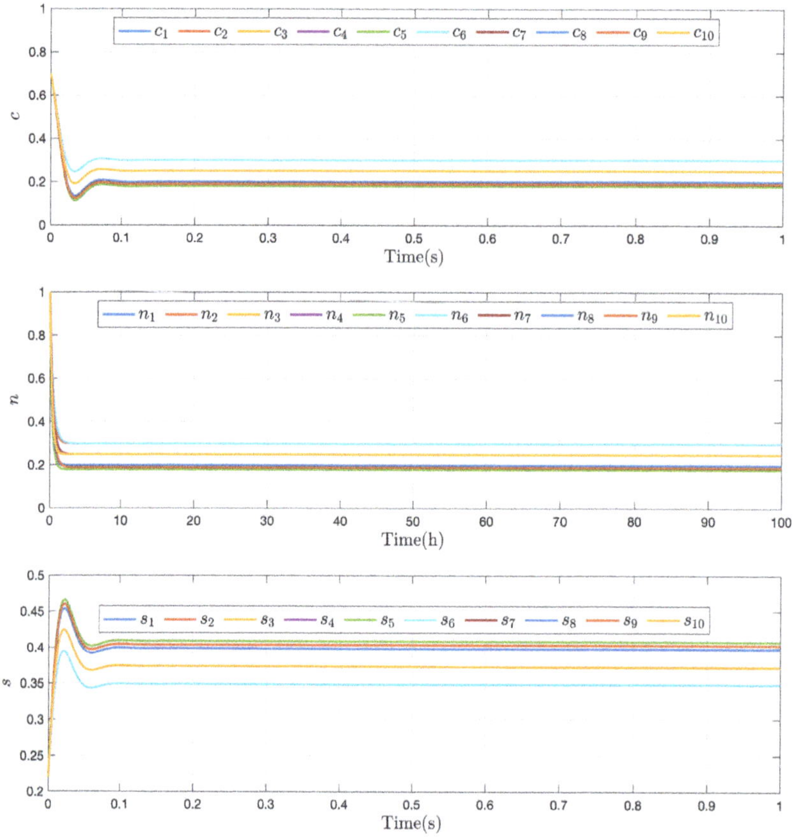

Figure 10. Scenario 2. Dynamics of Consumer Behavior, Motivation, and Social Intervention under Biospheric Value-Dominated Consumer.

These findings underscore the substantial impact of consumer motivations on system resource consumption and control costs while validating the controller's consistent performance across different consumer types. The comparative analysis highlights that optimizing consumer motivations not only reduces energy consumption but also achieves a balanced trade-off between control costs and sustainable operation, thereby enhancing the overall efficiency of microgrid systems.

Scenario 3: Effectiveness of the Control Strategy under Behavioral and Motivational Shifts

To assess the adaptability and robustness of the proposed control strategy under challenging conditions, a scenario was designed to simulate abrupt changes in consumer behavior and motivational parameters. This setup mimics the impact of external interventions, economic incentives, and social pressures on consumer decision-making and behavioral dynamics.

At $t = 0.5$, significant changes occur in the components D and E of the motivational model parameter n, as well as in the hedonic value w_{hed} and biospheric value w_{bio}. The updated parameters are defined as follows:

$$
\begin{aligned}
&D = 10^{-5} \times \text{diag}(1.4, 1.4, 1.6, 1.6, 1.44, 1.44, 1.6, 1.6, 1.48, 1.48), \\
&E = 10^{-5} \times \text{diag}(7.2, 6.3, 8.2, 6.8, 7.5, 6.5, 8.4, 7.8, 7.5, 8.6), \\
&w_{\text{hed}} = \begin{bmatrix} 0.45 & 0.5 & 0.35 & 0.5 & 0.45 & 0.4 & 0.3 & 0.5 & 0.3 & 0.5 \end{bmatrix}^T, \\
&w_{\text{bio}} = \begin{bmatrix} 0.3 & 0.5 & 0.45 & 0.5 & 0.3 & 0.4 & 0.3 & 0.5 & 0.45 & 0.5 \end{bmatrix}^T.
\end{aligned}
$$

At $t = 1$, the behavioral variable b abruptly shifts to 1, while the motivational model parameters D and E, along with the hedonic value w_{hed} and biospheric value w_{bio}, return to their initial values. This dual-phase scenario effectively captures dynamic shifts in consumer behavior and motivation.

The simulation results for Scenario 3, shown in Figures 11 and 12, demonstrate that the controller stabilizes voltage V_a near its target despite abrupt changes in consumer behavior and motivational parameters. Oscillations in current I_a are suppressed, and the system quickly converges to steady-state values. Power output P_vsc and load demand P_L also stabilize rapidly. The control strategy dynamically adjusts intervention levels based on the magnitude of disturbances, minimizing costs while maintaining efficiency. Consumer behavior and motivational variables converge to steady-state values without affecting system performance. Further results show that voltage V_x remains stable even with external interventions at $t = 1$. The controller effectively regulates voltage and manages load variations, guiding consumer behavior toward a steady state.

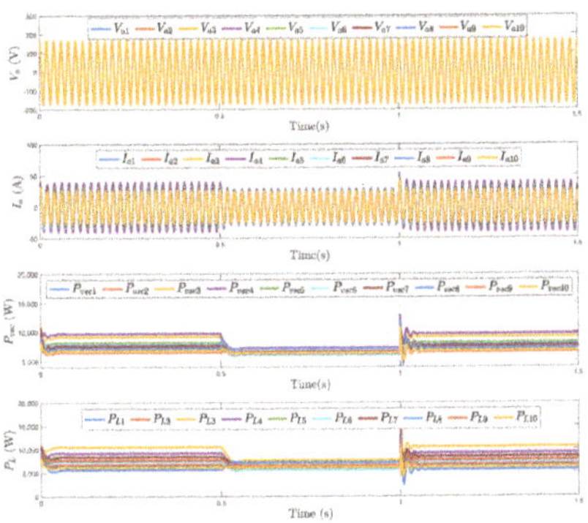

Figure 11. Scenario 3. Electrical Response of the System.

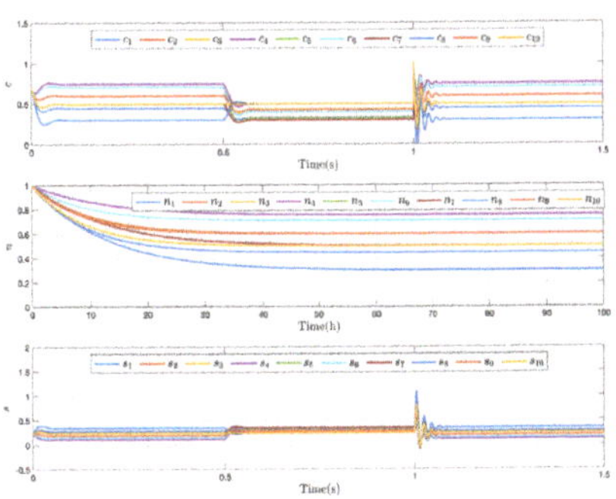

Figure 12. Scenario 3. Dynamics of Consumer Behavior, Motivation, and Social Intervention.

19. Tushar, W.; Saha, T.K.; Yuen, C.; Liddell, P.; Bean, R.; Poor, H.V. Peer-to-Peer Energy Trading with Sustainable User Participation: A Game Theoretic Approach. *IEEE Access* **2018**, *6*, 62932–62943. [CrossRef]
20. Cintuglu, M.H.; Mohammed, O.A. Behavior Modeling and Auction Architecture of Networked Microgrids for Frequency Support. *IEEE Trans. Ind. Inform.* **2017**, *13*, 1772–1782. [CrossRef]
21. O'Neill-Carrillo, E.; Mercado, E.; Luhring, O.; Jordán, I. Local Socio-Economic Development through Community-Based Distributed Energy Resources. In Proceedings of the 2018 IEEE International Symposium on Technology and Society (ISTAS), Washington, DC, USA, 13–14 November 2018; pp. 8–13.
22. Fragniere, E.; Sandoz, S.; Abdenadher, N.; Moussa, M.; Serugendo, G.D.M.; Glass, P. Fostering 'Energy Communities': An Ethnographic-SECI Approach to User-Centered Residential Micro-Smart Grid Adoption. In Proceedings of the 2023 11th International Conference on Smart Grid (icSmartGrid), Paris, France, 4–7 June 2023; pp. 1–5.
23. Cucuzzella, M.; Bouman, T.; Kosaraju, K.C.; Schuitema, G.; Lemmen, N.H.; Johnson-Zawadzki, S.; Fischione, C.; Steg, L.; Scherpen, J.M.A. Distributed Control of DC Grids: Integrating Prosumers' Motives. *IEEE Trans. Power Syst.* **2022**, *37*, 3299–3310. [CrossRef]
24. Feng, S.; Cucuzzella, M.; Bouman, T.; Steg, L.; Scherpen, J.M.A. An Integrated Human–Cyber–Physical Framework for Control of Microgrids. *IEEE Trans. Smart Grid* **2023**, *14*, 3388–3400. [CrossRef]
25. Doumen, S.C.; Hönen, J.; Nguyen, P.; Hurink, J.L.; Zwart, B.; Kok, K. Modeling and Demonstrating the Effect of Human Decisions on the Distribution Grid. In Proceedings of the 2023 IEEE Power & Energy Society Innovative Smart Grid Technologies Conference (ISGT), Washington, DC, USA, 16–19 January 2023; pp. 1–5.
26. Kulikov, A.L.; Sharygin, M.V. Automatic control system of power supply of active consumers. In Proceedings of the 2017 International Conference on Industrial Engineering, Applications and Manufacturing (ICIEAM), St. Petersburg, Russia, 16–19 May 2017; pp. 1–6.
27. Akter, M.; Rahman, M.A.; Munsi, M.S.; Siddique, A.B.; Sarkar, S.K.; Das, S.K. Multi-Objective Model Reference Modified Adaptive PID Framework to Islanded Microgrid Control under Various Load Conditions. In Proceedings of the 2018 International Conference on Advancement in Electrical and Electronic Engineering (ICAEEE), Gazipur, Bangladesh, 22–24 November 2018; pp. 1–4.
28. Patel, T.; Chauhan, P. Battery Storage Assisted Plug-and-Play of Wind Generator in Standalone Microgrid for Off-Grid Residential Consumers. *IEEE Trans. Consum. Electron.* **2024**, *70*, 460–470. [CrossRef]
29. Simpson-Porco, J.W.; Shafiee, Q.; Dörfler, F.; Vasquez, J.C.; Guerrero, J.M.; Bullo, F. Secondary Frequency and Voltage Control of Islanded Microgrids via Distributed Averaging. *IEEE Trans. Ind. Electron.* **2015**, *62*, 7025–7038. [CrossRef]
30. Liu, Z.; Wang, L.; Ma, L. A Transactive Energy Framework for Coordinated Energy Management of Networked Microgrids with Distributionally Robust Optimization. *IEEE Trans. Power Syst.* **2020**, *35*, 395–404. [CrossRef]
31. Fan, Z.; Fan, B.; Peng, J.; Liu, W. Operation Loss Minimization Targeted Distributed Optimal Control of DC Microgrids. *IEEE Syst. J.* **2021**, *15*, 5186–5196. [CrossRef]
32. Kokotović, P.; Khalil, H.K.; O'reilly, J. *Singular Perturbation Methods in Control: Analysis and Design*; Society for Industrial and Applied Mathematics: Philadelphia, PA, USA, 1999.
33. Litkouhi, B.; Khalil, H. Multirate and composite control of two-time-scale discrete-time systems. *IEEE Trans. Autom. Control.* **1985**, *30*, 645–651. [CrossRef]
34. Zhang, Y.; Ma, L.; Yang, C.; Wang, G.; Dai, W. Asynchronous Event-Triggered Control for Two-Time-Scale CPSs with Dual-Scale Channel: Dealing with Hybrid Attacks. *IEEE Trans. Control. Netw. Syst.* **2024**, *11*, 999–1011. [CrossRef]
35. Yu, T.-T.; Wang, Y.-W.; Lei, Y.; Liu, X.-K. Event-Triggered H∞ Control for Fuzzy Two-Time-Scale Systems. *IEEE Trans. Fuzzy Syst.* **2024**, *32*, 2211–2222. [CrossRef]
36. Doan, T.T. Nonlinear Two-Time-Scale Stochastic Approximation: Convergence and Finite-Time Performance. *IEEE Trans. Autom. Control* **2023**, *68*, 4695–4705. [CrossRef]
37. Yuan, Z.-P.; Xia, J.; Li, P. Two-Time-Scale Energy Management for Microgrids with Data-Based Day-Ahead Distributionally Robust Chance-Constrained Scheduling. *IEEE Trans. Smart Grid* **2021**, *12*, 4778–4787. [CrossRef]
38. Zhao, Z.; Yang, P.; Guerrero, J.M.; Xu, Z.; Green, T.C. Multiple-Time-Scales Hierarchical Frequency Stability Control Strategy of Medium-Voltage Isolated Microgrid. *IEEE Trans. Power Electron.* **2016**, *31*, 5974–5991. [CrossRef]
39. Bao, Z.; Zhou, Q.; Yang, Z.; Yang, Q.; Xu, L.; Wu, T. A Multi Time-Scale and Multi Energy-Type Coordinated Microgrid Scheduling Solution—Part I: Model and Methodology. *IEEE Trans. Power Syst.* **2015**, *30*, 2257–2266. [CrossRef]
40. Zhang, Z.; Dong, Z.Y.; Yue, D. Multiple Time-Scale Voltage Regulation for Active Distribution Networks Via Three-Level Coordinated Control. *IEEE Trans. Ind. Inform.* **2024**, *20*, 4429–4439. [CrossRef]

41. Merchán-Riveros, M.C.; Albea, C. Three Time-Scale Singular Perturbation Hybrid Control and Large-Signal Analysis Stability in AC-Microgrids. *IEEE Trans. Circuits Syst. I Regul. Pap.* **2023**, *70*, 3373–3386. [CrossRef]
42. Yu, X.; Wang, B.; Dong, H. A distributed algorithm based on KKT conditions for convex intersection computation. In Proceedings of the 2017 Chinese Automation Congress (CAC), Jinan, China, 20–22 October 2017; pp. 7676–7680.

electronics

MDPI

Article

Research on Bus Voltage Stability Control Technology of the DC Microgrid Based on the Disturbance Estimation Feedforward Compensation Strategy

Huiting Qiao, Hongyan Xin *, Kaiman Li and Zeyuan Yu

Energy Development Research Institute, China Southern Power Grid, Guangzhou 510530, China;
qiaoht@csg.cn (H.Q.); likm@csg.cn (K.L.); yuzy@csg.cn (Z.Y.)
* Correspondence: xinhongyan0109@163.com

Abstract: This paper proposes a control method for the voltage stability of DC microgrid buses based on a disturbance estimation feedforward compensation strategy, aiming to enhance the dynamic response characteristics of the system. A nonlinear disturbance observer is designed to estimate the load current, and the estimated value is applied to the control loop through the feedforward compensation strategy, effectively mitigating the impact of load disturbances on voltage. To simplify the complex dynamic characteristics of the nonlinear observer, a linear equivalent method is employed. This strategy significantly improves the dynamic regulation performance of the DC bus voltage, ensuring system stability and control accuracy.

Keywords: DC microgrid; voltage stability; disturbance estimation; feedforward compensation

1. Introduction

As an important part of the future power system, the DC microgrid has been widely used in the scenarios of new energy grid connection, distributed energy access, and frequent load changes [1–3]. Its high efficiency, flexibility, and compatibility with energy storage systems show great potential in terms of grid stability, energy efficiency, and environmental friendliness. However, the dynamic characteristics of DC microgrids are complex and susceptible to external disturbance, especially the rapid change of the load current, which makes it difficult to guarantee the dynamic stability of DC bus voltage. Therefore, it is particularly important to design an effective control strategy for bus voltage stability control of the DC microgrid [4–6].

The bus voltage stability control of the DC microgrid is an important research field that ensures the reliability of the system. With the increase in distributed energy and load fluctuations, the control strategies for bus voltage show a diversified trend [7,8]. At present, the main busbar voltage control methods include the sag control method [9,10], the hysteresis control method [11,12], the model predictive control method [13,14], and the disturbance observer and feedforward compensation control method [15]. The sag control method mainly realizes the regulation of bus voltage by adjusting the power distribution curve. Taye et al. [16] proposed a sag control method for distributed energy systems and studied its influence on power distribution. Lin et al. [17] introduced the concept of virtual impedance to improve the voltage regulation accuracy of sag control. Montegiglio et al. [18] conducted an in-depth analysis of the dynamic stability of sag control in microgrids and discussed the influence of load fluctuation on voltage stability. In addition, Du et al. [19] applied the droop control method in a distributed power generation system and verified its effectiveness in inverter control. Although sag control has advantages in distributed systems without communication, its voltage regulation accuracy is poor, and it cannot meet the dynamic performance requirements.

The hysteresis control method is widely used in power electronic equipment because of its superior dynamic response performance. Zhou et al. [20] studied the application of hysteresis control in boost converters and proved that it has good control performance when the load changes sharply. Li et al. [21] proposed a hysteresis current control strategy suitable for grid-connected inverters, which significantly improved the transient response of the system. At the same time, Tang et al. [22] designed a hysteresis current control method for single-phase inverters, which has excellent performance in dynamic responses. Behera et al. [23] discussed the application of hysteresis control in renewable energy microgrids and demonstrated its effect on voltage regulation. However, the switching frequency of the hysteresis control method is not fixed, which can easily cause an electromagnetic disturbance, which limits its popularization in practical applications.

Control strategies should be selected based on the specific equipment type used in the DC microgrid, such as the power sources, converters, and energy storage systems. For instance, photovoltaic power generation systems often face significant power fluctuations due to environmental conditions such as cloud cover and sunlight intensity. These fluctuations require more adaptive and accurate control strategies to ensure stable bus voltage regulation. In comparison, battery energy storage systems (BESS) have relatively stable power outputs, making them more suited for control strategies that focus on voltage regulation under varying load conditions. Additionally, fuel cells and distributed generation sources each have unique characteristics that must be considered when designing control systems for voltage stability.

Power converters, such as DC-DC converters and DC-AC inverters, also play a critical role in determining the best control method. DC-DC converters often require voltage and current regulation, especially when handling energy from sources like photovoltaic systems. Meanwhile, DC-AC inverters, which are frequently used for integrating renewable energy sources, often require more sophisticated control methods, such as model predictive control (MPC) or disturbance observers, due to their interaction with AC grids.

Model predictive control (MPC) has become a research hotspot in recent years because of its accurate prediction and high dynamic performance of complex systems. Jayan et al. [24] proposed a MPC method for multilevel inverters, which effectively improved the control accuracy of the system. Cheng et al. [25] further studied the application of MPC in power electronic devices and drive systems, demonstrating its adaptability to complex systems. Yu et al. [26] discussed the advantages and disadvantages of MPC in distributed energy systems through a comprehensive review of MPC in microgrids. Yu et al. [27] proposed a MPC strategy for power converters and drivers to improve their computational complexity and real-time control performance. However, the MPC method has a strong dependence on the mathematical model of the system and high computational complexity, which limits its application in real-time control.

Disturbance observer and feedforward compensation methods are particularly effective in DC microgrids with frequent and rapid load disturbances. These methods estimate the disturbance in the system, which allows for real-time compensation to maintain voltage stability. Alipour et al. [28] proposed a disturbance observer-based control method for power converters in DC microgrids to improve the dynamic response of the system. Wang et al. [29] designed a nonlinear disturbance observer for voltage regulation of the DC bus, which significantly improved the stability of the system. Cui et al. [30] proposed a novel feedforward compensation strategy to improve the bus voltage regulation effect through real-time observation of the load disturbance. Yu et al. [31] optimized the existing feedforward compensation control and proposed a compensation strategy suitable for scenarios with rapid load changes. In addition, Sarrafan et al. [32] designed a disturbance suppression control method using the observer to improve the voltage stability of the system by observing load disturbances. Hanzaei et al. [33] studied the control strategy based on disturbance observation in nonlinear systems and proposed a scheme to improve the voltage regulation performance. When selecting the most appropriate control strategy, the type of power converter and the energy storage system play an important role in determining the dynamic characteristics of the system and the level of disturbance it can handle. For example, systems relying heavily on photovoltaic generation and energy storage will benefit from control methods that can handle both steady-state regulation and dynamic disturbances. In contrast, systems utilizing inverter-based distributed generation may need more advanced control systems capable of real-time adaptation to rapidly changing load profiles.

However, the existing feedforward compensation control methods based on disturbance observation have some shortcomings. First of all, the design of a nonlinear disturbance observer is complicated, especially for the dynamic characteristics of the power generation system. The existing observer has a high computational cost and dynamic response delay in practical application. Secondly, when the load current changes rapidly, the compensation effect of the existing method is not ideal, and the compensation is not timely or the observation is not accurate, which affects the regulation accuracy of the bus voltage. To solve the above problems, this paper presents a method of bus voltage stability control based on a disturbance estimation feedforward compensation control strategy. The main contributions are as follows:

(1) By designing a nonlinear disturbance observer for the load current, an accurate estimation of the load current disturbance is realized.
(2) Combined with the linear equivalent processing method of the nonlinear observer, the complex dynamic characteristics of the nonlinear observer are simplified.
(3) The feedforward compensation control gain based on load current disturbance estimation is further designed, which significantly improves the dynamic regulation ability and voltage control accuracy of the system.

The rest of this paper is organized as follows. Section 2 introduces the power generation system model of a DC microgrid and the design of a voltage–current double closed-loop control strategy. In Section 3, the design process of the nonlinear disturbance observer and disturbance estimation feedforward compensation control strategy is introduced in detail. A comparative simulation verification experiment of the bus voltage dynamic characteristics of the DC microgrid is given to prove the effectiveness of the proposed method in Section 4. Finally, a conclusion is drawn in Section 5.

2. DC Microgrid Architecture and Design of the Voltage–Current Double Closed-Loop Control Strategy

The DC microgrid system architecture consists of a distribution network, photovoltaic power generation system, controllable distributed power sources, energy storage system, demand response loads, and conventional loads. All these components are ultimately controlled by the microgrid central controller. The structure of the DC microgrid system is shown in Figure 1 below. In the following sections, the power generation system model and the voltage–current dual closed-loop control strategy in the system will be analyzed and studied.

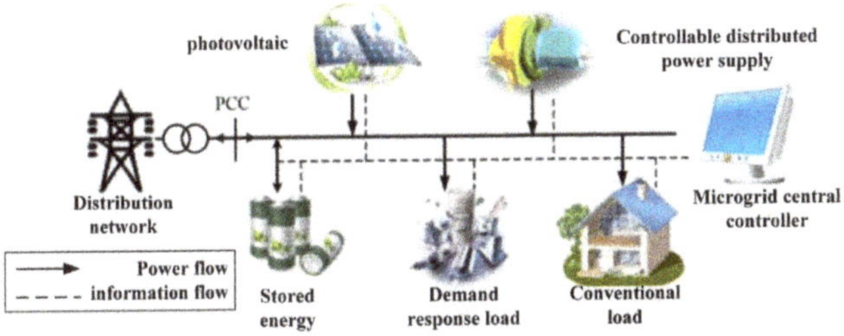

Figure 1. DC microgrid architecture.

The synchronous generator plays a critical role in the DC microgrid as a primary power source, ensuring voltage and frequency stability. It not only provides a stable power output but also offers necessary support during load fluctuations, renewable energy generation variations, or system faults. In the DC microgrid, as the load demand changes or the output from renewable energy sources like photovoltaic systems fluctuates, the synchronous generator adjusts its output power to maintain the balance of the system, ensuring the grid remains stable under sudden disturbances.

The power generation system of a DC microgrid consists of an engine and its speed regulation system, a synchronous generator, and a PWM-VSR rectifier. For the power generation system of a DC microgrid using the PWM-VSR rectification method, its dynamic characteristic mathematical model can be re-expressed as follows:

$$\begin{cases} L_q \frac{di_q}{dt} = -r_s i_q - \omega_r L_d i_d + \omega_r \lambda_m - u_q \\ L_d \frac{di_d}{dt} = -r_s i_d + \omega_r L_q i_q - u_d \\ C_{dc} \frac{dv_{dc}}{dt} = \frac{3(u_d i_d + u_q i_q)}{2v_{dc}} - i_L \end{cases} \quad (1)$$

In Equation (1), i_q and i_d represent the synchronous generator stator currents, with the direction of flow out of the motor taken as positive. u_q and u_d represent the synchronous generator stator voltages, and v_{dc} is the output direct current voltage. i_L is the unknown load current of unknown magnitude.

The synchronous generator exhibits strong dynamic response capabilities, quickly adjusting its output in response to load changes or system disturbances to maintain system stability. For example, during sudden load increases, photovoltaic generation variations, or energy storage system interventions, the synchronous generator helps regulate the voltage and frequency by adjusting its output, preventing overloads or voltage fluctuations from impacting the microgrid. Particularly in frequency regulation, the synchronous generator's response is critical for maintaining frequency stability and ensuring the grid's dynamic response capability under various disturbances. In this paper, a PWM-VSR dual closed-loop current decoupling control structure was designed for the voltage–current inner loop controller of the DC microgrid power generation system. The control system block diagram is shown in Figure 2.

The stator current of the DC microgrid power generation system was selected with a PI controller, and its control action can be expressed as follows:

$$\begin{cases} e_q^* = k_{pi}(i_q^* - i_q) + k_{ii} \int (i_q^* - i_q) dt \\ e_d^* = k_{pi}(i_d^* - i_d) + k_{ii} \int (i_d^* - i_d) dt \end{cases} \quad (2)$$

The expression of the decoupled controller function in Figure 2 is

$$\begin{cases} u_q^* = -\omega_r L_d i_d + \omega_r \lambda_m - e_q^* \\ u_d^* = \omega_r L_q i_q - e_d^* \end{cases} \tag{3}$$

By substituting Equations (2) and (3) into Equation (1), the closed-loop transfer function of the stator current can be obtained as follows:

$$G_i(s) = \frac{i_{qd}(s)}{i_{qd}^*(s)} = \frac{k_{pi}s + k_{ii}}{L_q s^2 + (r_s + k_{pi})s + k_{ii}} \tag{4}$$

It can be seen from Equation (4) that, under decoupling control, the stator current becomes a typical linear second-order system, and the controller parameters can be obtained by setting the damping ratio ξ and the natural frequency ω_n.

$$2\xi\omega_n = \frac{r_s + k_{pi}}{L_q} \tag{5}$$

$$\omega_n^2 = \frac{k_{ii}}{L_q} \tag{6}$$

The current control damping ratio of PWM-VSR in the DC microgrid power generation system is set to $\xi = 1.8$, and the natural frequency is set to $\omega_n = 110$ rad/s.

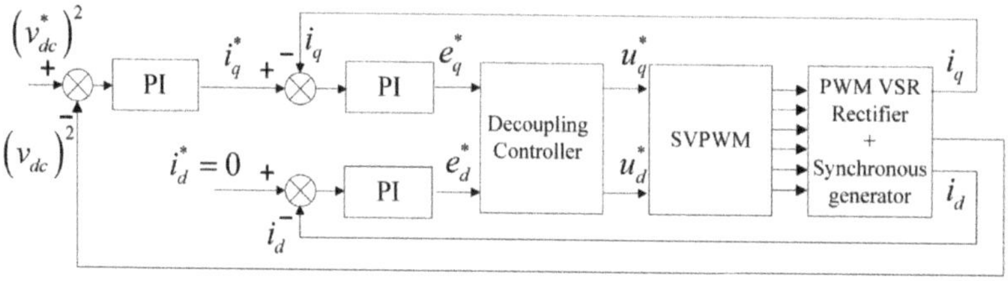

Figure 2. The control structure of the synchronous generator PWM−VSR dual−loop controller.

As can be seen from the PWM-VSR double closed-loop control structure diagram of the synchronous generator, the v_{dc}^2 control method is selected in this paper to adjust the bus voltage of the DC microgrid power generation system, and its closed-loop transfer function can be expressed as

$$G_v(s) = \frac{3U_q(sk_{pv} + k_{iv})}{C_{dc}s^2 + 3U_q k_{pv}s + 3U_q k_{iv}} \tag{7}$$

It can be seen from Equation (7) that the bus voltage v_{dc}^2 control loop is still a typical second-order system, and the damping ratio ξ and natural frequency ω_n of the system can also be set to obtain the parameters of the controller.

The synchronous generator works closely with other key components of the DC microgrid, such as photovoltaic systems, energy storage systems, and controllable distributed power sources. Through coordinated control strategies, these components work together to ensure power balance and system stability within the microgrid. For instance, when load fluctuations occur, the energy storage system can adjust its storage state to balance the power, while the synchronous generator provides necessary stability support to maintain the voltage and frequency of the grid. Furthermore, variations in the output of photovoltaic systems can affect the microgrid's stability and the synchronous generator coordinates with the photovoltaic systems to dynamically adjust the power generation to balance the power flow and stabilize the grid voltage.

3. Design of the Disturbance Estimation Feedforward Compensation Control Strategy for Bus Voltage in the DC Microgrid

It can be seen from Equation (1) that the change in the unknown load current will directly affect the dynamic characteristics of DC bus voltage. Therefore, a nonlinear disturbance observer for the load current is designed for the DC microgrid power generation system to estimate the load current of the DC microgrid power generation system.

3.1. Design of the Nonlinear Disturbance Observer for the Load Current

The nonlinear disturbance observer is designed to estimate the dynamic changes in the load current within the DC microgrid. The primary function of the observer is to accurately estimate the load current and predict future disturbances based on the observed changes, providing real-time information for the feedforward compensation control.

Before the design of the nonlinear disturbance observer for the DC microgrid power generation system, this paper first makes a general statement on the working principle of the nonlinear disturbance observer. Most of the existing nonlinear systems can be described by the following standard affine system model:

$$\begin{cases} \dot{x} = f(x) + g_1(x)u + g_2(x)d \\ y = h(x) \end{cases} \tag{8}$$

In Equation (8), x is the state variable of the system, u is the control input variable of the system, d is the unknown constant disturbance variable of the system, $f(x)$ is the system state function, $g_1(x)$ is the input channel function of the system, $g_2(x)$ is the disturbance channel function of the system, and $h(x)$ is the output function of the system.

According to the basic principle of nonlinear disturbance estimation, the unknown disturbance variable d in Equation (8) can be approximated by a nonlinear disturbance observer as follows:

$$\begin{cases} \dot{z} = -l(x)g_2(x)z - l(x)[g_2(x)p(x) + f(x) + g_1(x)u] \\ \hat{d} = z + p(x) \end{cases} \tag{9}$$

In Equation (9), $\hat{d}$ is the estimated value of the system disturbance, z is the intermediate variable of the disturbance observer, $p(x)$ is the nonlinear function to be designed in the disturbance observer, $l(x)$ is the gain of the disturbance observer, and $l(x)$ and $p(x)$ satisfy the following relation:

$$l(x) = \frac{\partial p(x)}{\partial x} \tag{10}$$

Therefore, this paper defines the estimated error of disturbance as

$$e_d = \hat{d} - d \tag{11}$$

According to Equation (8)–(11), the dynamic characteristics of the disturbance estimation error can be derived:

$$\begin{aligned} \dot{e}_d &= \dot{\hat{d}} - \dot{d} = \dot{z} + \tfrac{\partial p(x)}{\partial x}\dot{x} \\ &= -l(x)g_2(x)e_d \end{aligned} \tag{12}$$

If the system shown in Equation (12) is asymptotically stable, then the estimation error e_d of the disturbance variable will eventually converge to 0, and the disturbance estimate will eventually converge to the actual value of the disturbance, while the speed of convergence depends on the size of $l(x)g_2(x)$.

By using the nonlinear disturbance observer shown in Equation (9), the load current nonlinear disturbance observer can be designed and studied for the DC microgrid power generation system with PWM-VSR rectification control. The mathematical model of the DC microgrid power generation system can be rewritten in the standard form of nonlinear affine system as follows:

$$\begin{array}{ccccc} \begin{bmatrix} \dot{v}_{dc} \\ \dot{i}_q \\ \dot{i}_d \end{bmatrix} & = & \begin{bmatrix} 3(u_d i_d + u_q i_q)/(2C_{dc} v_{dc}) \\ (-r_s i_q - \omega_r L_d i_d + \omega_r \lambda_m - u_q)/L_q \\ (-r_s i_d + \omega_r L_q i_q - u_d)/L_d \end{bmatrix} & + & \begin{bmatrix} -1/C_{dc} \\ 0 \\ 0 \end{bmatrix} \cdot i_L \\ \dot{x} & = & f(x) + g_1(x) \cdot u & & g_2(x) \cdot d \end{array} \tag{13}$$

In Equation (13), the load current i_L is the disturbance quantity d to be estimated in the system, and the disturbance channel function $g_2(x)$ is

$$g_2(x) = \begin{bmatrix} -1/C_{dc} & 0 & 0 \end{bmatrix}^T \tag{14}$$

It can be seen from Equation (14) that the disturbance channel $g_2(x)$ of the power generation system of the DC microgrid has only one non-zero variable, so the disturbance observer gain $l(x)$ can be designed as

$$l(x) = \begin{bmatrix} l_1 & 0 & 0 \end{bmatrix} \tag{15}$$

According to Equations (10) and (15), the undetermined function $p(x)$ of the disturbance observer can be obtained as follows:

$$p(x) = l_1 v_{dc} \tag{16}$$

The dynamic characteristics of the estimation error of the nonlinear disturbance observer shown in Equation (12) can be simplified as

$$\dot{e}_d = -l(x) g_2(x) e_d = l_1/C_{dc} \cdot e_d \tag{17}$$

According to the nonlinear disturbance observer in Equation (9) and the affine mathematical model of the power generation system, the design results of the load current disturbance observer of the DC microgrid power generation system can be obtained as follows:

$$\begin{cases} \dot{z} = l_1/C_{dc} z + l_1^2 v_{dc}/C_{dc} - l_1 \cdot 3(u_d i_d + u_q i_q)/(2C_{dc} v_{dc}) \\ \hat{i}_L = z + l_1 v_{dc} \end{cases} \tag{18}$$

In Equation (18), $\hat{i}_L$ is the estimated load current of the DC microgrid power generation system, z is the intermediate variable of the disturbance observer, v_{dc} is the DC bus voltage, u_d and u_q are the stator voltages of the synchronous generator in the rotor coordinate system, i_d and i_q are the stator currents, and l_1 is the gain of the disturbance observer to be designed.

As can be seen from Equation (18), the designed disturbance observer needs information about the stator voltage of the synchronous generator, which will be troublesome to use. Therefore, the disturbance observer given in Equation (18) is further improved in this paper.

First, it is assumed that the estimated error convergence bandwidth of the designed nonlinear disturbance observer is much smaller than the stator current control bandwidth of the synchronous generator, so that the dynamic change process of the stator current in the synchronous generator can be ignored. According to the mathematical model of the stator voltage of the power generation system shown in Equation (1), it can be obtained:

$$\begin{cases} u_q = -r_s i_q - \omega_r L_d i_d + \omega_r \lambda_m \\ u_d = -r_s i_d + \omega_r L_q i_q \end{cases} \quad (19)$$

According to the conditions of the stator current ring $i_d^* = 0$ and the electron-generating resistance $r_s \approx 0$ of the power generation system, the mathematical model of the stator voltage of the power generation system Equation (19) can be further simplified as follows:

$$\begin{cases} u_q = \omega_r \lambda_m \\ u_d = \omega_r L_q i_q \end{cases} \quad (20)$$

By substituting the simplified stator voltage Equation (20) into the nonlinear disturbance observer Equation (18), the final design result of the nonlinear disturbance observer for the load current of the power generation system can be obtained:

$$\begin{cases} \dot{z} = l_1/C_{dc} z + l_1^2 v_{dc}/C_{dc} - l_1 \cdot (3\omega_r \lambda_m i_q)/(2C_{dc} v_{dc}) \\ \hat{i}_L = z + l_1 v_{dc} \end{cases} \quad (21)$$

where ω_r is the speed of the generator, and λ_m is the flux linkage of the synchronous motor.

It can be seen from Equation (17) that, as long as l_1/C_{dc} is less than zero, the disturbance observer estimation results of the power generation system will inevitably converge progressively to the actual load current value. Therefore, the dynamic relationship between the estimated load current $\hat{i}_L$ and the actual load current i_L can be further written in the form of a transfer function:

$$\hat{i}_L(s) = \frac{1}{\tau_{NDO} s + 1} i_L(s) \quad (22)$$

where $\tau_{NDO} = -C_{dc}/l_1$. From the above equation, it can be seen that disturbance observer is equivalent to a low-pass filter time constant for τ_{NDO}, τ_{NDO} is smaller, and the disturbance observer to estimate the load current convergence speed is faster.

Thus, the design of the nonlinear disturbance observer for the load current of the auxiliary power generation system is completed. In this paper, the estimated value of the load current will be used to carry out feedforward compensation for the bus voltage of the DC microgrid. The designed feedforward compensation controller structure principle of power generation system disturbance is shown in Figure 3.

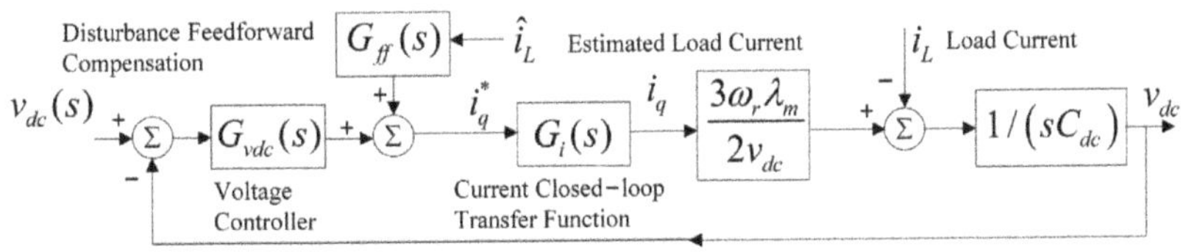

Figure 3. Disturbance feedforward compensation controller for the auxiliary generation system.

As can be seen from Figure 3, the load current $\hat{i}_L$ estimated by the disturbance observer is superimposed to the given value $G_{ff}(s)$ of the stator current of the generator via the feedforward compensation control gain i_q^*. The purpose is to use the value estimated by the nonlinear disturbance observation to offset the influence of the load current change on the DC bus voltage.

This observer is tightly integrated with the feedforward compensation control strategy. By using the load current estimates provided by the observer, we can compensate for disturbances in advance. Specifically, the estimated values from the observer are fed into the feedforward control loop to adjust the voltage before disturbances impact the system, thereby reducing the effect of disturbances on the bus voltage.

3.2. Design of the Disturbance Estimation Feedforward Compensation Control Strategy for the Load Current

In DC microgrids, maintaining the stability of the bus voltage is crucial to ensuring the safe and reliable operation of the system. To effectively address the impact of load disturbances on the bus voltage, this paper proposes a disturbance estimation-based feedforward compensation control strategy. By designing a nonlinear disturbance observer to estimate the load current, the estimated value is applied to the voltage control loop through feedforward compensation, significantly enhancing the system's dynamic response capability. This section first introduces the basic working principle of the feedforward compensation control strategy, then details the calculation process of the control gain and optimizes the differential component of the controller gain. Finally, this section presents an improved control method by combining the V-I droop control strategy, and the overall system working principle is illustrated through a control structure diagram.

As shown in Figure 3, the dynamic characteristics of bus voltage under the control of feedforward compensation for the disturbance estimation can be expressed as

$$\begin{aligned} v_{dc}(s) = & \frac{(1/sC_{dc})G_{vdc}(s)G_i(s)(3\omega_r\lambda_m/2v_{dc})}{1+(1/sC_{dc})G_{vdc}(s)G_i(s)(3\omega_r\lambda_m/2v_{dc})}v^*_{dc}(s) \\ & -\frac{1}{sC_{dc}+G_{vdc}(s)G_i(s)(3\omega_r\lambda_m/2v_{dc})}i_L(s) \\ & +\frac{G_{ff}(s)G_i(s)(3\omega_r\lambda_m/2v_{dc})}{sC_{dc}+G_{vdc}(s)G_i(s)(3\omega_r\lambda_m/2v_{dc})}\hat{i}_L(s) \end{aligned} \tag{23}$$

where $G_{vdc}(s)$ is the DC bus voltage controller, $G_i(s)$ is the closed-loop transfer function of the stator current, and $G_{ff}(s)$ is the gain of the disturbance feedforward compensation controller. As can be seen from Equation (23), in order to eliminate the influence of the load current i_L on the bus voltage v_{dc} of the DC microgrid power generation system, the gain $G_{ff}(s)$ of the disturbance feedforward compensation controller needs to meet the following relation:

$$G_{ff}(s)G_i(s)(3\omega_r\lambda_m/2v_{dc})\hat{i}_L(s) = i_L(s) \tag{24}$$

From the dynamic characteristics of the nonlinear disturbance observer, it can be seen that there is a low-pass filter $1/(\tau_{NDO}s+1)$ between the estimated load current $\hat{i}_L(s)$ and the actual current $i_L(s)$. Therefore, the gain of the disturbance feedforward compensation controller $G_{ff}(s)$ can be obtained by Equation (24), the magnitude of which is

$$G_{ff}(s) = \frac{\tau_{NDO}s+1}{G_i(s)(3\omega_r\lambda_m/2v_{dc})} \tag{25}$$

In this paper, by substituting Equation (4) of the closed-loop control transfer function $G_i(s)$ of the stator current of the generator into Equation (25) and further simplifying the controller gain of the disturbance feedforward compensation, the following can be obtained:

$$\begin{aligned} G_{ff}(s) &= \frac{\tau_{NDO}s+1}{3\omega_r\lambda_m/2v_{dc}}\cdot\left(\frac{L_qs+r_s}{k_{pi}s+k_{ii}}s+1\right) \\ &\approx \frac{\tau_{NDO}s+1}{(3\omega_r\lambda_m/2v_{dc})}\left(\frac{L_q}{k_{pi}}s+1\right) \\ &\approx \frac{1+\left(\tau_{NDO}+L_q/k_{pi}\right)s}{(3\omega_r\lambda_m/2v_{dc})} \end{aligned} \tag{26}$$

It can be seen from Equation (26) that the gain of the disturbance feedforward compensation controller $G_{ff}(s)$ contains a pure differential link s, which will introduce a high-frequency disturbance into the given value of the stator current i_q. Therefore, this paper uses the differential link containing a low-pass filter to replace it, and then, the final disturbance feedforward compensation controller gain $G_{ff}(s)$ is as follows:

$$G_{ff}(s) = \frac{1}{(3\omega_r\lambda_m/2v_{dc})} \cdot \left(1 + \frac{(\tau_{NDO} + L_q/k_{pi})s}{\tau_{LF}s + 1}\right) \tag{27}$$

In Equation (27), τ_{LF} is the low-pass filter time constant of the differential element in the gain $G_{ff}(s)$ of the disturbance feedforward compensation controller.

Therefore, this paper completes the design of a feedforward compensation control strategy for load current disturbance estimation of the DC microgrid power generation system. Combined with the V-I sag control strategy at the voltage end, an improved V-I sag control method for the DC microgrid power generation system based on disturbance estimation feedforward compensation is formed, and its control structure schematic diagram is shown in Figure 4.

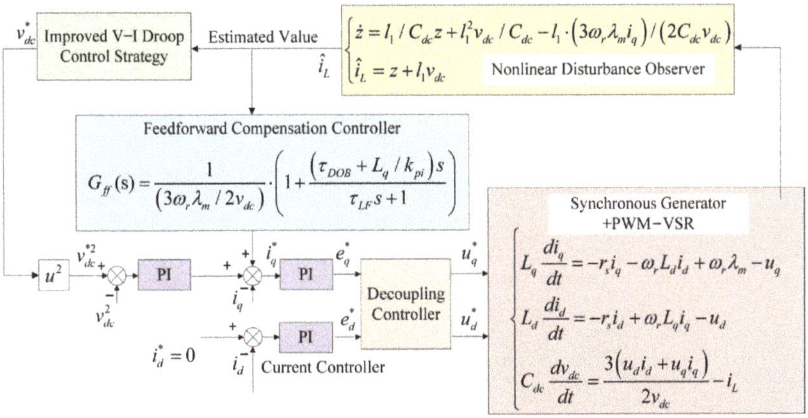

Figure 4. Structure of the modified droop controller for the DC microgrid power generation system based on disturbance estimation and feedforward compensation.

Based on the above content, in order to solve the contradiction between the dynamic regulation characteristics of DC bus voltage in the DC microgrid power generation system and system stability, a feedforward compensation control strategy for the power generation system disturbance estimation is proposed in this paper, and a nonlinear disturbance observer for the DC microgrid power generation system is designed to estimate the load disturbance current. Then, the disturbance feedforward compensation control method is given, and the disturbance feedforward compensation control gain is designed. The dynamic influence of load current change on the DC bus voltage is offset by the estimated disturbance current, and the contradiction between bus voltage stability and dynamic regulation characteristics of the original DC microgrid is solved.

4. Comparative Simulation and Verification of Bus Voltage Dynamic Characteristics of the DC Microgrid

In order to verify the effectiveness of the proposed feedforward compensation control strategy for disturbance estimation, this paper further conducts time–domain simulation analysis of the dynamic characteristics of the DC microgrid power generation system, as follows.

4.1. Effectiveness Simulation Verification of the Nonlinear Disturbance Observer Design for the DC Microgrid Power Generation System

In order to verify the effectiveness of the design of nonlinear disturbance observer for the load current of the DC microgrid power generation system, a simulation example is designed for the dynamic characteristics of the power generation system under the condition of load current variation. In this simulation example, the DC microgrid power generation system is connected to an equivalent constant power load, and the system only operates under double closed-loop voltage–current control.

The DC microgrid power generation system initially works in a no-load state, the prime mover speed is 376.9 rad/s, and the DC bus voltage is 5 kV. At 10 s, the constant power load connected to the DC microgrid power generation system changes from 0 MW to 2 MW, and the corresponding simulation results of the system dynamic characteristics are shown in Figure 5.

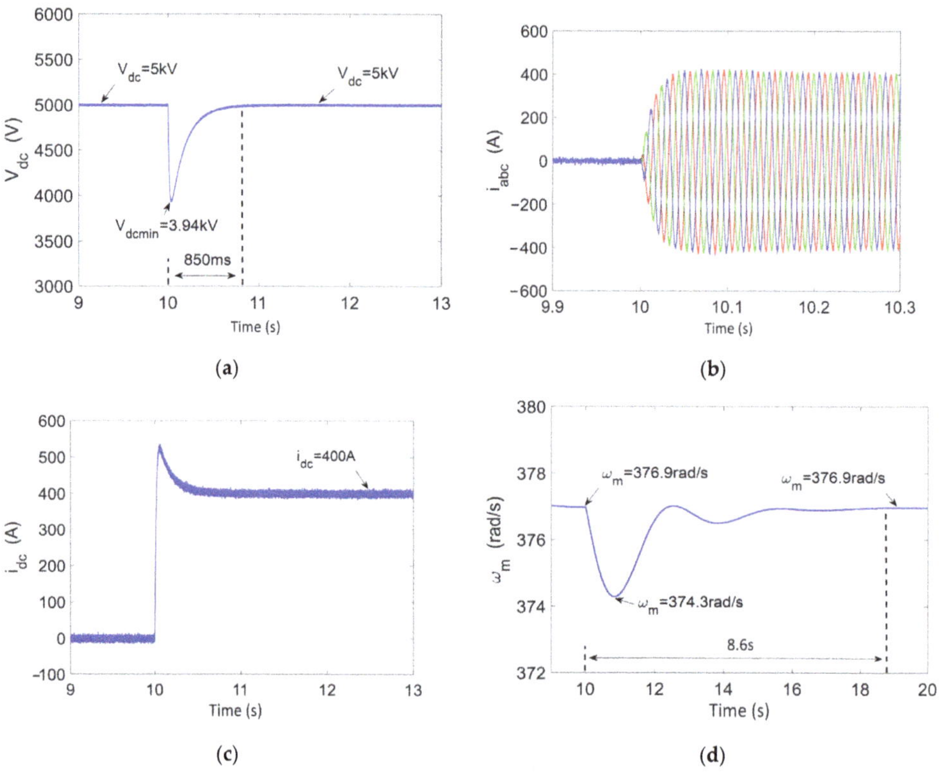

Figure 5. (**a**) DC bus voltage. (**b**) Stator current of the synchronous generator. (**c**) DC output current. (**d**) Generator speed.

As can be seen from the simulation curve of the dynamic response characteristics of DC bus voltage shown in Figure 5, the sudden increase in the load current will lead to rapid discharge of the bus capacitance, and the DC bus voltage will drop from the original 5 kV to 3.94 kV in less than 50 ms. Due to the voltage regulation of the DC microgrid power generation system, the stator current of the synchronous generator and the DC current output of the system will gradually rise, and the DC bus voltage will start to adjust back and return to the given 5 kV after about 850 ms.

As can be seen from the simulation results of the speed characteristics of the synchronous generator in Figure 5, the sudden increase in the constant power load will reduce the prime mover speed from 376.9 rad/s to 374.3 rad/s. Due to the adjustment and control of the prime mover speed regulation system, the original expected speed of 376.9 rad/s will be restored at 18.6 s. The comparison simulation curve between the dynamic characteristics of the load current of the DC microgrid power generation system and the tracking characteristics of the estimated value of the nonlinear disturbance observer is shown in Figure 6.

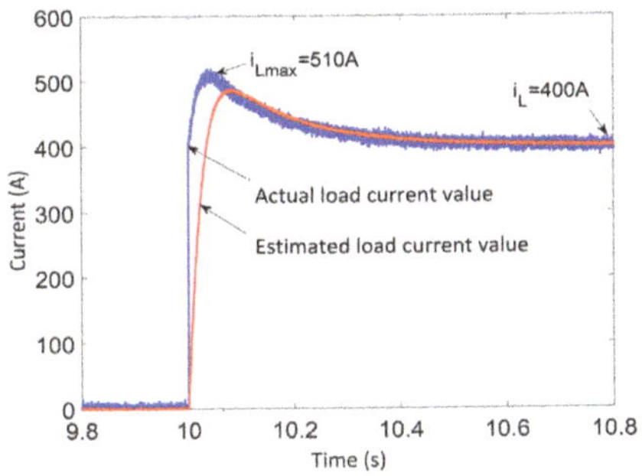

Figure 6. Simulation results for the actual load current and nonlinear disturbance observer estimations.

It can be seen from the simulation results that the load current estimate given by the nonlinear disturbance observer filters out the high-frequency noise in the original load current, which verifies that the load current estimate of the nonlinear disturbance observer has the characteristics of a low-pass filter and further verifies the validity of the theoretical derivation.

However, it should be noted that there is a certain overshoot of the constant power load current in the simulation results, which is caused by the voltage drop of the DC bus. When the DC bus voltage is adjusted back to 5 kV at 10.8 s, the constant power load current also stabilizes at 400 A, which is consistent with the given 2 MW load power.

In the simulation example, this paper verifies that the nonlinear disturbance observer has a good dynamic tracking effect and there is no static deviation between the estimated load current and the actual load current and fully verifies the validity of the design of the nonlinear disturbance observer for the DC microgrid power generation system through the time–domain simulation analysis of the system characteristics when the constant power load power surges.

4.2. Effectiveness Simulation Verification of the Disturbance Feedforward Compensation Control Method Design for the DC Microgrid Power Generation System

In this example, a comparative time–domain simulation study of the DC bus voltage dynamic response characteristics of the system before and after the disturbance feedforward compensation control strategy is enabled is carried out to verify the effectiveness of the disturbance feedforward compensation control strategy design.

In this simulation example, the constant power load suddenly increases from 2 MW to 3 MW at 20 s, and the corresponding DC bus voltage dynamic characteristics of the time–domain simulation results are shown in Figure 7.

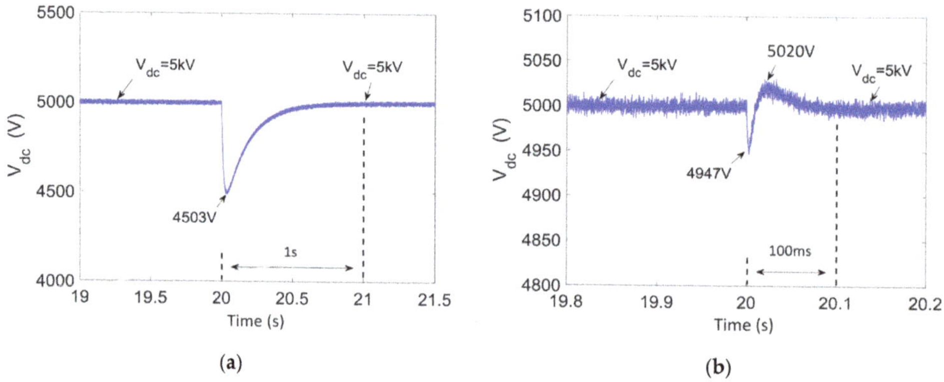

Figure 7. (**a**) Non-disturbance feedforward compensation control; (**b**) feedforward compensation control with disturbance.

As can be seen from Figure 7a, at 20s, when the constant power load suddenly increases from 2 MW to 3 MW, the simulation results of bus voltage dynamic characteristics of the DC microgrid generation system without the disturbance feedforward compensation control strategy show a serious voltage sag phenomenon. The lowest value of DC bus voltage is 4503 V, and its dynamic control deviation is close to 10%. It cannot meet the power quality requirements of the DC microgrid power generation system. Moreover, as can be seen from Figure 7, the callback speed of the DC bus voltage is very slow, and it takes about 1 s to slowly recover to the given standard voltage of 5 kV. The dynamic characteristics of DC bus voltage are extremely sensitive to the change in load power.

As can be seen from Figure 7b, in the constant power load, the mw spurted from 2 to 3 mw 20 s, the disturbance feedforward compensation control strategy of the DC microgrid power system bus voltage dynamic control effect was good, the lowest value of the DC bus voltage was 4947 v, and the maximum dynamic control deviation was only 1%. Although the DC bus voltage has an overshoot of about 20 V during the callback process, it only accounts for 0.4% of the 5 kV standard voltage. Moreover, the callback speed of the DC bus voltage is very fast, and it only takes about 100 ms to restore to the standard 5 kV, and the dynamic response characteristics of the voltage are excellent.

Comparing the time–domain simulation results of the system characteristics shown in Figure 7a,b, it can be clearly seen that the disturbance feedforward compensation control strategy proposed in this paper effectively offsets the influence of the constant power load current change on the dynamic characteristics of the DC bus voltage of the system, which reduces the instantaneous drop amplitude of the bus voltage and speeds up the callback speed of the DC bus voltage. The DC bus voltage dynamic response characteristics meet the power quality requirements of the DC microgrid power generation system.

In the simulation example, this paper verifies the effectiveness of the design of the DC bus voltage feedforward compensation control strategy by comparing the dynamic characteristics of the DC bus voltage before and after the disturbance feedforward compensation control strategy is enabled.

4.3. Analysis of the Influence of Disturbance Estimation Feedforward Compensation Controller Parameter Variation on the Dynamic Characteristics of DC Bus Voltage

In this example, the time–domain simulation study of system characteristics is carried out around the influence of disturbance estimation feedforward compensation controller parameter change on the DC bus voltage dynamic characteristics.

- A. Influence of the time constant τ_{LF} on the DC bus voltage dynamic characteristics

The first is the time–domain simulation analysis of the influence of the change of the time constant of the low-pass filter in the approximate differential link of the disturbance feedforward compensator on the dynamic characteristics of the DC bus voltage. A 2 MW constant power load is applied at 10 s, and the obtained time–domain simulation results of the system dynamic response characteristics are shown in Figure 8a, as the simulation results of the dynamic characteristics of DC bus voltage with the low-pass filter time constant $\tau_{LF} = 1/1000$ of the approximate differential link in the disturbance feedforward compensator. Figure 8b corresponds to the simulation results of the DC bus voltage dynamic characteristics at $\tau_{LF} = 1/200$.

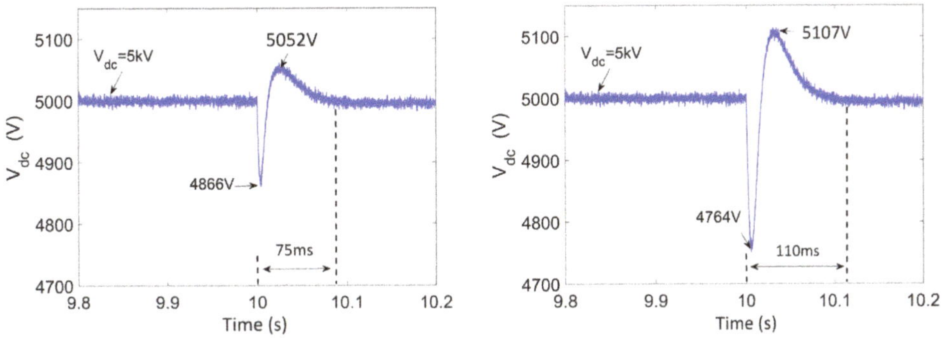

(a) (b)

Figure 8. (**a**) The time constant of the low-pass filter is 1/1000. (**b**) The low-pass filter time constant is 1/200.

It can be seen from the time–domain simulation analysis results of the above system that, the smaller the time constant τ_{LF} of the approximate differential link of the disturbance feedforward compensator, the smaller the corresponding DC bus voltage dynamic control deviation and the voltage callback time. When $\tau_{LF} = 1/1000$, the corresponding minimum DC bus voltage is 4866 V, the dynamic control deviation of the voltage is 2.8%, the maximum voltage is 5052 V, the overshoot is about 1%, and the voltage callback time is about 75 ms. When $\tau_{LF} = 1/200$, the corresponding DC bus voltage minimum value is 4764 V, the dynamic control deviation is about 5%, the maximum voltage is 5107 V, the overshoot is about 2%, and the voltage callback time is about 110 ms.

In summary, the selection of the low-pass filter time constant τ_{LF} of the approximate differential link in the disturbance feedforward compensator needs careful consideration: if τ_{LF} is too small, the high-frequency noise suppression ability of the approximate differential link is insufficient. If τ_{LF} is too large, it will have an adverse effect on the dynamic characteristics of DC bus voltage. In this paper, through the time–domain simulation of the system, the high-frequency noise suppression ability and voltage dynamic characteristics of the approximate differential link are considered, and the time constant of the low-pass filter in the approximate differential link is determined to be $\tau_{LF} = 1/1000$.

- B. Influence of time constant τ_{NDO} on the DC bus voltage dynamic characteristics

In the simulation example, the system load power suddenly increases from 0 MW to 2 MW at 10 s, and the corresponding time–domain simulation results of the system dynamic characteristics are shown in Figure 9. Among them, Figure 9a,b correspond to the time–domain simulation results of the system characteristics when the time constant of the nonlinear disturbance observer is $\tau_{NDO} = 20$ ms. Figure 9c,d correspond to the time–domain simulation results of the $\tau_{NDO} = 200$ ms system characteristics.

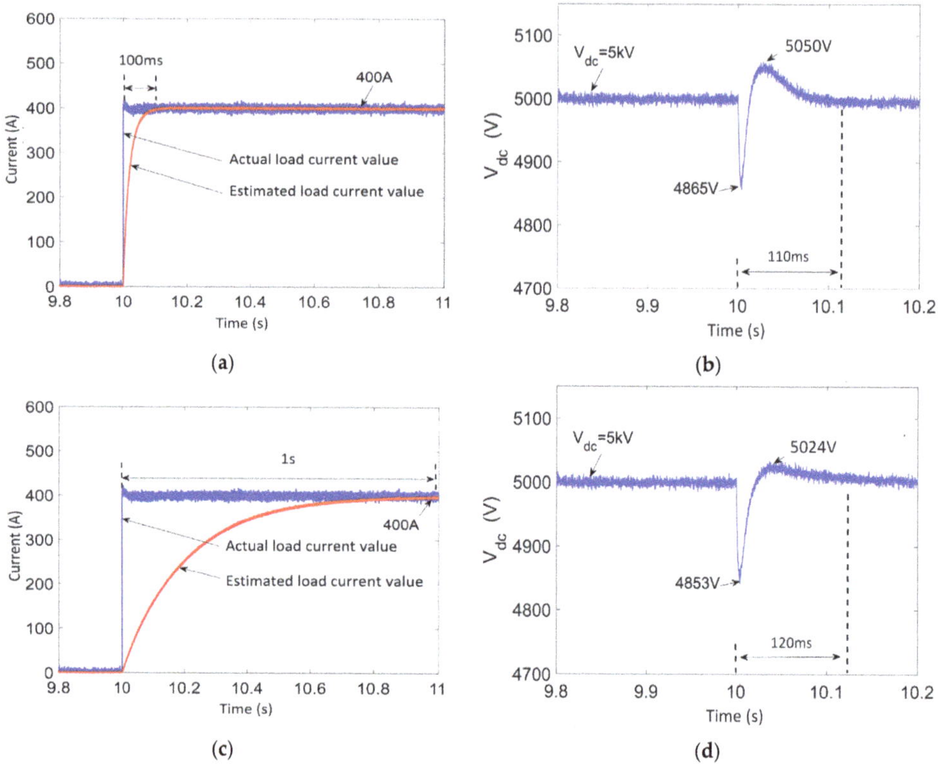

Figure 9. (**a**) Load current and disturbance estimation when τ_{NDO} is 20 ms. (**b**) The DC bus voltage when τ_{NDO} is 20 ms; (**c**) load current and disturbance estimation when τ_{NDO} is 200 ms; (**d**) τ_{NDO} is the DC bus voltage at 200 ms.

As can be seen from Figure 9a, with the nonlinear disturbance observer for the $\tau_{NDO} = 20$ ms time constant, the nonlinear disturbance observer presents a good dynamic load current estimate of the tracking effect, and about a 100 ms estimate of the size of the load current can be realized without static error tracking. As shown in Figure 9b, the minimum value of the DC bus voltage is 4865 V, the dynamic control deviation is about 3%, the maximum voltage is 5050 V, the overshot is about 1%, and the callback time of the DC bus voltage is about 110 ms.

As you can see from Figure 9c, when the nonlinear disturbance observer time constant is $\tau_{NDO} = 200$ ms, estimates of the nonlinear disturbance observer given for the load current can be realized without static error tracking, but its tracking speed is slower, an intermediate process taking about 1 s or so. From the simulation results of the DC bus voltage dynamic regulation characteristics shown in Figure 9d, it can be seen that the minimum DC bus voltage corresponding to $\tau_{NDO} = 200$ ms is 4853 V, which corresponds to a dynamic control deviation of about 3%, which is not much different from 4865 V corresponding to Figure 9b. The voltage callback time is also about 120 ms, and the maximum voltage is 5024 V. On the contrary, it is slightly smaller than 5050 V in Figure 9b.

From the above system characteristic of the time–domain simulation results, it can be found that the nonlinear disturbance observer time constant τ_{NDO} change will impact the dynamic tracking characteristics of the disturbance estimation, and the smaller the τ_{NDO}, the better the dynamic effect of the load current estimates. However, the change in τ_{NDO} has no significant effect on the dynamic characteristics of the DC bus voltage regulation, because the dynamic characteristics of the disturbance observer have been taken into account and compensated for in the design process of the disturbance feed-forward compensation controller. Therefore, the change in the time constant τ_{NDO} of the nonlinear disturbance observer does not affect the dynamic characteristics of the DC bus voltage, which also verifies the validity of the design results of the disturbance feedforward compensation controller.

5. Conclusions

This paper presents a feedforward compensation control strategy for disturbance estimation of the DC microgrid power generation system, which solves the contradiction between the dynamic response characteristics and stability of the bus voltage of the DC microgrid power generation system. A nonlinear disturbance observer for the DC microgrid power generation system is designed, and the load current estimate is given by using the observer. The disturbance feedforward compensation controller is designed, and the influence of load current on the dynamic characteristics of DC bus voltage is offset by the estimation of the disturbance observer. A time–domain simulation experiment of dynamic characteristics of the DC microgrid power generation system is designed to verify the effectiveness of the proposed feedforward compensation control strategy for disturbance estimation, and the influence of the approximate differential low-pass filter parameters in the feedforward compensation controller for disturbance estimation and the time constant of the nonlinear disturbance observer on the dynamic characteristics of the system are simulated. The effectiveness of the feedforward compensation control strategy for disturbance estimation of the power generation system is verified.

Although the proposed feedforward compensation control strategy significantly improves the dynamic response of the DC microgrid's bus voltage, it is not without limitations. The performance of the disturbance estimation is sensitive to the accuracy of the load current measurement and the nonlinear disturbance observer's parameters. For large disturbances or highly dynamic load changes, the effectiveness of the control may degrade, and the system may not maintain optimal voltage regulation. Further studies will focus on evaluating the strategy under more extreme conditions and refining the estimation and compensation approach for better robustness.

Author Contributions: Conceptualization, H.Q. and H.X.; methodology, H.Q. and K.L.; software, H.X. and Z.Y.; validation, H.X., K.L. and Z.Y.; formal analysis, H.Q. and Z.Y.; data curation, K.L. and Z.Y.; writing—original draft, H.Q., H.X. and K.L.; writing—review and editing, H.Q., H.X. and K.L.; supervision, Z.Y. All authors have read and agreed to the published version of the manuscript.

Funding: This work was supported in part by the National Natural Science Foundation of China (NSFC) under grant 52271313, in part by the Natural Science Foundation of Shannxi Province under grant 2023JCQN0751, and in part by the Fundamental Research Funds for the Central Universities under grant 3072024GH0405.

Data Availability Statement: The original contributions presented in the study are included in the article. Further inquiries can be directed to the corresponding author.

Conflicts of Interest: The authors declare no conflicts of interest.

References

1. Xu, Q.; Zhang, C.; Wen, C. A Novel Composite Nonlinear Controller for Stabilization of Constant Power Load in DC Microgrid. *IEEE Trans. Smart Grid* **2019**, *10*, 752–761. [CrossRef]

2. Meng, J.; Zhang, Y.; Wang, Y. Flexible virtual capacitance control strategy for a DC microgrid with multiple constraints. *IET Renew. Power Gener.* **2020**, *14*, 3469–3478. [CrossRef]
3. Gao, F.; Bozhko, S.; Costabeber, A. Comparative Stability Analysis of Droop Control Approaches in Voltage-Source-Converter-Based DC Microgrids. *IEEE Trans. Power Electr.* **2017**, *32*, 2395–2415. [CrossRef]
4. Gui, Y.; Han, R.; Guerrero, J. Large-Signal Stability Improvement of DC-DC Converters in DC Microgrid. *IEEE Trans. Energy Conver.* **2021**, *36*, 2534–2544. [CrossRef]
5. Jiang, J.; Liu, F.; Pan, S. A Conservatism-Free Large Signal Stability Analysis Method for DC Microgrid Based on Mixed Potential Theory. *IEEE Trans. Power Electr.* **2019**, *34*, 11342–11351. [CrossRef]
6. Raghavendra, N.K.; Bhooshan, R.; Arghya, M. Adaptive energy management strategy for sustainable voltage control of PV-hydro-battery integrated DC microgrid. *J. Clean. Prod.* **2021**, *315*, 128102.
7. Dragicevic, T.; Lu, X.; Vasquez, J. DC Microgrids-Part II: A Review of Power Architectures, Applications, and Standardization Issues. *IEEE Trans. Power Electr.* **2016**, *31*, 3528–3549. [CrossRef]
8. Raghavendra, N.K.; Bhooshan, R.; Arghya, M. Assessment of energy management technique for achieving the sustainable voltage level during grid outage of hydro generator interfaced DC Micro-Grid. *Sustain. Energy Technol. Assess.* **2021**, *46*, 101231.
9. Chen, X.; Zhou, J.; Shi, M. A Novel Virtual Resistor and Capacitor Droop Control for HESS in Medium-Voltage DC System. *IEEE Trans. Power Syst.* **2019**, *34*, 2518–2527. [CrossRef]
10. Ni, J.; Zhao, B.; Goudarzi, A. A Dispatchable Droop Control Method for PV Systems in DC Microgrids. *IEEE Access* **2023**, *11*, 7588–7598. [CrossRef]
11. Lin, G.; Ma, J.; Li, Y. A Virtual Inertia and Damping Control to Suppress Voltage Oscillation in Islanded DC Microgrid. *IEEE Trans. Energy Conver.* **2021**, *36*, 1711–1721. [CrossRef]
12. Zhu, X.; Xie, Z.; Jing, S. Distributed Virtual Inertia Control and Stability Analysis of DC Microgrid. *IET Gen. Transm. Distrib.* **2018**, *12*, 3477–3486. [CrossRef]
13. Zhi, N.; Ding, K.; Du, L. An SOC-Based Virtual DC Machine Control for Distributed Storage Systems in DC Microgrids. *IEEE Trans. Energy Conver.* **2020**, *35*, 1411–1420. [CrossRef]
14. Wang, Y.; Wang, Z.; Lei, M. Analysis and Control of DC Voltage Dynamics Based on a Practical Reduced-Order Model of Droop-Controlled VSC-MTDC System in DC Voltage Control Timescale. *IEEE Trans. Power Deliv.* **2020**, *39*, 1031–1039. [CrossRef]
15. Zhi, N.; Ming, X.; Ding, Y. Power-Loop-Free Virtual DC Machine Control with Differential Compensation. *IEEE Trans. Ind. Appl.* **2022**, *58*, 413–422. [CrossRef]
16. Taye, B.; Choudhury, N. A Dynamic Droop Control for a DC Microgrid to Enhance Voltage Profile and Proportional Current Sharing. *Electr. Pow. Syst. Res.* **2023**, *221*, 109438. [CrossRef]
17. Lin, X.; Yu, R.; Yu, J. Dual-frames-impedance-based Analysis of Dynamic Phase Difference Effect on Grid-forming Converter with Different Power Synchronization Controls. *Int. J. Electr. Power Energy Syst.* **2024**, *158*, 109928. [CrossRef]
18. Montegiglio, P.; Acciani, G.; Dicorato, M. A Decentralized Power and Bus Voltage Regulation Approach for DC Microgrids. *IEEE Trans. Ind. Appl.* **2023**, *59*, 4773–4785. [CrossRef]
19. Du, W.; Nguyen, Q.; Wang, S. Positive-Sequence Modeling of Droop-Controlled Grid-Forming Inverters for Transient Stability Simulation of Transmission Systems. *IEEE Trans. Power Deliv.* **2024**, *39*, 1736–1748. [CrossRef]
20. Zhou, Y.; Zhu, W.; Tong, G. A Core Loss Calculation Method for DC/DC Power Converters Based on Sinusoidal Losses. *IEEE Trans. Power Electr.* **2023**, *38*, 692–702.
21. Li, X.; Liu, Y.; Zhang, H. Hybrid-Modulation Hysteresis Scheme Based Decoupled Power Control of Grid-Connected Inverter. *IEEE J. Em. Sel. Top. Power Electron.* **2023**, *11*, 276–287. [CrossRef]
22. Tang, Y.; Zhang, C.; Guo, Y. Optimization of Zero-Crossing Distortion for Unipolar BCM Grid-Tied Inverter. *IEEE J. Em. Sel. Top. Power Electron.* **2023**, *11*, 3680–3691. [CrossRef]
23. Behera, M.; Saikia, L. A Novel Spontaneous Control for Autonomous Microgrid VSC System Using BPF Droop and Improved Hysteresis Band Control Scheme. *Electr. Pow. Syst. Res.* **2023**, *220*, 109262. [CrossRef]
24. Jayan, V.; Ghias, A. Computationally-Efficient Model Predictive Control of Dual-Output Multilevel Converter in Hybrid Microgrid. *IEEE Trans. Power Electr.* **2023**, *38*, 5898–5910. [CrossRef]
25. Cheng, L.; Qiu, L.; Wu, W. Virtual Voltage Vector-Based Sequential Model-Free Predictive Control for Multi-paralleled NPC Inverters. *IEEE J. Em. Sel. Top. Power Electron.* **2023**, *11*, 3108–3116. [CrossRef]
26. Yu, Y.; Liu, G.; Hu, W. Coordinated Distributed Predictive Control for Voltage Regulation of DC Microgrids with Communication Delays and Data Loss. *IEEE Trans. Smart Grid* **2023**, *14*, 1708–1722. [CrossRef]
27. Yu, Y.; Liu, G.; Huang, Y. Coordinated Predictive Secondary Control for DC Microgrids Based on High-Order Fully Actuated System Approaches. *IEEE Trans. Smart Grid* **2024**, *15*, 19–33. [CrossRef]
28. Alipour, M.; Zarei, J.; Razavi-Far, R. Observer-Based Backstepping Sliding Mode Control Design for Microgrids Feeding a Constant Power Load. *IEEE Trans. Ind. Electron.* **2023**, *70*, 465–473. [CrossRef]

29. Wang, Z.; Hu, J. Composite Sliding Mode Control for Converters with Unknown Constant Power Load. *IEEE Trans. Ind. Appl.* **2024**, *60*, 4098–4108. [CrossRef]
30. Cui, Y.; Yin, Z.; Bai, C. Robust Control of IPMSM Based on Doubly-Fed Differential Compensation Linear Active Disturbance Rejection Controller. *IEEE J. Em. Sel. Top. Power Electron.* **2024**, *12*, 936–949. [CrossRef]
31. Yu, D.; Zhang, W.; Chu, Z. An optimized synchronous approach to DC/DC droop backstepping control considering voltage compensation in DC microgrids. *Electr. Pow. Syst. Res.* **2024**, *228*, 110014. [CrossRef]
32. Sarrafan, N.; Zarei, J.; Horiyat, N. A Novel Fast Fixed-Time Backstepping Control of DC Microgrids Feeding Constant Power Loads. *IEEE Trans. Ind. Electron.* **2023**, *70*, 5917–5926. [CrossRef]
33. Hanzaei, S.; Korki, M.; Zhang, X. Distributed Cooperative Voltage Mode Control for DC-isolated Microgrids Powered by Renewable Energy Sources. *Int. J. Electr. Power Energy Syst.* **2023**, *152*, 109175. [CrossRef]

electronics

MDPI

Article

Dynamic Characteristic Analysis of Multi-Virtual Synchronous Generator Systems Considering Line Impedance in Multi-Node Microgrid

Wei Xie [1], Liangzi Li [1], Weihao Kong [1], Zheng Peng [1], Xiaogang Li [1], Dandan Jiao [1], Chenyi Xu [2,*] and Zebin Yang [2]

[1] Henan Jiuyu Enpai Power Technol Co., Ltd., Zhengzhou 450052, China; xnyjcz@163.com (W.X.); chengzi_llz@sina.com (L.L.); kongweihao1996@hotmail.com (W.K.); 18240585107@163.com (Z.P.); lxg622@163.com (X.L.); 13939069667@163.com (D.J.)

[2] State Key Laboratory of Electrical Insulation and Power Equipment, Xi'an 710045, China; yangzbin@xjtu.edu.cn

* Correspondence: 1471608065@stu.xjtu.edu.cn

Abstract: With the increasing integration of distributed energy resources into modern power systems, virtual synchronous generators (VSGs) have been a promising approach to imitate the inertial response of synchronous generators, thereby enhancing microgrid stability in a dynamic state. When many VSGs are integrated into microgrids, the dynamic characteristics of the system become increasingly complex. Current studies typically assume that different VSGs are connected to a common coupling point, focusing on analyzing the interaction characteristics, which may overlook the widely distributed line impedances in microgrids with distance between different facilities. This may lead to incomplete understanding of the interaction dynamics when VSGs are distributed over long feeder lines. Therefore, this paper proposes and investigates a multi-node, multi-VSG model incorporating line impedances among different nodes, establishing transfer function models for multi-node load disturbances and the frequency responses of individual VSGs. The study explores the dynamic response characteristics of VSGs under varying parameter influences and proposes principles for designing VSG port impedance and inertia parameters to optimize system dynamic frequency characteristics. The findings, validated through simulations in PSCAD v46, provide insights for enhancing the flexibility and reliability of grids incorporating VSGs.

Keywords: VSG; multi-node; dynamic frequency characteristic; microgrids; frequency regulation

1. Introduction

Grid-connected inverters are essential for linking renewable energy sources and energy storage with the microgrid [1–4]. However, with the increased installation of distributed power sources, the reduction in synchronous generators leads to decreased system inertia and increased voltage fluctuations [5–7].

The Grid-Following (GFL) control strategy, as the mainstream inverter control strategy currently applied in power grids, has been successfully utilized for the grid connection of renewable energy sources [2,8]. GFL control utilizes a phase-locked loop (PLL) to track the phase angle of the grid voltage [3,9]. This synchronization method allows for rapid dynamic adjustment of the output current, which enhances grid flexibility and optimizes the use of renewable resources such as wind and solar power [4]. Nonetheless, in the weak grid, the synchronization loop may have problems in stability, leading to potential disconnection of inverters [10]. Additionally, when grid power is mainly supplied by inverters using GFL control, the lack of inertia can cause substantial frequency deviations due to changes in generation or load, resulting in system instability [8]. Therefore, as grid stability continues to weaken, strategies like the virtual synchronous generator (VSG), which provide voltage/frequency support and improved power regulation [3], become increasingly valuable. These strategies can provide relatively stable voltage and frequency when facing disturbance from both the source and load sides [11].

The VSG control method is designed based on the mechanical equations of a synchronous generator (SG) to develop an active power control loop, thereby establishing frequency and phase references [5,12]. To adapt the VSG control to many different situations, including filter type, sensor type and location, and dc-link power source type, many different VSG methods are proposed [13–15]. The critical parts of them include active and reactive power control, which relate to the frequency and voltage control, respectively [16,17]. By controlling the power of energy storage on the inverter's dc side, the VSG provides rotational kinetic energy to simulate the inertia of SG, offering damping and inertia after frequency variations. Moreover, reactive power control, like virtual excitation control or direct reactive power droop control [5], is investigated for optimizing control strategies to enhance voltage performance in the dynamic state [6].

Correspondingly, the performance of VSGs in microgrids is investigated in many studies. Usually, single-machine systems are analyzed in which a single VSG connects to a voltage source in series with line impedance. In these studies, small-signal models have been developed for a single-VSG system, and frequency domain analysis is used to explore the stability boundaries considering virtual inertia and other control parameters [18–20]. However, the conclusion of single-VSG analysis neglects the interaction among different inverters, which may not be used in situations with multiple distributed sources, and the related dynamic performance is not considered [21]. Therefore, some researchers study the interaction among different VSGs, where stability issues and transient behaviors are studied under various conditions. In [22], a method based on the Lyapunov method is proposed to control the oscillations of VSGs around the grid's center-of-inertia frequency. In [23], a state-space model of two VSGs is developed to analyze the stability and dynamic performance. In [24], a master–slave control method with instantaneous frequency is proposed in islanded microgrids. In summary, this research focuses on the stability of multiple VSGs in parallel at a Point of Common Coupling (PCC), either in an islanded or a grid-connected microgrid. These studies do not fully consider the dynamic interaction and the frequency response features.

Usually, the line impedance between different nodes is simplified and not considered [25]. The interactions between multiple VSG units introduce further complexity, particularly when connected across different nodes. In [26], a parameter-adaptive VSG controller for multi-terminal direct-current systems is proposed to improve the damping of the low-frequency oscillation problem and mitigate the impact of VSGs on dc voltage stability. In [27,28], a transfer function model of VSGs is proposed to analyze the frequency and phase response of an ac bus. Interactive frequency performance is improved with VSG control methods. However, this paper also uses the PCC to describe the interaction among different inverters, without considering the impedance among different nodes. Moreover, in the following study [29], this model is further developed to describe the power interaction with the same weakness. Some papers consider virtual damping in improving the dynamic performance. In [30], the influence of damping is analyzed after considering changing of the reference, where the transient power is also studied. Other studies may consider impedance in relation to both resistance and inductance. For example, in both [25,30], the resistance in line impedance is considered for performance analysis. However, in all of these studies, they use the PCC as a base and all converters connect to the PCC. Simulation-based analysis is also considered in some studies. This technology is usually applied to power grids, which mainly focus on specific configurations. While simulations serve as a valuable analytical tool, their inherent constraints highlight the need for mathematical modeling to enhance the understanding of multi-VSG dynamics. Therefore, in the situation that the line is long and the impedance between different nodes cannot be neglected, the performance may not be precise enough. It is necessary to develop models that incorporate the influence of line impedance between different nodes to better determine the performance of VSG-based systems.

This paper investigates the influence of line impedance parameters between different nodes and load disturbances with various spatial distributions on the configuration of inertia and port impedance in multi-VSG parallel systems under a multi-node network topology. The dynamic characteristics are analyzed using a transfer function model and validated through simulation results. This study provides insights for improving the frequency dynamic characteristics of multi-VSG systems, aiming to enhance frequency stability, flexibility, and reliability in distribution networks, especially in systems where VSGs are widely dispersed. The main contributions of this paper are summarized as follows:

(1) A parallel model of multi-VSGs is proposed. Compared with existing models, this model considers that the VSGs are not connected to a PCC, where the line impedance between different nodes is considered.
(2) With the proposed model, different parameters of VSGs from the front node to end node are considered. Therefore, the VSG's parameters in different locations can be further studied to improve the transient performance, which was often neglected in previous research.
(3) In the parameter analysis, both the inertia and impedance are considered in many different scenarios, which can provide a substantial reference for choosing reasonable parameters whilst considering the location of VSGs.

2. VSG Control Strategies and Dynamic Characteristic Analysis

2.1. The Principle of the VSG Control Strategy

In a typical weak grid, considering factors such as LCL-type filter impedance, inverter virtual impedance, and transformer equivalent impedance, the AC port characteristics of inverters are nearly inductive. Therefore, the phase difference is related to the active power among different nodes, where the *P-f* droop characteristic is adopted to realize frequency and active power regulation. The active power control strategy employed in this paper for a VSG is centered around the rotor mechanical equations and frequency control equations of synchronous generators. The relationship between torque and power is illustrated in Equation (1). Here, T_{m} represents the mechanical torque and T_{e} represents the electromagnetic torque. In this paper, T_{m} in a VSG is determined by the droop control loop, which is illustrated in Equation (2), replacing the traditional governor control loop for primary frequency regulation and simplifying the control loop.

$$J\frac{\mathrm{d}\omega}{\mathrm{d}t} = T_{\mathrm{m}} - T_{\mathrm{e}} - D_{\mathrm{p}}(\omega - \omega_{\mathrm{g}}) \tag{1}$$

where J is the rotational inertia; D_{p} is the damping coefficient; ω is the synchronous machine rotational frequency; ω_{g} is frequency at the grid connection point.

$$T_{\mathrm{m}} = \frac{P_{\mathrm{ref}} + k_{\mathrm{p}}(\omega_{\mathrm{ref}} - \omega)}{\omega} \approx \frac{P_{\mathrm{ref}} + k_{\mathrm{p}}(\omega_{\mathrm{ref}} - \omega)}{\omega_0} \tag{2}$$

where k_{p} is the droop coefficient; ω_{ref} and P_{ref} are the reference of frequency and power; ω is the actual frequency in a VSG. For analytical convenience, the rated frequency ω_0 is used in Equation (2) to replace the actual frequency in the power and torque relationship.

By combining Equations (1) and (2) and neglecting the effect of the damping coefficient, we obtain Equation (3).

$$J\omega_0\frac{\mathrm{d}\omega}{\mathrm{d}t} = P_{\mathrm{ref}} - P_{\mathrm{out}} + k_{\mathrm{P}}(\omega_{\mathrm{ref}} - \omega) \tag{3}$$

2.2. The Complete VSG Control Strategy

With the analysis, the diagram of the complete control loop of VSG is shown in Figure 1. The active power control of a VSG is from Equations (1) and (3). The reactive power loop of the VSG is same as the regular Q-V droop control. In addition, the inner dual loop control is applied, which includes both the voltage and current control loops with a PI regulator.

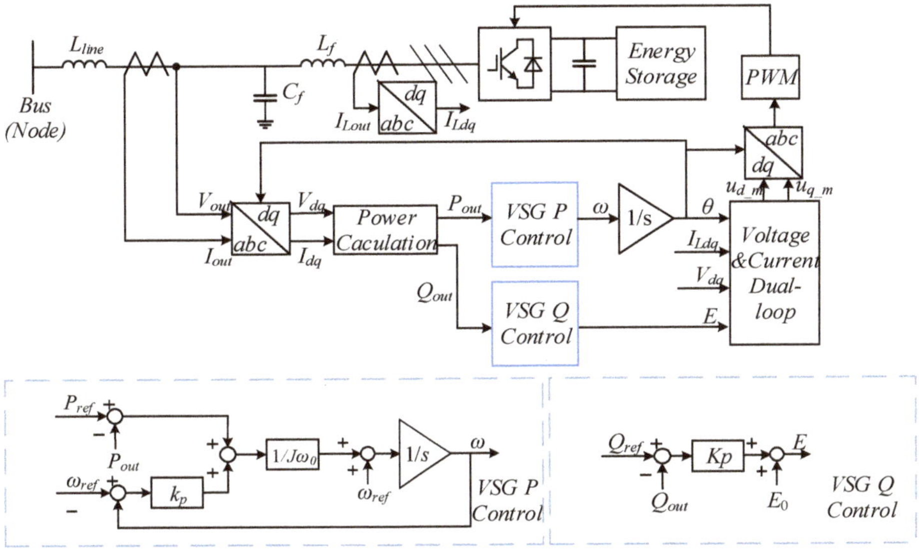

Figure 1. The diagram of complete VSG control strategy.

2.3. *Dynamic Characteristic Analysis of Single Grid-Connected VSG*

To analyze the impact of the frequency dynamics of the VSG control strategy on load power disturbance, we consider the case of a single grid-connected VSG operating in parallel with a load. Ignoring the frequency characteristics of the load and line losses, when the load changes, the power should be balanced among the grid, VSG and load, which is shown in (4).

$$\Delta P_g + \Delta P_{\mathrm{vsg}} = \Delta P_{\mathrm{load}} \tag{4}$$

We extract the relationship between frequency variation and power variation and perform the Laplace transform in Equation (3), which allows us to assess the impact of changes in inverter output power on the variation in inverter output frequency, as shown in Equation (5). The equation indicates that changes in inverter output power affect the deviation of the steady-state frequency. Adjusting the inertia influences the transient frequency variation.

$$\Delta\omega_{\mathrm{vsg}} = -\frac{\Delta P_{\mathrm{vsg}}}{J\omega_0 s + k_{\mathrm{p}}} \tag{5}$$

When the output port of the VSG is predominantly inductive with reactance X_{L}, the relationship between active power and phase can be expressed as Equation (6). This expression is one of the theoretical foundations for frequency control in a VSG.

$$\Delta P_{\mathrm{vsg}} = \frac{3EU_{\mathrm{s}}}{2X_{\mathrm{L}}}\Delta\delta = k_{\mathrm{vsg}}(\Delta\theta_{vsg} - \Delta\theta_1) \tag{6}$$

where E is the equivalent electromotive force at the output port of the VSG; U_{s} is the voltage at the node; k_{vsg} is the line coefficient under the assumption of there being no voltage magnitude variation, which equals $3EU_{\mathrm{s}}/2X_{\mathrm{L}}$; θ_1 is the voltage phase at the node.

Similarly, for the power grid, a similar relationship between active power output and phase angle is shown in Equation (7).

$$\Delta P_{\mathrm{g}} = \frac{3U_{g}U_{\mathrm{s}}}{2X_{\mathrm{g}}}\Delta\delta_{g} = k_{\mathrm{g}}(\Delta\theta_{g} - \Delta\theta_{1}) \tag{7}$$

where U_{g} is the voltage at the grid, k_{g} is the line coefficient under the assumption of minus voltage magnitude variation, and θ_{g} is the voltage phase at the grid. Here, the grid phase is set as the reference axis of phase.

Combining Equations (4)–(7), for a single grid-connected VSG operating in parallel with a load, the impact of load power variation on the phase of the bus is given by Equation (8). It can be observed that a change in load power causes an instantaneous phase shift at the bus, and the subsequent frequency variation is influenced by the rotational inertia and the impedance of the VSG's output port.

$$\frac{\Delta\theta_{1}}{\Delta P_{\mathrm{Load}}} = -\frac{J\omega_{0}s^{2} + k_{\mathrm{p}}s + k_{\mathrm{d}}}{(k_{\mathrm{d}} + k_{g})(J\omega_{0}s^{2} + k_{\mathrm{p}}s) + k_{\mathrm{d}}k_{g}} \tag{8}$$

It can also be observed that the VSG emulates the inertia characteristics of synchronous generators, improving the dynamic frequency response of the grid, reducing the rate of change in frequency, and providing inertial support for the grid frequency. This paper further develops the models with the impact of line impedance between different nodes, thereby enhancing the frequency performance analysis for multiple VSGs in microgrids.

3. Analysis and Modeling of Multi-VSG Parallel Systems

In a multi-VSG parallel system with a PCC, all inverters deliver power to the same node. This section uses a three-machine system as an example to establish the mathematical models for both the single-node multi-machine parallel topology with a common coupling point and the multi-node multi-machine parallel topology. Based on these models, a comparative analysis is performed to investigate the impact of line impedances between nodes on the system's frequency response.

3.1. *Transfer Function Model of a Single-Node Multi-Machine Grid-Connected System*

Firstly, by further developing a model based on the model in Section 2 into a dual-machine system, the transfer function for the node frequency response can be easily obtained. After the load changes, the voltage phase at the node will experience a sudden shift, and the subsequent frequency variation will be influenced by the rotational inertia set for both VSGs (J_1 and J_2) and the magnitude of the droop coefficient.

$$\frac{\Delta\theta_{1}}{\Delta P_{\mathrm{load}}} = -\left(\frac{1}{k_{g} + \frac{1}{\frac{1}{k_{\mathrm{d1}}} + \frac{1}{J_{1}\omega_{0}s^{2} + k_{\mathrm{p1}}s}} + \frac{1}{\frac{1}{k_{\mathrm{d2}}} + \frac{1}{J_{2}\omega_{0}s^{2} + k_{\mathrm{p2}}s}}}\right) \tag{9}$$

By comparing the forms of Equations (8) and (9), the phase deviation at the connected node caused by a load step change for each VSG can be described by a similar expression in a single-node multi-machine system. The expression can also be derived from Equations (5) and (6), as shown in Equation (10). This expression is defined as the additional active power compensation provided by the VSG when a load step occurs at the connected node, denoted in this paper as $k_{\mathrm{m}}(\mathrm{s})$, which equals $\Delta P_{\mathrm{vsg\_i}}/\Delta\theta_{1}$.

$$k_{\mathrm{m}}(s) = \frac{1}{\frac{1}{k_{\mathrm{d}}} + \frac{1}{J\omega_{0}s^{2} + k_{\mathrm{p}}s}} \tag{10}$$

At this point, the single-node two-VSG system and extended single-node n-VSG grid-connected systems corresponding to Equations (8) and (9) can all be represented by a single expression, as shown in Equation (11). For the single-node three-VSG system in Figure 2, n = 3 in Equation (11).

$$\frac{\Delta\theta_1}{\Delta P_{\text{load}}} = -\left(\frac{1}{K_g + \sum_{i=1}^{n} K_{mi}(s)}\right) \tag{11}$$

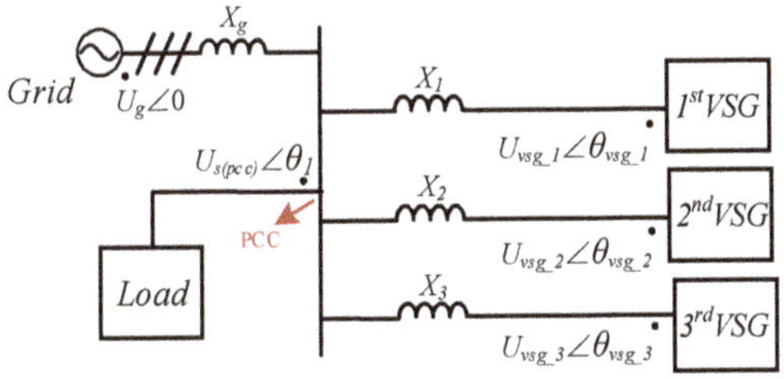

Figure 2. The topology of a single-node three-machine grid-connected system.

If the VSG's control parameters and port impedances are similar, the system can be approximately analyzed as a single-machine system. In this case, considering the dynamic frequency response of nodes in a multi-machine system has its limitations. Therefore, this paper analyzes a multi-VSG topology that considers the line impedance between nodes.

3.2. Transfer Function Model of a Multi-Node Multi-Machine Grid-Connected System

To illustrate the general characteristics of frequency response under impedance effects, this paper investigates the transfer function model in a three-node three-VSG grid structure, as shown in Figure 3, to obtain the principle that can be further used in the system with more VSGs. Firstly, due to the presence of line impedance between nodes, spatial differences arise among the nodes. It is necessary to distinguish between each node, the connected VSGs, and the loads. To clearly express the relationships between quantities in a multi-node system, the state variables are defined and represented in vector form.

$$\begin{cases} \Delta\omega_{\text{vsg}} = \left[\Delta\omega_{\text{vsg\_1}} \quad \Delta\omega_{\text{vsg\_2}} \quad \Delta\omega_{\text{vsg\_3}}\right]^{\text{T}} \\ \Delta\theta_{\text{node}} = \left[\Delta\theta_{\text{node\_1}} \quad \Delta\theta_{\text{node\_2}} \quad \Delta\theta_{\text{node\_3}}\right]^{\text{T}} \\ \Delta P_{\text{vsg}} = \left[\Delta P_{\text{vsg\_1}} \quad \Delta P_{\text{vsg\_2}} \quad \Delta P_{\text{vsg\_3}}\right]^{\text{T}} \\ \Delta P_{\text{load}} = \left[\Delta P_{\text{load\_1}} \quad \Delta P_{\text{load\_2}} \quad \Delta P_{\text{load\_3}}\right]^{\text{T}} \end{cases} \tag{12}$$

where θ_{node} is the voltage phase at the node and subscript i is used to distinguish each node, along with the VSG and load connected to it.

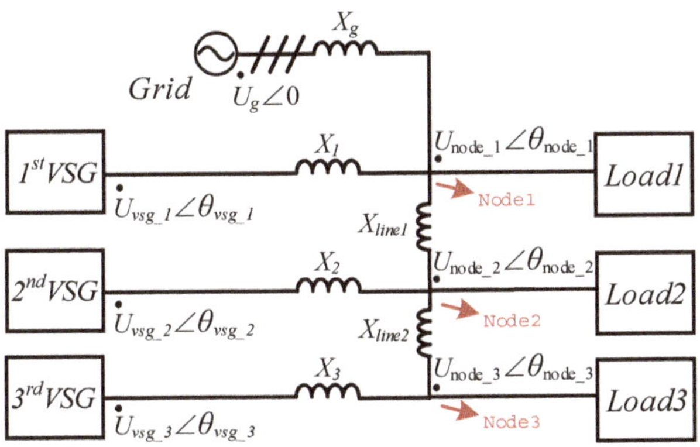

Figure 3. The topology of a three-node three-machine grid-connected system.

Combining the content from Section 2, Equation (13) holds.

$$\Delta P_g + \sum_{i=1}^{3} \Delta P_{\text{vsg}\_i} = \sum_{i=1}^{3} \Delta P_{\text{load}\_i} \tag{13}$$

Considering the relationship of active power between nodes, Equation (14) holds.

$$\frac{3U_{\text{node}\_i}U_{\text{node}\_i+1}}{2X_{\text{line}\_i}}\Delta\delta_{\text{node}\_i} = k_{\text{L}\_i}(\Delta\theta_{\text{node}\_i} - \Delta\theta_{\text{node}\_i+1}) \tag{14}$$

where U_{node} is the voltage at the node; k_{L} is the line coefficient under the assumption of there being no voltage magnitude variation. Considering the power flow relationships in the three-node system, Equation (15) holds.

$$\begin{cases} K_{\text{L}\_1}(\Delta\theta_{\text{node}\_1} - \Delta\theta_{\text{node}\_2}) = \Delta P_{\text{load}\_2} + \Delta P_{\text{load}\_3} - \Delta P_{\text{vsg}\_\text{out}\_2} - \Delta P_{\text{vsg}\_\text{out}\_3} \\ K_{\text{L}\_2}(\Delta\theta_{\text{node}\_2} - \Delta\theta_{\text{node}\_3}) = \Delta P_{\text{load}\_3} - \Delta P_{\text{vsg}\_\text{out}\_3} \end{cases} \tag{15}$$

Combining Equations (5)–(10) and (13)–(15) and expressing them using the vector form in Equation (12), the following relationship holds.

$$\begin{cases} \Delta\omega_{\text{vsg}} = \text{A}_1\Delta P_{\text{out}} \\ \Delta P_{\text{out}} = \text{A}_2\Delta\theta_{\text{node}} \\ \text{A}_3\Delta P_{\text{out}} + \text{A}_4\Delta\theta_{\text{node}} = \text{A}_5\Delta P_{\text{load}} \end{cases} \tag{16}$$

where

$$\text{A}_1 = \begin{bmatrix} -\frac{1}{J_i\omega_0 s + k_{\text{p}\_i}} & 0 & 0 \\ 0 & -\frac{1}{J_i\omega_0 s + k_{\text{p}\_i}} & 0 \\ 0 & 0 & -\frac{1}{J_i\omega_0 s + k_{\text{p}\_i}} \end{bmatrix}$$

$$\text{A}_2 = \begin{bmatrix} -k_{\text{m}\_1}(s) & 0 & 0 \\ 0 & -k_{\text{m}\_2}(s) & 0 \\ 0 & 0 & -k_{\text{m}\_3}(s) \end{bmatrix}$$

$$\text{A}_3 = \text{A}_5 = \begin{bmatrix} 1 & 1 & 1 \\ 0 & 1 & 1 \\ 0 & 0 & 1 \end{bmatrix}, \text{A}_4 = \begin{bmatrix} -k_g & 0 & 0 \\ k_{\text{L}\_1} & -k_{\text{L}\_1} & 0 \\ 0 & k_{\text{L}\_2} & -k_{\text{L}\_2} \end{bmatrix}$$

Performing matrix operations on this equation yields the following result:

$$\Delta\omega_{\text{vsg}} = \text{A}_1\text{A}_2(\text{A}_2\text{A}_3 + \text{A}_4)^{-1}\text{A}_5\Delta P_{\text{load}} = G\Delta P_{\text{load}} \tag{17}$$

where

$$G = \begin{bmatrix} G_{\omega 11}(s) & G_{\omega 12}(s) & G_{\omega 13}(s) \\ G_{\omega 21}(s) & G_{\omega 22}(s) & G_{\omega 23}(s) \\ G_{\omega 31}(s) & G_{\omega 32}(s) & G_{\omega 33}(s) \end{bmatrix}$$

where G represents the relationship matrix between the frequency of each VSG and the load disturbances. Using this relationship, Equation (18) can be derived.

$$\Delta\omega_{\text{vsg}\_1} = G_{\omega 11}\Delta P_{\text{load}\_1} + G_{\omega 12}\Delta P_{\text{load}\_2} + G_{\omega 13}\Delta P_{\text{load}\_3} \tag{18}$$

And $G_{\omega 1i}$(s) represents the transfer function of the frequency of the *VSG* connected at node 1 with respect to the load power change at node i. From Equation (18), it can be observed that in the three-node model, the impact of load disturbances at each node on the frequency of any *VSG* varies. Simplifying these load disturbances as if they all occur at a single node in a single-node model will introduce discrepancies in the dynamic response analysis. The subsequent analysis of the three-node, three-*VSG* model is performed utilizing this transfer function model.

4. Dynamic Frequency Characteristics of Three-Node Three-VSG System

Using the transfer function presented in Section 3, the dynamic frequency responses of each *VSG* can be obtained under load step changes at different nodes. By analyzing the maximum frequency deviation (MFD), rate of change of frequency (RoCoF) and adjustment time to reach a steady state under disturbances, the dynamic characteristics of each *VSG* can be assessed.

This section examines the effects of load disturbance locations, line impedance between nodes, and the virtual inertia (J) and port impedance of each VSG on the frequency dynamic characteristics of VSGs. Sections 4.1 and 4.2 focus on analyzing how variations in practical conditions—such as load disturbance locations and line impedance—affect the dynamic frequency at each node. Conversely, Sections 4.3 and 4.4 investigate how to optimally configure parameters and allocate resources under these varying conditions to achieve an optimized performance in practical scenarios. Specifically, this section identifies the patterns of changes in VSG frequency dynamics influenced by these factors and, based on these observations, proposes and validates optimization strategies for frequency dynamics within a three-node, three-VSG grid.

4.1. Dynamic Frequency Characteristics with Load Disturbances at Different Locations

First, the impact of load disturbance location is analyzed to differentiate between single-node and multi-node models, particularly regarding the consideration of spatial differences between nodes. The findings highlight the comprehensiveness and practicality of multi-node models in capturing dynamic characteristics more effectively compared to single-node models.

The specific parameters are outlined in Table 1. To reflect the differences caused by disturbance location, disturbances are concentrated at each of the three nodes. The frequency characteristics of the VSGs at the initial and terminal nodes are observed to derive the underlying patterns. The results are shown in Figure 4.

Table 1. Parameter of the transfer function model of three-node three-VSG system.

Parameters	Values
E_{vsg_i} (i = 1, 2, 3)	220 V
U_{node_i} (i = 1, 2, 3)	220 V
J_i	1 kg·m^2
ω_0	314 rad/s
L_g L_i (i = 1, 2, 3)	0.4 mH
L_{XLine_i} (i = 1, 2)	0.4 mH
Kp	50 × 10^3 w·s/rad

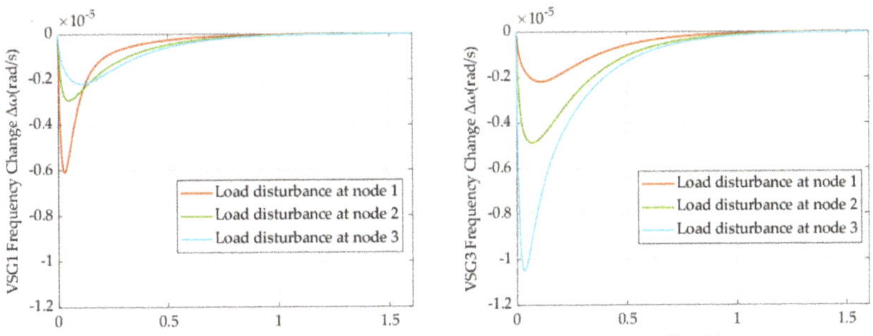

(**a**) VSG1(initial) (**b**) VSG3(terminal)

Figure 4. Step response of VSG frequency at different load positions.

For the terminal VSG3, as the disturbance distance increases, the MFD decreases and RoCoF reduces, and the same applies for the initial VSG1. However, the terminal VSG experiences a greater frequency deviation at shorter disturbance distances compared to the initial VSG1, indicating that the dynamic frequency performance of the terminal VSG3 is significantly weaker than that of the initial VSG1. Therefore, to optimize the dynamic frequency characteristics of VSGs in a multi-node model, it is essential to first enhance the dynamic performance of the terminal VSG. Additionally, balancing the load distribution to reduce the burden on the terminal VSG can optimize the overall system's frequency dynamic characteristics.

4.2. Dynamic Frequency Characteristics Under Different Line Impedences

To assess the impact of varying line impedances on the dynamic frequency characteristics of VSGs under load disturbances within this structure, four scenarios are considered in this section. The specific parameters are outlined in Table 2, and other parameters are the same as those listed in Table 1. The results are shown in Figures 5–8.

Table 2. Parameter of the transfer function model of the three-node three-VSG system.

Scenario	L_{Xline_1} (Lx1)	L_{Xline_2} (Lx2)	Load Set
1	0.1/0.2/0.4/0.6/0.8 mH	0.4 mH	Evenly
2	0.1/0.2/0.4/0.6/0.8 mH	0.4 mH	Node 1
3	0.1/0.2/0.4/0.6/0.8 mH	0.4 mH	Node 3
4	0.1/0.2/0.4/0.6/0.8 mH	0.1/0.2/0.4/0.6/0.8 mH	Node 1

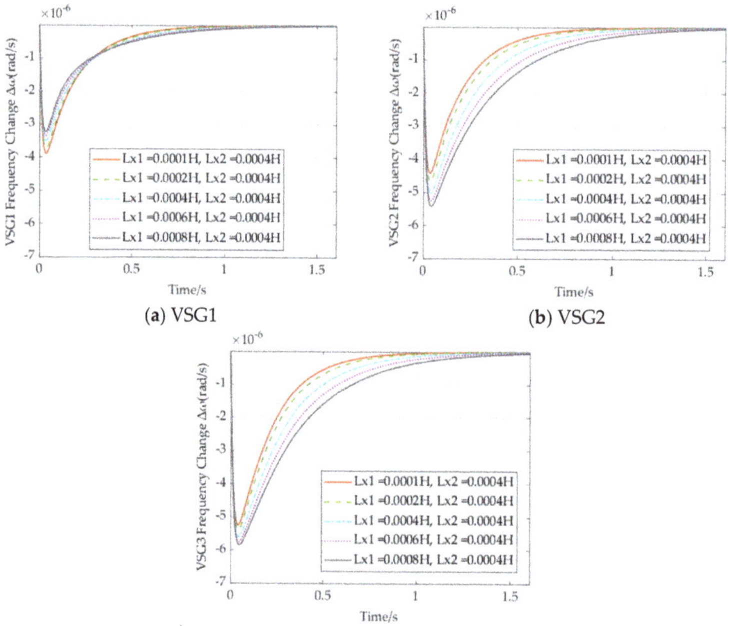

(**a**) VSG1 (**b**) VSG2 (**c**) VSG3

Figure 5. Step response of VSG frequency in scenario 1.

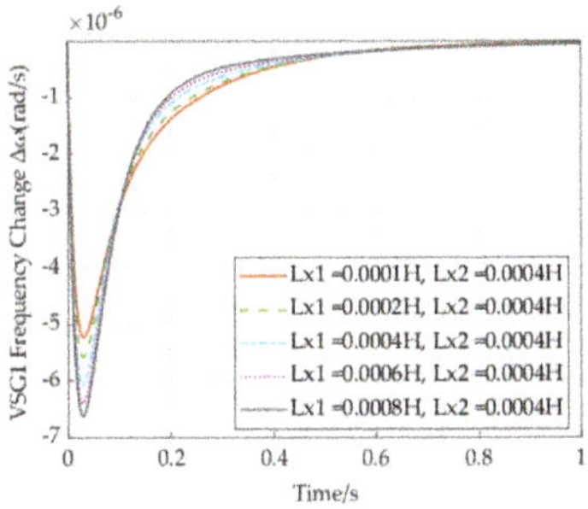

(**a**) VSG1

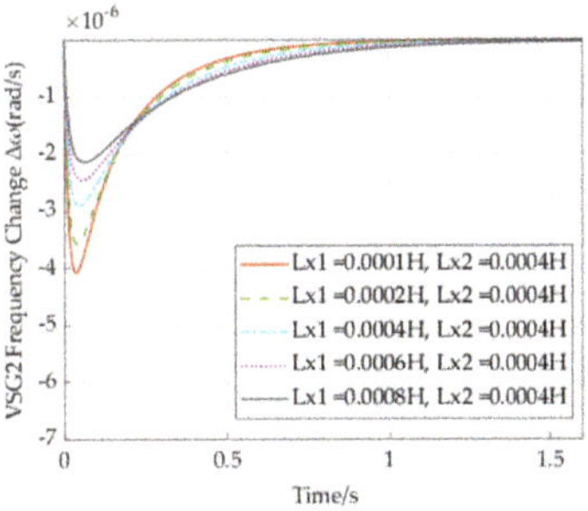

(**b**) VSG2

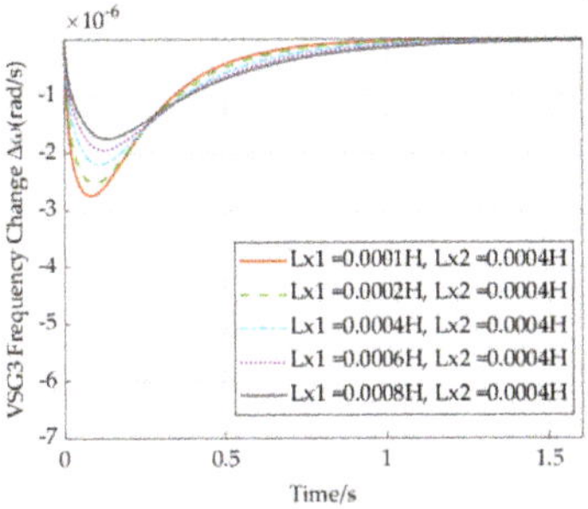

(**c**) VSG3

Figure 6. Step response of VSG frequency in scenario 2.

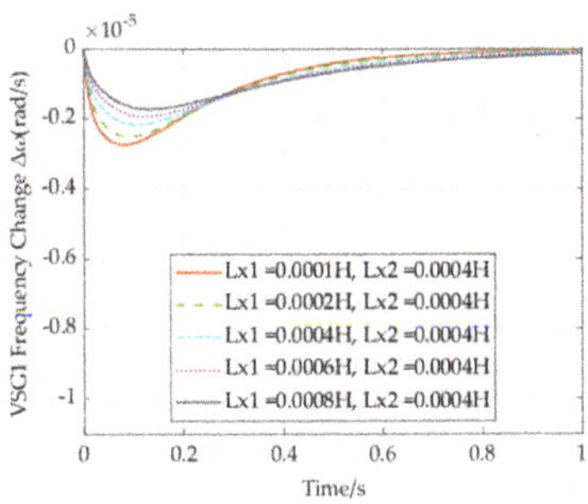

(**a**) VSG1

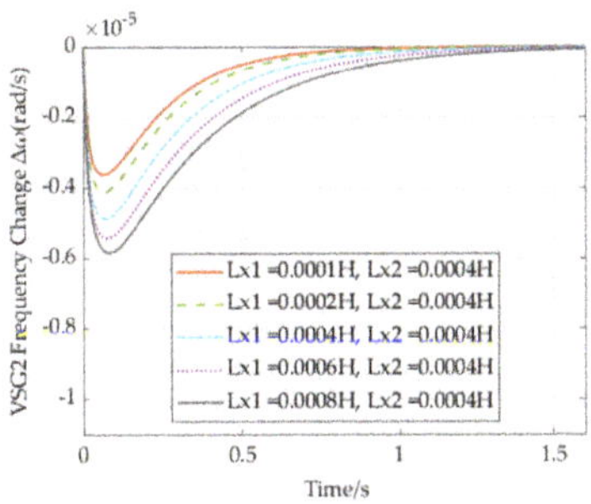

(**b**) VSG2

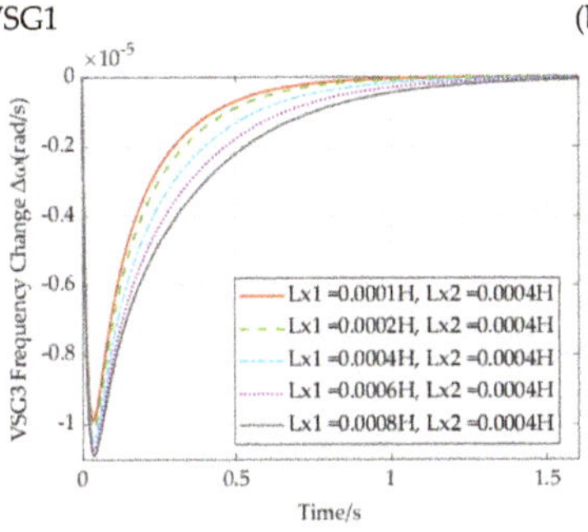

(**c**) VSG3

Figure 7. Step response of VSG frequency in scenario 3.

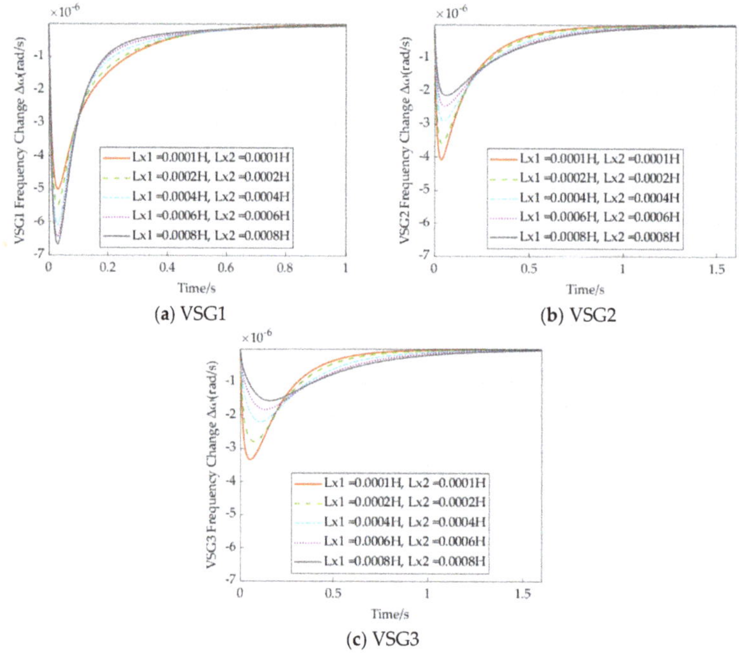

Figure 8. Step response of VSG frequency in scenario 4.

Scenario 1: The line impedance $L_{\mathrm{Xline\_1}}$ between Nodes 1 and 2 is varied, while the line impedance $L_{\mathrm{Xline\_2}}$ between Nodes 2 and 3 is kept constant. The dynamic frequency responses of the three VSGs are analyzed after the load disturbance is evenly distributed across the three nodes.

Scenario 2: The line impedance $L_{\mathrm{Xline\_1}}$ between Nodes 1 and 2 is varied, while the line impedance $L_{\mathrm{Xline\_2}}$ between Nodes 2 and 3 is kept constant. The dynamic frequency responses of the three VSGs are analyzed following a load disturbance at Node 1.

Scenario 3: The line impedance $L_{\mathrm{Xline\_1}}$ between Nodes 1 and 2 is varied, while the line impedance $L_{\mathrm{Xline\_2}}$ between Nodes 2 and 3 is kept constant. The dynamic frequency responses of the three VSGs are analyzed following a load disturbance at Node 3.

Scenario 4: Both the line impedance $L_{\mathrm{Xline\_1}}$ between Nodes 1 and 2 and the line impedance $L_{\mathrm{Xline\_2}}$ between Nodes 2 and 3 are varied. The dynamic frequency responses of the three VSGs are analyzed following a load disturbance at Node 1.

As can be observed in Figure 5, when load disturbances are reasonably distributed across each node, increasing the line impedance between Node 1 and Node 2 results in a reduction in the MFD of VSG1, along with an increase in the adjustment time, effectively improving the frequency dynamic response in the microgrid context. However, the MFD of the VSGs connected at Node 2 and Node 3 increases, leading to a certain loss in dynamic performance at those nodes. Furthermore, as shown in Figure 5b,c, both the MFD and RoCoF increase as the VSG is closer to the terminal, demonstrating that the dynamic performance of the frequency deteriorates when the node approaches the end of the line.

From Figure 6, when load disturbances are primarily concentrated at Node 1 (the initial node), the frequency variation of VSGs closer to Node 1 is larger, indicating that load disturbances primarily rely on local VSGs for inertia support. The impact on VSGs further from the disturbance is smaller. Even at the terminal, where VSG3 is more susceptible to disturbances, the frequency deviation and rate of change are significantly smaller compared to VSG1 and VSG2, which are closer to the initial node. Increasing the line impedance between Node 1 and Node 2 causes load disturbances to rely even more on the inertia support provided by local VSGs. Consequently, the frequency deviation of VSG1 increases, adjustment time decreases, and dynamic performance deteriorates. In contrast, VSGs 2 and 3, separated by the impedance between Nodes 1 and 2, experience a reduction in frequency deviation and rate of change. This suggests that increased line impedance acts as a barrier to disturbances, making local disturbances more reliant on local VSG inertia support and consequently reducing the dynamic frequency characteristics of the local VSG.

When disturbances are concentrated at Node 3 (the terminal) in Figure 7, increasing the line impedance between Node 1 and Node 2 results in a decrease in the dynamic performance of VSG2 and VSG3 at Nodes 2 and 3, while the dynamic performance of VSG1 at Node 1 improves. This demonstrates that line impedance acts as a barrier to load disturbances, with the impact of the disturbances being increasingly supported by VSG2 and VSG3, which are separated by the impedance.

As the line impedance between nodes increases in Figure 8, it further limits the propagation of load disturbances, resulting in greater support for the disturbances from the local VSG1. The impact on VSG2 and VSG3 is reduced, which benefits the optimization of the dynamic frequency characteristics of VSG2 and VSG3.

4.3. Dynamic Frequency Characteristics Under Different Inertia

VSGs require additional capacity to provide virtual inertia, with larger virtual rotational inertia necessitating greater investment. Therefore, optimizing the allocation of virtual inertia is crucial from an economic perspective. In this section, the impact of virtual inertia allocation is analyzed using a three-node, three-VSG model. To evaluate the effect of varying inertia on the dynamic frequency response of VSGs under load disturbances, three scenarios are considered. The parameters are detailed in Table 3, and the results are presented in Figures 9–14.

Table 3. Parameter of the transfer function model of three-node three-VSG system.

Scenario	J_1	$X_{\text{line\_1}}/X_{\text{line\_2}}$	Load Set
1	0.5/1/3/5/8 kg·m^2	0.4 mH	Evenly
2	0.5/1/3/5/8 kg·m^2	0.4 mH	Node 1
3	0.5/1/3/5/8 kg·m^2	0.4 mH	Node 3

When load disturbances are evenly distributed across each node and the line impedance between nodes is set to 0.4 mH, the MFD and RoCoF are larger for VSG3, which is closer to the terminal. This further confirms that the dynamic frequency performance of VSGs deteriorates as they approach the terminal. Additionally, as the inertia (J) of the initial VSG1 increases, the dynamic frequency characteristics of VSG1 are optimized, while those of VSG2 and VSG3 deteriorate. Notably, the MFD of VSG3, closer to the terminal, is greater than that of VSG2. This indicates that when load disturbances are evenly distributed, the optimization of dynamic performance at the initial end results in a further improvement in dynamic performance at the terminal end.

Figure 10 shows that when the load is evenly distributed across the three nodes, both the MFD and RoCoF of VSG1 exhibit a decreasing trend as J_1 varies, leading to improved dynamic frequency characteristics. Specifically, the rise in RoCoF, indicating a slower rate of frequency decline, contributes to the optimization of the frequency's dynamic response. In contrast, the MFD of VSG2 and VSG3, positioned at terminal nodes, displays a more substantial variation, with this trend becoming more pronounced as they are located closer to the network's edge.

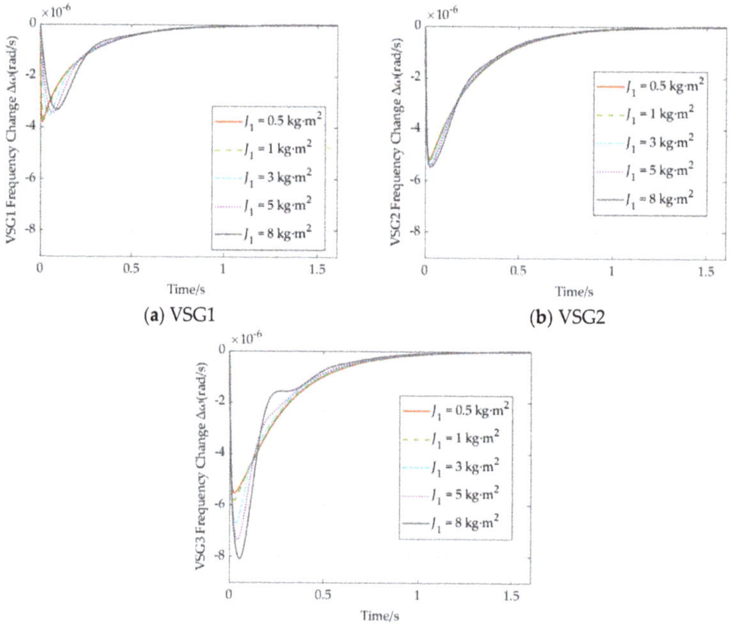

(**a**) VSG1 (**b**) VSG2 (**c**) VSG3

Figure 9. Step response of VSG frequency in scenario 1.

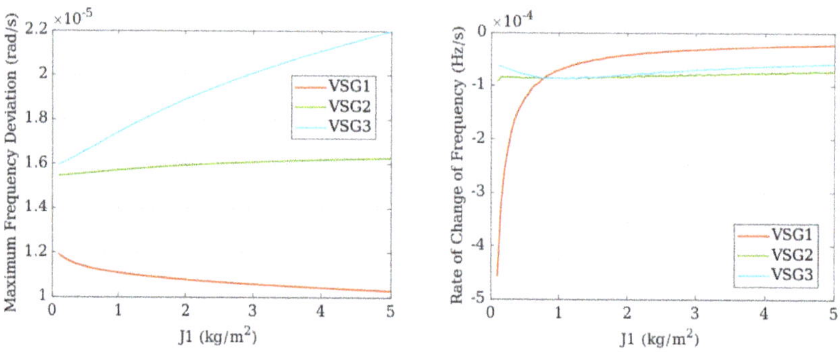

(**a**) MFD (**b**) RoCoF

Figure 10. MFD and RoCoF in scenario 1.

Figure 11 shows that when load disturbances are primarily concentrated at Node 1, increasing the virtual inertia (J) of VSG1 improves its dynamic frequency characteristics. Additionally, this increase also optimizes the dynamic frequency performance of VSG2 and VSG3, which are closer to the terminal. This indicates that when load disturbances are concentrated, increasing the inertia (J) of the local VSG not only enhances the dynamic frequency characteristics of the local VSG but also has a beneficial effect on nearby VSGs.

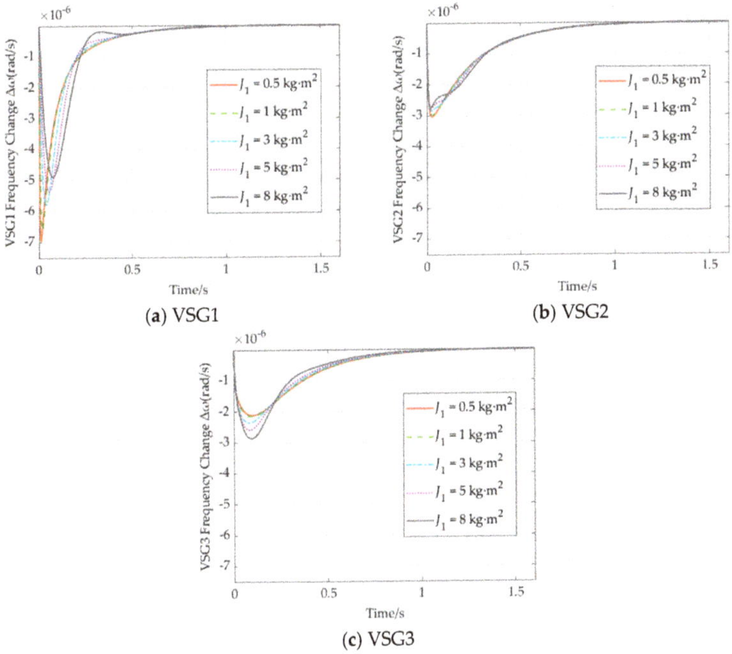

Figure 11. Step response of VSG frequency in scenario 2.

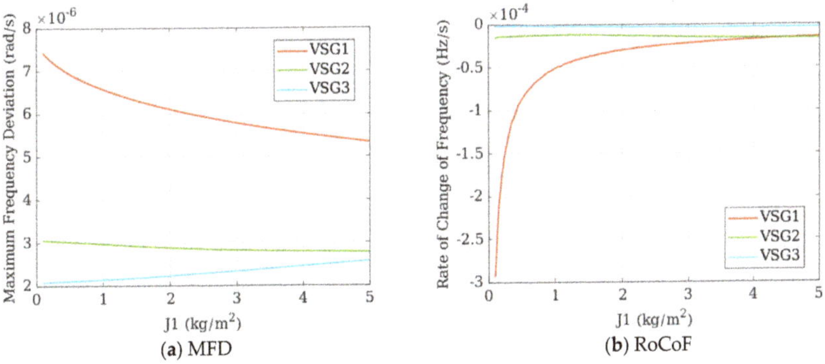

Figure 12. MFD and RoCoF in scenario 2.

From Figure 12, it can be clearly observed that as the value of J_1 varies, both the MFD and RoCoF of VSG1 exhibit a more sensitive change compared to VSGs at other nodes, indicating a significant optimization effect. This also demonstrates that the inertia parameter has a greater impact on the local VSG connected to the node where the load disturbance occurs.

When load disturbances are primarily concentrated at the terminal (Node 3) in Figure 13, increasing the inertia (J) of VSG1 at the initial node results in a decrease in RoCoF but an increase in the maximum frequency deviation of VSG1. The maximum frequency deviation of VSG2 also increases, and the deviation for VSG3 becomes larger compared to VSG2, with increased fluctuations during the recovery process. In this case, increasing the virtual inertia of VSG1 is detrimental to the overall dynamic frequency performance of the system.

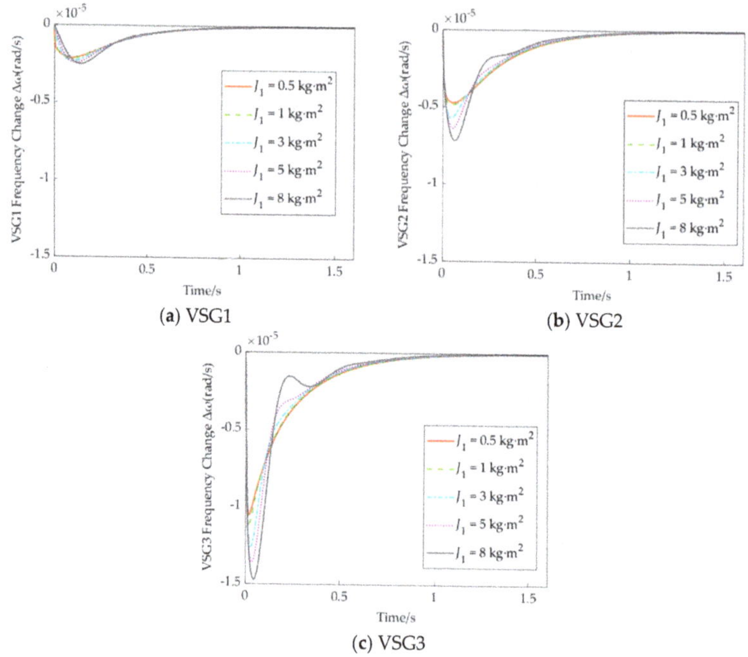

(**a**) VSG1 (**b**) VSG2 (**c**) VSG3

Figure 13. Step response of VSG frequency in scenario 3.

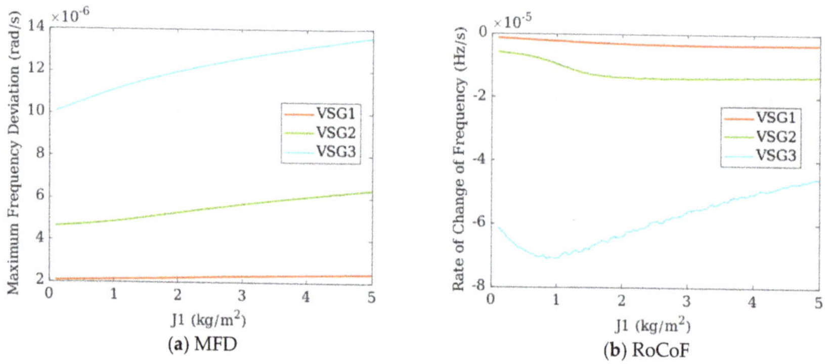

(**a**) MFD (**b**) RoCoF

Figure 14. MFD and RoCoF in scenario 3.

From Figure 14, it can be clearly observed that when the load is primarily concentrated at Node 3, the MFD of VSG2 and VSG3 shows a significantly increasing trend as J_1 varies, while VSG1 is not as sensitive. A comparison with Figures 10 and 12 reveals a strong correlation between the setting of inertia (J) and the location where the load is concentrated.

However, it is worth noting that in Figure 10b, as J_1 varies, the RoCoF of VSG3 initially improves and then deteriorates. This indicates that when the inertia parameter is not well matched with the load disturbance, increasing inertia (J) can lead to an overall rise in system MFD, but it also enhances the RoCoF of the VSG at the load location.

4.4. Dynamic Frequency Characteristics Under Different Port Impedances

The output port impedance of the inverter, including line impedance, virtual impedance, and transformer equivalent impedance, can be flexibly adjusted by modifying the virtual impedance. Meanwhile, the line impedance between nodes, as a grid parameter, is generally treated as a fixed value. Thus, studying the impact of VSG port impedance on the frequency dynamic response of the inverter provides another approach to optimizing dynamic performance, which will be discussed in this section.

Considering the conclusion drawn in Section 4.2, when the line impedance between a node and its neighboring nodes decreases, the reliance of the local VSG on frequency support for mitigating local load disturbances is reduced, leading to improved dynamic frequency characteristics for the local VSG, and vice versa. Based on this conclusion, a hypothesis can be proposed: when the impedance at the VSG port increases, the load disturbance at the connected node will reduce its reliance on the frequency support of the corresponding VSG (i.e., the local VSG), thereby improving the local VSG's dynamic frequency characteristics, but at the cost of deteriorating the dynamic frequency performance of other VSGs in the system.

To validate the aforementioned hypothesis and explore the influence of inertia within this context, this section establishes two scenarios for discussion. The specific parameters are outlined in Table 4 and the results are shown in Figures 15–18.

Table 4. Parameter of the transfer function model of three-node three-VSG system.

Scenario	$J_1/J_2/J_3$	L_2	Load Set
1	2/2/2 kg·m^2	0.1/0.2/0.4/0.8/1 mH	Node 2
2	1/1/1 kg·m^2	0.1/0.2/0.4/0.8/1 mH	Node 2

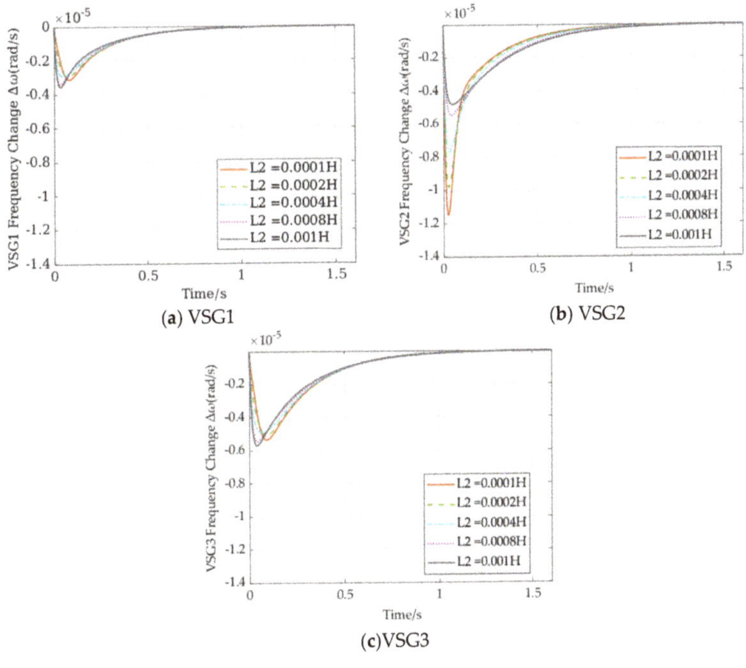

Figure 15. Step response of VSG frequency in scenario 1.

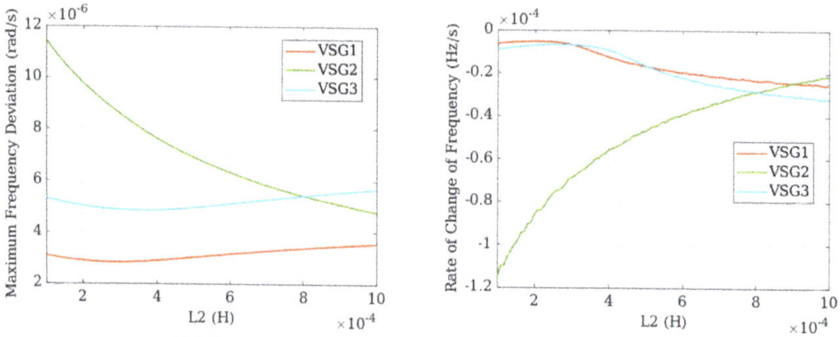

(**a**) MFD (**b**) RoCoF

Figure 16. MFD and RoCoF in scenario 1.

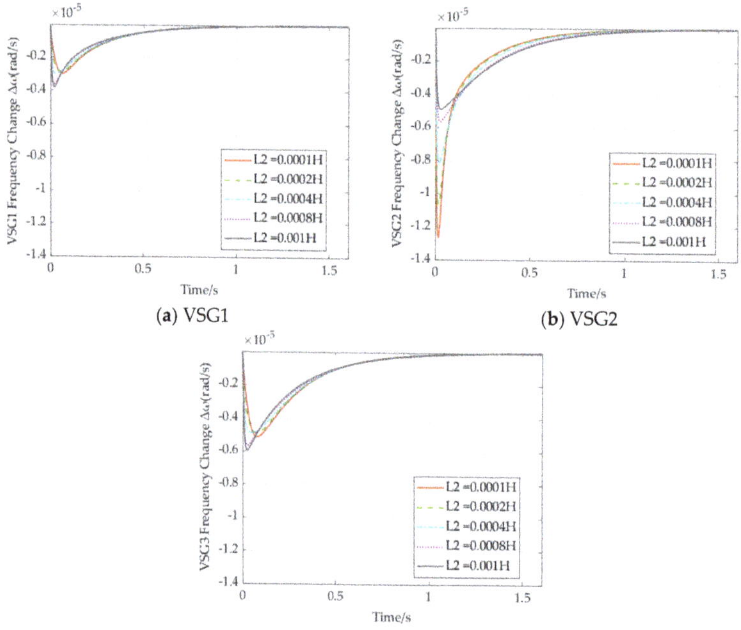

(**c**) VSG3

Figure 17. Step response of VSG frequency in scenario 2.

In Figure 15, when the load disturbance is primarily concentrated at Node 2, increasing the port impedance of VSG2 results in a significant reduction in the frequency deviation and an increase in the recovery time of VSG2. This indicates that the dynamic frequency characteristics of VSG2 are optimized as the port impedance increases, thus validating the hypothesis that increasing port impedance benefits the dynamic frequency characteristics of VSGs.

At the same time, the frequency response of neighboring VSGs, VSG1 and VSG3, follows a similar pattern. As the port impedance L2 increases to 1 mH, the frequency deviation range and rate reach their maximum, causing a deterioration in frequency characteristics. When L2 is reduced gradually to 0.2 mH, the frequency deviation decreases and reaches its minimum value. However, further decreasing L2 to 0.1 mH results in an increase in frequency deviation range.

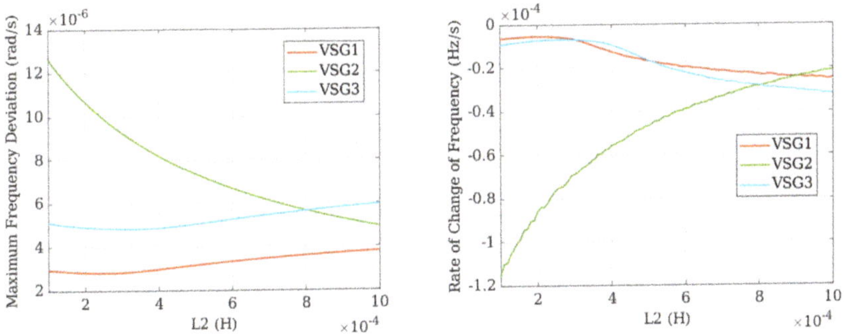

(a) MFD (b) RoCoF

Figure 18. MFD and RoCoF in scenario 2.

From Figure 16, it can also be observed that as the port impedance L2 increases, both the MFD and RoCoF of VSG2 show a significant improvement trend. However, the frequency responses of the other two VSGs deteriorate correspondingly.

This demonstrates that increasing port impedance can negatively impact the dynamic characteristics of VSGs at other nodes. Additionally, it is observed that the frequency response deviation of VSG3 is notably larger than that of VSG1, corroborating the conclusion in Section 4.1.

In Case 2, when the virtual inertia (J) of all three VSGs is simultaneously reduced, the frequency response characteristics follow a pattern similar to that observed in Case 1. Specifically, the frequency deviation range of VSG2 increases significantly when the port impedance is small, while the frequency deviations of VSG1 and VSG3 remain relatively unchanged. As the port impedance increases to 1 mH, the frequency deviation range of VSG2 shows minimal variation.

This indicates that as the port impedance decreases, the impact of the virtual inertia set on the VSGs becomes more pronounced compared to when the port impedance is large. Therefore, while considering the efficiency of virtual inertia usage, a larger port impedance for VSGs is not always beneficial. From the comparison between Figures 16 and 18, it can be seen that changing the value of inertia (J) for the three VSGs only affects the magnitude of the values, without impacting the overall trend of variation. Therefore, integrating the conclusions from the above discussion with the experimental results, under the conditions of this section, a port impedance value for L2 in the range of 0.4–0.8 mH is suitable.

4.5. Optimization Strategies and Validation of Dynamic Frequency

From the analysis of the results on the dynamic frequency characteristics of VSGs under various individual and combined influencing factors discussed in Sections 4.1–4.4, the following qualitative conclusions and dynamic performance optimization strategies can be derived:

Sections 4.1 and 4.2 focus on the impact of actual load disturbance locations and line impedance between nodes within the circuit topology on the dynamic frequency characteristics of each VSG. The key qualitative conclusions are as follows:

1. The dynamic frequency characteristics of each VSG vary depending on the location of the load disturbance. The closer the disturbance occurs to the connected node, the greater the MFD and RoCoF during the frequency response process. Additionally, for each VSG, differences arise due to the position of the connected node. Compared to terminal nodes, VSGs located closer to the grid exhibit superior dynamic performance, especially when load disturbances occur at the connected node, where terminal VSGs show poorer performance.

2. Increasing the line impedance between nodes has a detrimental effect on frequency response during unbalanced load conditions. When load disturbances occur at one end of

the line impedance, the VSG at the opposite end will participate less in active power regulation during frequency response, thereby improving its dynamic frequency characteristics. Conversely, the dynamic frequency characteristics of the VSG on the load disturbance side will deteriorate.

Sections 4.3 and 4.4 discuss the impact of parameter J and port impedance under specific conditions of node impedance and disturbance locations. The key qualitative conclusions are as follows:

1. When a disturbance occurs at the connected node, increasing the local VSG's inertia can significantly improve its dynamic performance. However, this can lead to a deterioration in the dynamic performance of VSGs at other nodes, with the degree of impact varying depending on the specific location of the disturbance.

2. Increasing the port impedance can improve the dynamic frequency performance of the VSG at the connected node, but it also results in a reduction in performance at other nodes. Increasing the overall system inertia level does not change this trend.

Based on the analysis of the effects of practical influencing factors, as well as the summarized impacts of virtual inertia J and port impedance on the dynamic frequency characteristics of VSGs under different conditions, the following optimization conclusions are derived:

1. The dynamic frequency performance of the VSG at the initial node is better than that at the terminal node. Therefore, optimization should focus on the terminal VSG. The line impedance between connected nodes impedes active power interaction between the two nodes, which in turn worsens the dynamic frequency characteristics on the side where the load disturbance occurs. Thus, attention should be given to the combined impact of load disturbance location and the magnitude of line impedance between nodes on the dynamic performance at each node, in order to assess the strength of dynamic frequency characteristics for each VSG.

2. The inertia of each VSG should be reasonably set according to the distribution of load disturbances, with increased inertia allocated to areas where load disturbances are concentrated. If the load disturbance does not match the inertia parameter, it is likely to result in a deterioration of the frequency characteristics at the disturbance node, although it may also lead to an improvement in the RoCoF of the VSG at the node connected to the load disturbance.

3. Increasing the port impedance will improve the dynamic frequency characteristics of the VSG at that node, but it will also lead to a deterioration in the dynamic characteristics of VSGs at other nodes. This further necessitates that the port impedance be well matched to the load disturbance location.

To theoretically validate the correctness of the optimization strategies, a set of parameters for the three-node model is considered. Under the guidance of the dynamic frequency optimization strategies, optimization is performed, and the dynamic frequency characteristics of the initial and terminal VSGs are compared before and after the optimization. The parameters before and after optimization are shown in Table 5. And the results are shown in Figure 19.

Table 5. Parameters before and after adjustment.

Fixed Parameters		**Values**	
L_{XLine_i} **(i = 1, 2)**		**0.4 mH/0.2 mH**	
Adjustment	**J1/J2/J3**	**L1/L2/L3**	**Load Set**
Before	3/1/1 kg·m^2	0.4/0.6/0.2 mH	10 kw in Node 3
After	1/1/3 kg·m^2	0.2/0.4/0.6 mH	10 kw in Node 3

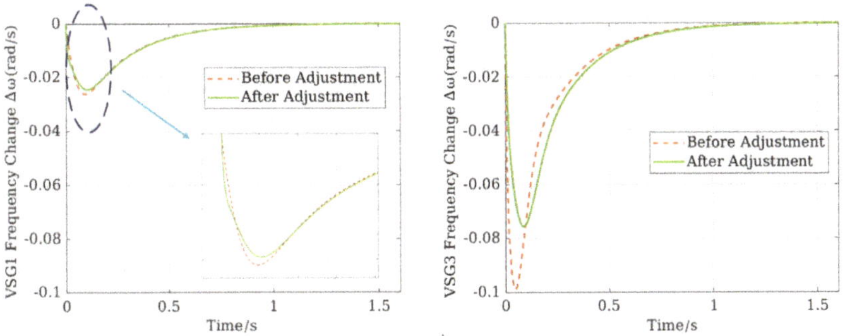

(**a**) VSG1 (**b**) VSG3

Figure 19. Step response of VSG frequency before and after adjustment.

From Figure 19, it can be observed that when the load disturbance occurs at the end, specifically at Node 3, increasing the port impedance and virtual inertia at VSG3, while reducing the port impedance of VSG2 and VSG1, effectively optimizes the dynamic frequency response of VSG3. Furthermore, the dynamic frequency response of VSG1, located at the front end, shows no significant changes, demonstrating the effectiveness of this optimization strategy.

5. Simulation Validation

To further verify the validity of the model and the correctness of the frequency dynamic performance optimization strategy, the topology described in this paper is simulated using PSCAD v46. The simulation model uses the circuit structure shown in Figure 3.

5.1. Validation of the Transfer Function Model

In a three-node three-VSG system, when an external load undergoes a step change at Node 1, the frequency response of three VSGs is shown in Figure 15. And the parameters of the simulation and theoretical experiment are shown in Table 6. It should be noted that in the simulation, to eliminate the influence of inner loop parameters on the dynamic frequency response of each VSG, the inner loop parameters are set to the same values.

Table 6. Parameter of three-node three-VSG system.

Parameters	Values
E_{vsg_i} (i = 1, 2, 3)	220 V
U_{node_i} (i = 1, 2, 3)	220 V
J_i (i = 1, 2, 3)	1 kg m^2
ω_0	314.15 rad/s
Lg L_i (i = 1, 2, 3)	0.4 mH
L_{XLine_i} (i = 1, 2)	0.4 mH
Lf/Cf	2 mH/40 µF
Kp	50 × 10^3 w s/rad
ΔP_{load}	300 kW
Voltage Loop P/I	50/2
Current Loop P/I	0.13/2

As shown in Figure 20, under the same parameter settings and disturbance values, the simulation and theoretical frequency response results exhibit a similar trend, though there is a certain degree of numerical error. This discrepancy is due to the theoretical calculation neglecting the variation in voltage amplitude at the nodes. The oscillations in frequency during the recovery process are determined by the bandwidth difference of the VSG P control loop and the dual-loop bandwidth in the simulation, but this does not affect the correctness of the trend observed in both results. Therefore, it can be concluded that the theoretical model has been validated.

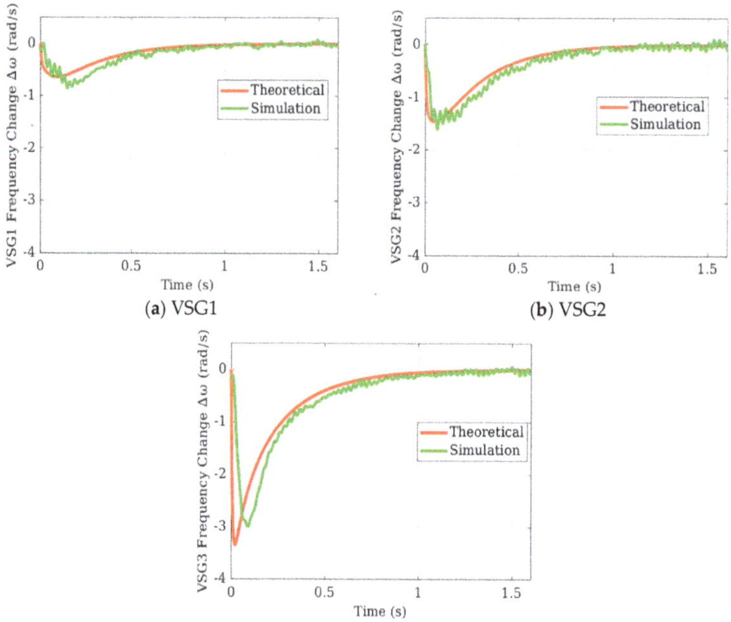

(**a**) VSG1 (**b**) VSG2 (**c**) VSG3

Figure 20. Theoretical and PSCAD/EMTDC results.

5.2. Validation of the Dynamic Performance Optimization Strategy

Next, the simulation model is used to validate the optimization strategy discussed in Section 4.5, with the specific parameters selected being the same as those in Table 4. And the load disturbance is set as 300 kW. To clearly demonstrate the correctness of the strategy, the results are also validated by analyzing the changes in the VSG port impedance and virtual inertia to eliminate the impact caused by the overlap of the two factors. The specific experimental parameters are shown in Table 7.

Table 7. The simulation experiment parameters.

Scenario	Attribute	J1/J2/J3	L1/L2/L3
1	Reference	3/1/1 kg m^2	0.4/0.6/0.2 mH
2	Vary L	3/1/1 kg m^2	0.2/0.4/0.6 mH
3	Vary J	1/1/3 kg m^2	0.4/0.6/0.2 mH
4	Vary Both	1/1/3 kg m^2	0.2/0.4/0.6 mH

In Figure 21, with only the port impedance changed, both the MFD and RoCoF of VSG3 are significantly improved, while the RoCoF of VSG1 shows a slight reduction. However, the MFD remains similar, and the overall system dynamic frequency performance is enhanced. This indicates that the adjustment of variable port impedance has effectively optimized the frequency dynamic response.

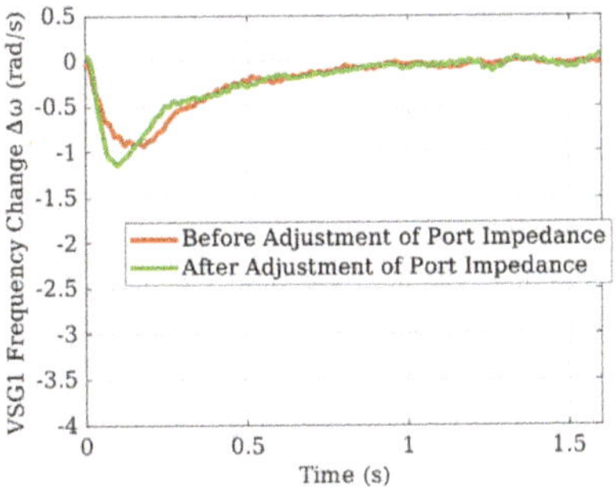

(**a**) VSG1

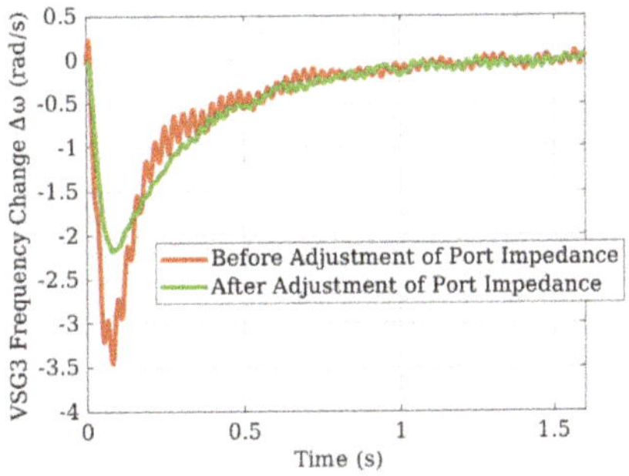

(**b**) VSG3

Figure 21. Simulation validation results of varying port impedance.

In Figure 22, similarly, when only the inertia is changed without altering the port impedance, the RoCoF of VSG3 decreases, but the MFD of VSG3 shows little variation. Meanwhile, the RoCoF of VSG1, located at the initial end, remains nearly constant. Although the overall system dynamic frequency performance improves, the effect is not pronounced, which indicates that when the port impedance is not optimally configured, the frequency optimization benefits from increased inertia are limited.

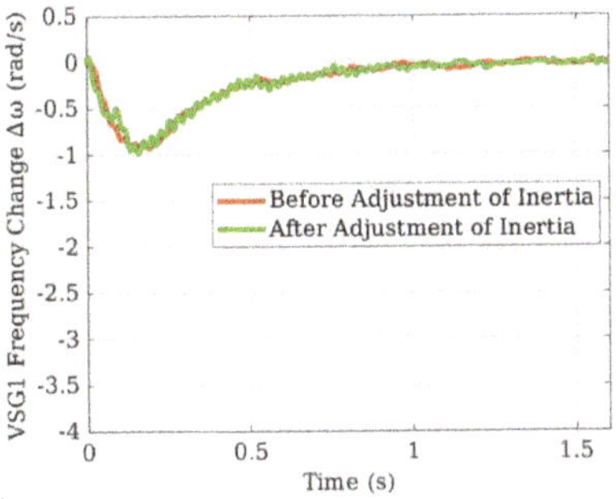

(**a**) VSG1

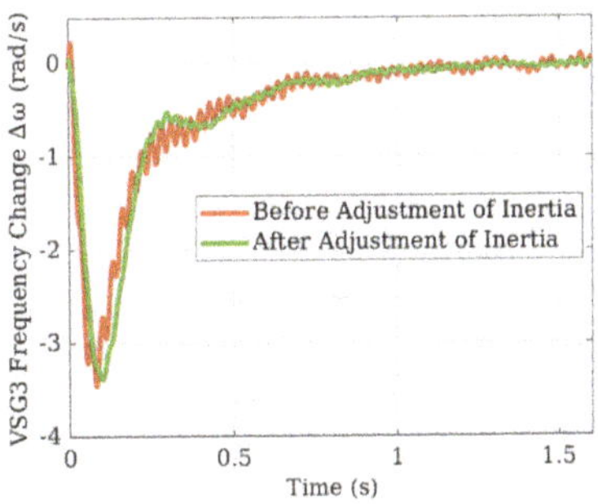

(**b**) VSG3

Figure 22. Simulation validation results of varying inertia.

In Figure 23, when both factors are altered, the dynamic performance of VSG3 shows significant improvement, while the RoCoF of VSG1 increases further, although the MFD remains relatively stable. This indicates a certain degree of enhancement in the system's frequency characteristics, thereby demonstrating the effectiveness of the frequency optimization strategy. And the specific experimental results are shown in the tables below.

Table 8 presents the MFD and RoCoF of each VSG during the frequency response process under each experimental condition, along with percentage comparisons to the reference group. From Table 8, a comparison between groups 1 and 4 clearly shows that under the dynamic frequency optimization strategy proposed in this paper, the MFD of the terminal VSG3 at the disturbance location significantly decreases, reaching only 62% of its value before adjustment. Although the MFDs of VSG1 and VSG2 both increase, Table 9 shows that the system's average MFD is only 86.1% of the original, demonstrating the effectiveness of the optimization strategy.

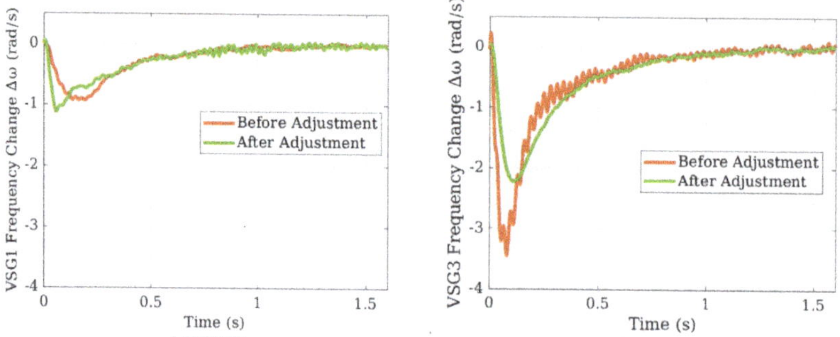

(**a**) VSG1 (**b**) VSG3

Figure 23. Simulation validation results of varying both factors.

Table 8. The first set of specific experimental results.

Scenario	VSG1		VSG2		VSG3	
	MFD (rad/s)	RoCoF (Hz/s)	MFD (rad/s)	RoCoF (Hz/s)	MFD (rad/s)	RoCoF (Hz/s)
1 (Reference)	0.788777	−0.154971	1.279293	−0.210996	3.069388	−0.158310
2	0.937706 1.19 × 100%	−0.128584 1.17 × 100%	1.517007 1.19 × 100%	−0.053582 1.75 × 100%	1.973153 0.64 × 100%	−0.138003 0.87 × 100%
3	0.908160 1.15 × 100%	−0.159783 1.03 × 100%	1.411547 1.10 × 100%	−0.274579 1.30 × 100%	2.723192 0.89 × 100%	−0.325253 2.06 × 100%
4	0.896958 1.14 × 100%	−0.189921 1.23 × 100%	1.618510 1.27 × 100%	−0.082723 1.61 × 100%	1.908000 0.62 × 100%	−0.240437 1.52 × 100%

Table 9. The second set of specific experimental results.

Scenario	Average of Three VSGs	
	MFD (rad/s)	RoCoF (Hz/s)
1 (Reference)	1.712486	−0.174759
2	1.475289 0.861 × 100%	−0.106723 0.611 × 100%
3	1.680300 0.981 × 100%	−0.253205 1.449 × 100%
4	1.474489 0.861 × 100%	−0.171027 0.979 × 100%

However, it is noteworthy that the comparison between scenarios 2 and 4 in Table 9 reveals that, despite similar average MFDs, reducing J1 and increasing J3 leads to a deterioration in the system's average RoCoF. The data in Table 8 indicate that the RoCoF change of the terminal VSG3 is most pronounced, consistent with the trend shown in Figure 14b in Section 4.3. This also highlights that when considering inertia configuration, it is insufficient to match the inertia only to the load; the impact of inertia reduction at non-disturbance nodes on the RoCoF at the disturbance node must also be considered.

Overall, the frequency optimization strategy proposed in this paper proves to be effective.

6. Conclusions

Considering a practical microgrid structure, this paper proposes a multi-node, multi-machine model that takes into account the line impedance between nodes and establishes a transfer function model from load disturbances to VSG frequency response. By analyzing the impact of inter-node line impedance and load disturbance location on dynamic frequency, as well as studying the effects of configuring parameters such as virtual inertia and port impedance under different disturbance conditions, a strategy is ultimately proposed to improve the dynamic frequency response of VSGs in multi-node microgrids. This approach explicitly considers spatial differences between nodes and provides a more effective strategy for improving frequency response in multi-node microgrid systems. The main conclusions are as follows:

1. Compared with traditional single-node analysis methods, the proposed strategy significantly enhances the dynamic frequency characteristics of the system under different load disturbance conditions. By appropriately adjusting virtual inertia and port impedance, the MFD and RoCoF of the system are effectively reduced, resulting in greater resistance to frequency disturbances.

2. The proposed method is suitable for complex, practical microgrid systems, taking into account the effects of inter-node line impedance and spatial differences, which improves the consistency of frequency response. At the same time, the strategy improves resource utilization efficiency and reduces the need for over-dimensioned hardware, although it also increases the complexity of modeling and parameter tuning, which may pose challenges for simpler systems.

Future research will focus on addressing the issue that optimizing frequency characteristics at specific nodes may lead to reduced performance at other nodes, in order to achieve balanced frequency response across all nodes. In addition, more advanced optimization algorithms will be explored to achieve deeper performance improvements, ultimately enabling optimal control solutions.

Author Contributions: Conceptualization, L.L.; Methodology, W.X. and W.K.; Software, Z.P.; Validation, X.L.; Formal analysis, W.X.; Resources, D.J.; Writing—original draft, W.X.; Writing—review & editing, D.J. and C.X.; Visualization, Z.Y. All authors have read and agreed to the published version of the manuscript.

Funding: The results of this paper is based on the project JYEPJS20230253 (Research of VSG for energy storage).

Data Availability Statement: The original contributions presented in the study are included in the article, further inquiries can be directed to the corresponding author.

Conflicts of Interest: Authors Wei Xie, Liangzi Li, Weihao Kong, Zheng Peng, Xiaogang Li and Dandan Jiao were employed by Henan Jiuyu Enpai Power Technol Co., Ltd. The remaining authors declare that the research was conducted in the absence of any commercial or financial relationships that could be construed as a potential conflict of interest.

References

1. Tian, J.; Wang, F.; Zhuo, F.; Wang, Y.; Wang, H.; Li, Y. A zero-backflow-power EPS control scheme with multiobjective coupled-relationship optimization in DAB-based converter. *IEEE J. Emerg. Sel. Top. Power Electron.* **2021**, *10*, 4128–4145. [CrossRef]
2. Zhu, Z.; Sun, S.; Huang, S. ILADRC resonance suppression control strategy for multiple parallel photovoltaic energy storage GFL VSG microgrid. *Electr. Eng.* **2024**, 1–14. [CrossRef]
3. Guan, M.; Pan, W.; Zhang, J.; Hao, Q.; Cheng, J.; Zheng, X. Synchronous generator emulation control strategy for voltage source converter (VSC) stations. *IEEE Trans. Power Syst.* **2015**, *30*, 3093–3101. [CrossRef]
4. Li, Y.; Gu, Y.; Green, T.C. Revisiting grid-forming and grid-following inverters: A duality theory. *IEEE Trans. Power Syst.* **2022**, *37*, 4541–4554. [CrossRef]
5. Wang, Z.; Yi, H.; Zhuo, F.; Sun, L.; Pei, Y.; Zhai, H.; Wu, J. A hardware structure of virtual synchronous generator in photovoltaic microgrid and its dynamic performance analysis. *Proc. CSEE* **2017**, *37*, 444–453.
6. Liu, J.; Miura, Y.; Ise, T. Comparison of dynamic characteristics between virtual synchronous generator and droop control in inverter-based distributed generators. *IEEE Trans. Power Electron.* **2015**, *31*, 3600–3611. [CrossRef]
7. Torres, M.; Lopes, L.A. Virtual synchronous generator control in autonomous wind-diesel power systems. In Proceedings of the 2009 IEEE Electrical Power & Energy Conference (EPEC), Montreal, QC, Canada, 22–23 October 2009; pp. 1–6.
8. Hu, B.; Cao, W.; Yang, M.; Lu, Y. The Optimal Ratio of GFM-Converters to GFL-Converters for Transient Voltage Regulation with Weak Grid Condition. In Proceedings of the 2023 IEEE 7th Conference on Energy Internet and Energy System Integration (EI2), Hangzhou, China, 15–18 December 2023; pp. 1993–1998.

9. Zhou, S.; Zou, X.; Zhu, D.; Tong, L.; Zhao, Y.; Kang, Y.; Yuan, X. An improved design of current controller for LCL-type grid-connected converter to reduce negative effect of PLL in weak grid. *IEEE J. Emerg. Sel. Top. Power Electron.* **2017**, *6*, 648–663. [CrossRef]
10. Askarian, A.; Park, J.; Salapaka, S. Enhanced Grid Following Inverter (E-GFL): A Unified Control Framework for Stiff and Weak Grids. *IEEE Trans. Power Electron.* **2024**, *39*, 5089–5107. [CrossRef]
11. Alipoor, J.; Miura, Y.; Ise, T. Stability assessment and optimization methods for microgrid with multiple VSG units. *IEEE Trans. Smart Grid* **2016**, *9*, 1462–1471. [CrossRef]
12. Driesen, J.; Visscher, K. Virtual synchronous generators. In Proceedings of the 2008 IEEE Power and Energy Society General Meeting-Conversion and Delivery of Electrical Energy in the 21st Century, Pittsburgh, PA, USA, 20–24 July 2008; pp. 1–3.
13. Sakimoto, K.; Miura, Y.; Ise, T. Stabilization of a power system including inverter-type distributed generators by a virtual synchronous generator. *Electr. Eng. Jpn.* **2014**, *187*, 7–17. [CrossRef]
14. Chen, Y.; Hesse, R.; Turschner, D.; Beck, H.P. Improving the grid power quality using virtual synchronous machines. In Proceedings of the 2011 International Conference on Power Engineering. Energy and Electrical Drives, Malaga, Spain, 11–13 May 2011; pp. 1–6.
15. Bevrani, H.; Ise, T.; Miura, Y. Virtual synchronous generators: A survey and new perspectives. *Int. J. Electr. Power Energy Syst.* **2014**, *54*, 244–254. [CrossRef]
16. Rongliang, S.; Xing, Z.; Fang, L.; Haizhen, X.; Yong, Y. Control technologies of multi-energy complementary microgrid operation based on virtual synchronous generator. *Trans. China Electrotech. Soc.* **2016**, *31*, 170–180.
17. Xie, N.; Liu, J.; Wang, Y.; Yin, Z.; Chen, C.; Zhao, Y. Combination analysis of VSG control converter and other converters. In Proceedings of the 2024 3rd International Conference on Energy Power and Electrical Technology (ICEPET), Chengdu, China, 17–19 May 2024; pp. 1868–1875.
18. Wang, G.; Fu, L.; Hu, Q.; Ma, F.; Liu, C.; Lin, Y. Low frequency oscillation analysis of VSG grid-connected system. In Proceedings of the 2021 3rd Asia Energy and Electrical Engineering Symposium (AEEES), Chengdu, China, 26–29 March 2021; pp. 631–637.
19. Sun, D.; Liu, H.; Wu, L.; Song, P.; Wang, X.; Li, D. Modeling and Characteristic Analysis of Influence of Virtual Synchronous Generatoron Low Frequency Oscillation. *Autom. Electr. Power Syst.* **2020**, *44*, 134–144.
20. Li, D.; Zhu, Q.; Lin, S.; Bian, X.Y. A self-adaptive inertia and damping combination control of VSG to support frequency stability. *IEEE Trans. Energy Convers.* **2016**, *2*, 397–398. [CrossRef]
21. Qi, Y.; Wu, H.; Pei, X.; Yang, R.; Zheng, Y. A Bandwidth-orienting Simplified and Design Method for Multiple VSG Parallel System. In Proceedings of the 2024 9th Asia Conference on Power and Electrical Engineering (ACPEE), Shanghai, China, 11–13 April 2024; pp. 2624–2629.
22. Choopani, M.; Hosseinian, S.H.; Vahidi, B. New transient stability and LVRT improvement of multi-VSG grids using the frequency of the center of inertia. *IEEE Trans. Power Syst.* **2019**, *35*, 527–538. [CrossRef]
23. Zhang, B.; Yan, X.; Huang, Y. Stability control and inertia matching method of multi-parallel virtual synchronous generators. *Trans. China Electrotech. Soc.* **2017**, *32*, 42–52.
24. Rasool, A.; Yan, X.; Rasool, H.; Guo, H. Correlation between multiple VSG sources for enhancing the power allocation in microgrid. In Proceedings of the 2018 IEEE Electrical Power and Energy Conference (EPEC), Toronto, ON, Canada, 10–11 October 2018; pp. 1–6.
25. Fu, S.; Sun, Y.; Li, L.; Liu, Z.; Han, H.; Su, M. Power oscillation suppression in multi-VSG grid by adaptive virtual impedance control. *IEEE Syst. J.* **2022**, *16*, 4744–4755. [CrossRef]
26. Wang, W.; Jiang, L.; Cao, Y.; Li, Y. A parameter alternating VSG controller of VSC-MTDC systems for low frequency oscillation damping. *IEEE Trans. Power Syst.* **2020**, *35*, 4609–4621. [CrossRef]
27. Wang, Z.; Zhuo, F.; Yi, H.; Wu, J.; Wang, F.; Zeng, Z. Analysis of Dynamic Frequency Performance Among Voltage-Controlled Inverters Considering Virtual Inertia Interaction in Microgrid. *IEEE Trans. Ind. Appl.* **2019**, *55*, 4135–4144. [CrossRef]
28. Sun, P.; Yao, J.; Zhao, Y.; Fang, X.; Cao, J. Stability Assessment and Damping Optimization Control of Multiple Grid-connected Virtual Synchronous Generators. *IEEE Trans. Energy Convers.* **2021**, *36*, 3555–3567. [CrossRef]
29. Wang, Z.; Yi, H.; Zhuo, F.; Wu, J.; Zhu, C. Analysis of Parameter Influence on Transient Active Power Circulation Among Different Generation Units in Microgrid. *IEEE Trans. Ind. Electron.* **2020**, *68*, 248–257. [CrossRef]
30. Chen, M.; Zhou, D.; Wu, C.; Blaabjerg, F. Characteristics of Parallel Inverters Applying Virtual Synchronous Generator Control. *IEEE Trans. Smart Grid* **2021**, *12*, 4690–4701. [CrossRef]

electronics

MDPI

Article

Simulation-Based Hybrid Energy Storage Composite-Target Planning with Power Quality Improvements for Integrated Energy Systems in Large-Building Microgrids

Chunguang He [1], Xiaolin Tan [1], Zixuan Liu [2], Jiakun An [1,*], Xuejun Li [3], Gengfeng Li [2] and Runfan Zhang [2]

1 State Grid Hebei Economic Research Institute, No. 27 Fuqiang Street, Yuhua District, Shijiazhuang 050081, China; gwhbjyyhcg@he.sgcc.com.cn (C.H.); gwhbjyytxl@he.sgcc.com.cn (X.T.)
2 The School of Electrical Engineering, Xi'an Jiaotong University, Xi'an 710049, China; lzx18323472768@stu.xjtu.edu.cn (Z.L.); gengfengli@xjtu.edu.cn (G.L.); r.zhang@xjtu.edu.cn (R.Z.)
3 State Grid Handan Electric Power Supply Company, Handan 056000, China; gwhbhdlxj@he.sgcc.com.cn
* Correspondence: gwhbjyyajk@he.sgcc.com.cn

Abstract: In this paper, we present an optimization planning method for enhancing power quality in integrated energy systems in large-building microgrids by adjusting the sizing and deployment of hybrid energy storage systems. These integrated energy systems incorporate wind and solar power, natural gas supply, and interactions with electric vehicles and the main power grid. In the optimization planning method developed, the objectives of cost-effective and low-carbon operation, the lifecycle cost of hybrid energy storage, power quality improvements, and renewable energy utilization are targeted and coordinated by using utility fusion theory. Our planning method addresses multiple energy forms—cooling, heating, electricity, natural gas, and renewable energies—which are integrated through a combined cooling, heating, and power system and a natural gas turbine. The hybrid energy storage system incorporates batteries and compressed-air energy storage systems to handle fast and slow variations in power demand, respectively. A sensitivity matrix between the output power of the energy sources and the voltage is modeled by using the power flow method in DistFlow, reflecting the improvements in power quality and the respective constraints. The method proposed is validated by simulating various typical scenarios on the modified IEEE 13-node distribution network topology. The novelty of this paper lies in its focus on the application of integrated energy systems within large buildings and its approach to hybrid energy storage system planning in multiple dimensions, including making co-location and capacity sizing decisions. Other innovative aspects include the coordination of hybrid energy storage combinations, simultaneous siting and sizing decisions, lifecycle cost calculations, and optimization for power quality enhancement. As part of these design considerations, microgrid-related technologies are integrated with cutting-edge nearly zero-energy building designs, representing a pioneering attempt within this field. Our results indicate that this multi-objective, multi-dimensional, utility fusion-based optimization method for hybrid energy storage significantly enhances the economic efficiency and quality of the operation of integrated energy systems in large-building microgrids in building-level energy distribution planning.

Keywords: integrated energy systems; hybrid energy storage; utility fusion theory; power quality improvements; large-building microgrids; multiple energy sources

1. Introduction

In recent years, integrated energy systems (IESs) have emerged as efficient energy supply models combining multiple forms of energy, such as cooling, heating, electricity, and gas, for unified planning and dispatch [1–3]. Incorporating this kind of design into the building sector, which involves major energy consumption, can facilitate the creation of nearly zero-energy buildings and support the transition to new power systems [4–6]. In this context, combined cooling, heating, and power (CCHP) systems with microgrids extend the energy supply capabilities of traditional microgrids. This integration not only fuses the functions of cooling, heating, and providing electricity but also promotes the full utilization of renewable energy sources [7,8]. However, several challenges are encountered in planning IESs in large buildings, including conflicts in system operation efficiency, the impact of uncertainties, high peak-load supply pressure, and the low absorption rates of wind and solar energy [9]. Numerous studies have proven that well-designed energy storage systems can act as buffers in ensuring efficient and stable system operation, making hybrid energy storage planning a crucial element for the stable, eco-friendly, and cost-effective development of IESs in microgrids for large buildings [10–14].

IES planning has been optimized widely in research, namely, by constructing multi-dimensional optimization objectives to evaluate that each aligns with the characteristics of the system models [15]. While some studies have considered multiple types of economic costs, such as operation and maintenance costs, equipment construction costs, and energy transaction costs [16–19], other studies have begun to consider their optimization objectives from multiple perspectives. These include minimizing the investment and operational costs, maximizing efficiency, and reducing carbon emissions under various uncertainties [20]. In some approaches, planning is modeled as a deterministic bi-objective optimization problem in which the investment and operational costs are targeted alongside robustness [21]. For integrated energy systems equipped with carbon capture, optimization models focus on minimizing the investment costs, energy purchase costs, energy sales revenue, and carbon costs [22]. In coordinating integrated energy system planning and designing capacity allocation schemes for scenarios involving high renewable energy penetration, both environmental penalties and subsidies for generating renewable energy are accounted for in the operational costs [23]. Another model proposed for a regional integrated energy system, which includes electric, thermal, and cooling buses, has also considered cost-effective and low-carbon operation of the system [24]. On the basis of these studies, it is evident that in addition to economic goals, multiple dimensions, including carbon emissions and system robustness, have been considered in research in order to meet the requirements of various scenarios.

Energy storage devices play the key bridging role of energy hubs in integrated energy systems. Indeed, researchers have utilized various types of energy storage systems to smooth the output power of stochastically volatile renewable energy systems, such as those incorporating wind and solar power [25,26], as well as to harmonize and compensate for fluctuating changes in customer loads, thereby improving their operational efficiency [27]. It has been demonstrated that introducing an optimal energy storage model into an integrated energy system can also enhance its operational flexibility [28]. Additionally, in simulating and comparing wind turbines with and without energy storage, the inclusion of energy storage was verified as an effective improvement in the efficiency of renewable energy utilization [29]. Thus, planning energy storage systems in designing IESs has been the subject of substantial research attention. Considering objectives such as battery lifespan degradation, minimizing the annual operational and storage investment costs, and economic efficiency, studies have explored how to optimize regional integrated energy systems (RIESs) by using hybrid battery/pumped storage, multi-energy storage systems in IESs, and hybrid energy storage planning [30–32]. Some research has also focused on a joint planning model for integrated energy systems with shared storage, aiming to plan the optimal construction timeline for storage equipment based on the total investment and operational costs [33]. However, despite these contributions, most of the existing studies on planning energy storage in IESs have focused primarily on the construction costs, the lifespan of the centralized storage systems, and the economic benefits to system operation while often overlooking the design of distributed hybrid energy storage systems at the operational level. Choosing the optimal siting can further enhance system performance, improve power quality, and increase the flexibility and safety of the configuration. With the current policy developments, higher renewable energy integration can also yield additional benefits, such as securing government subsidies and "green power certificates". Therefore, renewable energy utilization rates should be considered part of the optimization objectives. However, existing research on microgrid IEEs for large buildings has not fully leveraged the advantages of storing energy by using distributed siting and hybrid configurations to promote high-quality operation and renewable energy utilization, and such limitations have hindered the potential of energy storage systems in this context.

Motivated by the aforementioned studies, this paper uses a linearized DistFlow model to examine the underlying grid configuration and operational constraints of an IES for a large-building microgrid. Multiple optimization objectives are incorporated, including its cost-effective and low-carbon operation, lifecycle construction costs, system power quality improvements, and renewable energy absorption. By integrating these factors with utility fusion theory, we achieve optimal hybrid energy storage planning for IESs in large-building microgrids, ensuring low-carbon, cost-effective, and efficient renewable energy utilization while maintaining optimized system operation.

The salient features of this study are as follows:

- This study establishes a DistFlow power flow model of an IES for a large-building microgrid, improving power quality by considering the sensitivity of the node voltages and power injected into the microgrid. It also incorporates renewable energy, CCHP, electric vehicle charging interactions, and hybrid energy storage systems, facilitating the optimization of IESs that make use of multiple energy sources.
- The planning method proposed addresses how to manage fast and slow variations in power demand for hybrid energy storage on the basis of multi-cycle variations in power. This comprehensive model lays the foundation for optimizing the location and capacity of hybrid energy storage planned in IESs in large-building microgrids.
- Considering multiple dimensions beyond economic and low-carbon factors, including power quality optimization, new energy consumption, and the lifecycle cost of hybrid energy storage systems, this study coordinates all of these different aspects by using utility fusion theory, resulting in comprehensive and reliable planning under various scenarios.

The remainder of the paper organized as follows: Section 2 establishes the underlying model of the IES in a large-building microgrid. In Section 3, we design the multi-dimensional optimization objectives and formulate the operational constraints of the IES in a large-building microgrid. In Section 4, we demonstrate the effectiveness and performance of the planning model based on case studies. Section 5 discusses and summarizes the research content and value of this study. For the reader's convenience, Tables A1 and A2 list the variable and parameter symbols used in this paper, respectively.

2. Analysis of Integrated Energy Systems in Building Microgrids

2.1. Mathematical Models of Integrated Energy Systems in Large Buildings

Figure 1 illustrates the IES used in large-building microgrids discussed in this paper. To fully explore the potential of large buildings to accommodate renewable energy and the development of new power systems, this system takes into account the impact of wind and solar power generation. Additionally, this system features multi-energy complementarity, allowing users to switch between energy sources flexibly based on the electricity pricing at the time of use, facilitating a comprehensive response to demand [34]. Moreover, the system accounts for how charging stations in buildings can optimize the charging process to improve the energy supply–demand balance while still meeting users' needs. However, due to the uncertainty of electric vehicle charging and its limited scale, this aspect has a limited impact on energy systems involving numerous energy devices and complex operations. Therefore, the IES is configured with a more flexible and reliable hybrid energy storage system to accommodate fluctuations in the renewable energy output and load. The optimization planning of hybrid energy storage is at the core of designing an cost-effective, high-quality, operational IES for a large building. Specifically, the CCHP system established consists of electric chillers, electric heaters, microturbines, natural gas boilers, and lithium bromide absorption chillers. The microturbines simultaneously provide both electricity and heat. Meanwhile, the energy storage system, in combination with the electric heating and cooling units, meets the power, heating, and cooling demands of the IES by storing excess electricity during periods of low demand and releasing it with peak demand, thereby smoothing any demand spikes. The natural gas boilers can provide supplementary heating when electric heating is not feasible, and the absorption chillers can utilize any excess heat for cooling. By linking these components through a distribution network power flow model, the CCHP system achieves the gradient utilization of energy and the coordinated complementation of various energy forms.

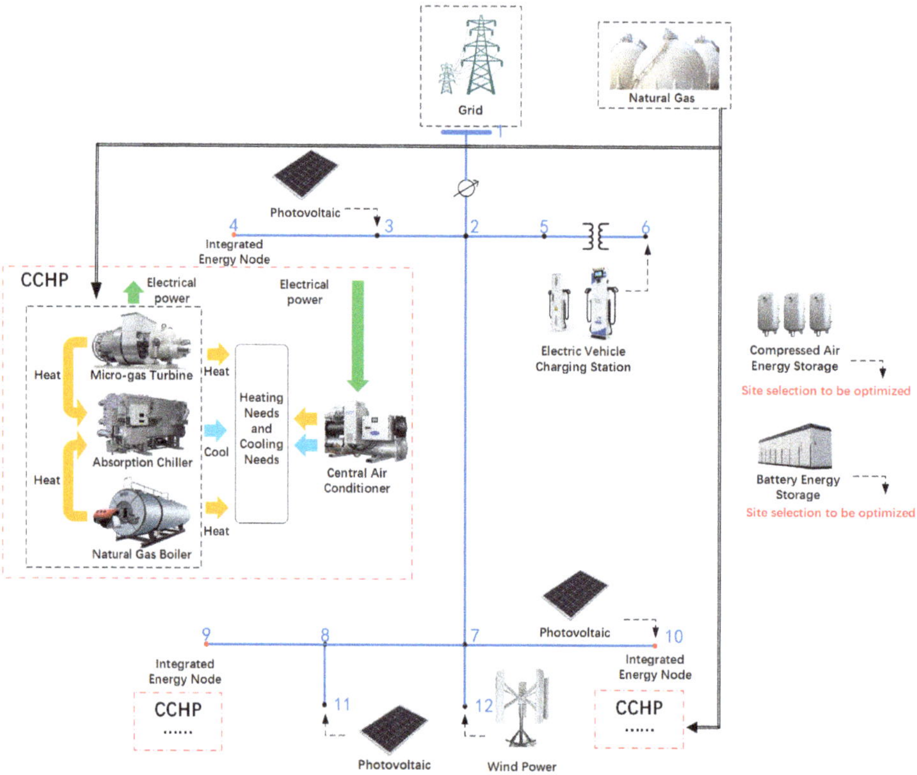

Figure 1. Distributed integrated energy system.

2.1.1. The Microturbine

The relationship between the heat generated by the microturbine and its electrical output power is [23]

$$q_{i,n}^{\mathrm{MT}}(t) = \frac{p_{i,n}^{\mathrm{MT}}(t)\left(1 - \eta_{\mathrm{e}}^{\mathrm{MT}} - \eta_{\mathrm{l}}^{\mathrm{MT}}\right)}{\eta_{\mathrm{e}}^{\mathrm{MT}}} \quad (1)$$

where $q_{i,n}^{\mathrm{MT}}$ is its exhaust heat at node i at time t in the nth scenario; $p_{i,n}^{\mathrm{MT}}$ is its electrical output power at node i at time t in the nth scenario; $\eta_{\mathrm{e}}^{\mathrm{MT}}$ is its electrical efficiency, set to 0.3; and $\eta_{\mathrm{l}}^{\mathrm{MT}}$ is its heat loss coefficient, set to 2%.

The natural gas consumed by the microturbine can be expressed as

$$v_{i,n}^{\mathrm{MT}}(t) = \frac{p_{i,n}^{\mathrm{MT}}(t)}{q_{i,n}^{\mathrm{MT}} L_{\mathrm{ng}}} \quad (2)$$

where $v_{i,n}^{\mathrm{MT}}$ is the amount of natural gas consumed by the microturbine per unit time; L_{ng} is the lower heating value of natural gas, set to 9.78 (kW·h)/m³; and ΔT is the length of the time step.

2.1.2. The Natural Gas Boiler

The heat generated by the natural gas boiler is [2]

$$q_{i,n}^{\mathrm{BLR}}(t) = n_{\mathrm{h}} L_{\mathrm{ng}} v_{i,n}^{\mathrm{BLR}}(t) \tag{3}$$

where $q_{i,n}^{\mathrm{BLR}}$ is the heat output of the natural gas boiler, n_{h} is the boiler's efficiency, and $v_{i,n}^{\mathrm{BLR}}$ is the natural gas consumption of the boiler per unit time.

2.1.3. The Electric Heater

The heat generated by the electric heater is

$$q_{i,n}^{\mathrm{EH}}(t) = \mathrm{C}_{\mathrm{op}}^{\mathrm{EH}} p_{i,n}^{\mathrm{EH}}(t) \tag{4}$$

where $q_{i,n}^{\mathrm{EH}}$ is the heat output of the electric heater, $\mathrm{C}_{\mathrm{op}}^{\mathrm{EH}}$ is its coefficient of performance (COP), and $p_{i,n}^{\mathrm{EH}}$ is its electrical power consumption.

2.1.4. The Electric Chiller

The cooling capacity of the electric chiller is

$$r_{i,n}^{\mathrm{EC}}(t) = \mathrm{C}_{\mathrm{op}}^{\mathrm{EC}} p_{i,n}^{\mathrm{EC}}(t) \tag{5}$$

where $r_{i,n}^{\mathrm{EC}}$ is the cooling capacity of the electric chiller, $\mathrm{C}_{\mathrm{op}}^{\mathrm{EC}}$ is its COP, and $p_{i,n}^{\mathrm{EC}}$ is its electrical power consumption.

2.1.5. The Absorption Chiller

Using heat to drive the cooling cycle, the cooling capacity of the absorption chiller is [2]

$$r_{i,n}^{\mathrm{ACH}}(t) = \mathrm{C}_{\mathrm{op}}^{\mathrm{ACH}} q_{i,n}^{\mathrm{ACH}}(t) \tag{6}$$

where $r_{i,n}^{\mathrm{ACH}}$ is the cooling output of the absorption chiller; $\mathrm{C}_{\mathrm{op}}^{\mathrm{ACH}}$ is its COP, set to 1.38; and $q_{i,n}^{\mathrm{ACH}}$ is its heat input at time t.

2.1.6. Hybrid Energy Storage

With the aim of achieving energy supply–demand balance and voltage support, this paper incorporates designing the parameters of hybrid energy storage devices into decision making for the optimization of IESs in microgrids for large buildings. By considering the characteristics of different energy storage technologies, these systems can be tailored to meet diverse needs [35,36]. Battery energy storage systems (BESSs) offer high flexibility and fast responses to energy or load fluctuations. In contrast, compressed-air energy storage (CAES) systems feature long lifespans, low loss, and low self-discharge rates but slower response times, making them suited to energy peak shifting. Therefore, this paper proposes a hybrid energy storage system that combines batteries and CAES for better adaptation to IESs in large-building microgrids, enhancing the flexibility of their design and economic efficiency. In terms of the differences in the charge–discharge response speed, power, and capacity costs between the two, CAES systems have a longer power update cycle, making them well equipped to handle shifts in renewable energy peaks and real-time electricity price utilization. Conversely, batteries require faster power update rates to absorb the volatility of renewable energy systems and the randomness of their short-term output. Currently, industrial facilities with high and fluctuating energy demands are among the types of buildings suited to installing CAES systems, where CAES can be used to balance their energy consumption. Additionally, large commercial buildings and data centers with sufficient space and continuous energy needs may also consider the use of CAES for backup power and load balancing, while remote or isolated facilities with limited grid access can use CAES as a long-term energy storage solution.

Batteries typically have a higher energy conversion efficiency and smaller discharge depth compared with CAES. On the other hand, CAES systems generally have lower energy conversion efficiency but can achieve larger charge–discharge depth. Therefore, when designing a hybrid energy storage system, it is essential to consider efficiency during the charge–discharge processes to ensure the corresponding energy-level calculations are accurate. The specific model of the energy storage system is built as follows [8]:

$$\begin{cases} p_{i,n}^{\mathrm{cCAES}}({}^{l}T_{\mathrm{CAES}}) = p_{i,n}^{\mathrm{cCAES}}({}^{l}T_{\mathrm{CAES}}+T_{\mathrm{BATT}}) = \cdots \\ = p_{i,n}^{\mathrm{cCAES}}({}^{l}T_{\mathrm{CAES}}+(N_{\mathrm{T}}-1)T_{\mathrm{BATT}}) \\ p_{i,n}^{\mathrm{dCAES}}(nN_{\mathrm{T}}T_{\mathrm{BATT}}) = p_{i,n}^{\mathrm{dCAES}}(nN_{\mathrm{T}}T_{\mathrm{BATT}}+T_{\mathrm{BATT}}) = \cdots \\ = p_{i,n}^{\mathrm{dCAES}}(nN_{\mathrm{T}}T_{\mathrm{BATT}}+(N_{\mathrm{T}}-1)T_{\mathrm{BATT}}) \\ e_{i,n}^{\mathrm{CAES}}(t) = E_{i,n}^{\mathrm{CAES0}} + \sum\limits_{s=1}^{t} \Delta T\left[n_{\mathrm{c}}^{\mathrm{CAES}} p_{i,n}^{\mathrm{cCAES}}(s) - p_{i,n}^{\mathrm{dCAES}}(s) \Big/ n_{\mathrm{d}}^{\mathrm{CAES}}\right] \\ e_{i,n}^{\mathrm{BATT}}(t) = E_{i,n}^{\mathrm{BATT0}} + \sum\limits_{s=1}^{t} \Delta T\left[n_{\mathrm{c}}^{\mathrm{BATT}} p_{i,n}^{\mathrm{cBATT}}(s) - p_{i,n}^{\mathrm{dBATT}}(s) \Big/ n_{\mathrm{d}}^{\mathrm{BATT}}\right] \end{cases} \tag{7}$$

where T_{BATT} is the power variation cycle of the BESS; ${}^{l}T_{\mathrm{CAES}}$ is the lth power variation cycle of the CAES system, ranging from 0 to $T/N_{\mathrm{T}} - 1$; N_{T} is the ratio of the power variation cycle of the CAES system to that of the BESS; $p_{i,n}^{\mathrm{dCAES}}$ is the discharge power of the CAES system; $p_{i,n}^{\mathrm{dBATT}}$ is the discharge power of the BESS; $p_{i,n}^{\mathrm{cCAES}}$ is the charge power of the CAES system; $p_{i,n}^{\mathrm{cBATT}}$ is the charge power of the BESS; $e_{i,n}^{\mathrm{CAES}}$ is the remaining energy of the CAES system; $E_{i,n}^{\mathrm{CAES0}}$ is the initial energy of the CAES system; $n_{\mathrm{c}}^{\mathrm{CAES}}$ and $n_{\mathrm{d}}^{\mathrm{CAES}}$ are the charging and discharging efficiency of the CAES system; $e_{i,n}^{\mathrm{BATT}}$ is the remaining energy of the BESS; $E_{i,n}^{\mathrm{BATT0}}$ is the initial energy of the BESS; and $n_{\mathrm{c}}^{\mathrm{BATT}}$ and $n_{\mathrm{d}}^{\mathrm{BATT}}$ are the charging and discharging efficiency of the BESS.

2.1.7. Electric Vehicle Charging Stations

Just as hybrid energy storage systems have a buffering effect on the electricity supply, EVs provide a limited response to demand. We modeled user charging behavior by utilizing variables such as user access to charging stations, the expected charging time, and the target battery level and integrated it into the model, prioritizing the need for electric vehicles to reach the user's expected battery level by the projected charging completion time. During the vehicle charging process, the charging load can be dynamically adjusted based on the real-time energy supply and demand, ensuring that the building's core load requirements are not compromised.

$$e_{i,n}^{\mathrm{EV}}(t) = E_{i,n}^{\mathrm{EV0}} + \sum_{s=1}^{t} \Delta T\left[n_{\mathrm{c}}^{\mathrm{EVCS}} p_{i,n}^{\mathrm{cEVCS}}(s) - p_{i,n}^{\mathrm{dEVCS}}(s) \Big/ n_{\mathrm{d}}^{\mathrm{EVCS}}\right] \tag{8}$$

where $p_{i,n}^{\mathrm{cEVCS}}$ is the charging power provided by the charging station to the electric vehicle, $p_{i,n}^{\mathrm{dEVCS}}$ is the discharging power drawn from the electric vehicle by the charging station, $e_{i,n}^{\mathrm{EV}}$ is the remaining energy (state of charge) of the electric vehicle, $E_{i,n}^{\mathrm{EV0}}$ is the initial energy (initial state of charge) of the electric vehicle, and $n_{\mathrm{c}}^{\mathrm{EVCS}}$ and $n_{\mathrm{d}}^{\mathrm{EVCS}}$ are the charging and discharging efficiency of the EV charging station.

2.2. The DistFlow Power Flow Model for an IES in a Large-Building Microgrid

Once the hybrid energy storage system is integrated, the system is able to balance electricity purchases from the main grid with the renewable energy output to meet the load demands and improve voltage quality. Basic power flow constraints, such as the power balance at each node and the voltage limits at each node, must be adhered to considering the formation of the underlying distribution network. The DistFlow model is adopted as the power flow model for the microgrid. For computational convenience, the DistFlow model is linearly approximated to establish the operating constraints of the underlying distribution network, and then a power flow model of this network is established.

Let $P_{i,n}$ and $Q_{i,n}$ represent the active and reactive power flows into node i in the underlying distribution network. Given that the underlying distribution network has a tree topology with the distributed generation located at the root node, each node has only one incoming power flow. The active power and reactive power balance at each node in the underlying distribution network can be written as

$$\begin{cases} \sum\limits_{j \in \mathcal{S}_i} P_{j,n} = P_{i,n} + p_{i,n}, & \forall i \in \mathcal{N} \\ \sum\limits_{j \in \mathcal{S}_i} Q_{j,n} = Q_{i,n} + q_{i,n}, & \forall i \in \mathcal{N} \end{cases} \tag{9}$$

where $\mathcal{S}_i$ denotes the set of child nodes for grid node i; $P_{j,n}$ and $Q_{j,n}$ are the active and reactive power flows from node i to node j in the nth scenario; $p_{i,n}$ and $q_{i,n}$ denotes the injected power of all the devices accessing node i in the nth scenario, including the contributions from the grid, photovoltaic systems, wind power systems, energy storage systems, and electrical loads; and $\mathcal{N}$ denotes the set of all the nodes in the microgrid. $p_{i,n}$ and $q_{i,n}$ are in the following form:

$$\begin{array}{l} p_{i,n} = p_{i,n}^{\mathrm{GP}}(t) + p_{i,n}^{\mathrm{MT}}(t) + p_{i,n}^{\mathrm{PV}}(t) + p_{i,n}^{\mathrm{WTG}}(t) + p_{i,n}^{\mathrm{dCAES}}(t) + p_{i,n}^{\mathrm{dBATT}}(t) + p_{i,n}^{\mathrm{dEVCS}}(t) \\ -p_{i,n}^{\mathrm{cCAES}}(t) - p_{i,n}^{\mathrm{cBATT}}(t) - p_{i,n}^{\mathrm{cEVCS}}(t) - p_{i,n}^{\mathrm{LOAD}}(t) - p_{i,n}^{\mathrm{EH}}(t) - p_{i,n}^{\mathrm{EC}}(t) \end{array} \tag{10}$$

where $p_{i,n}^{\mathrm{GP}}$ is the power purchased from the grid, $p_{i,n}^{\mathrm{PV}}$ is the output power from the photovoltaic system, $p_{i,n}^{\mathrm{WTG}}$ is the output power from the wind power system, and $p_{i,n}^{\mathrm{LOAD}}$ is the electrical load.

Considering the voltage constraints at the nodes in the system, the voltage at the grid connection node will be set as the reference value. According to the linearized DistFlow model, the voltage at the other nodes in the grid can be expressed as follows:

$$V_{i,n} = V_{j,n} - \frac{r_i P_{i,n} + x_i Q_{i,n}}{V_0}, j = \theta(i), i \in \overline{\mathcal{N}} \tag{11}$$

where $V_{i,n}$ is the voltage magnitude at node i in the nth scenario, $\theta(i)$ represents the previous node connected to node i, r_i is the line resistance with node i as the endpoint, x_i is the line reactance with node i as the endpoint, and V_0 is the reference voltage for the entire microgrid.

3. Configuration of Distributed Hybrid Energy Storage System's Capacity and Its Operational Optimization

The variables for optimization include the hourly quantities of electricity and gas purchased for each day of the integrated energy system's operation, the operational power values in each part of the hybrid energy storage system, and the design capacity and rated power of each energy storage device based on the operational conditions. The aim of the optimization process is to minimize the total lifecycle costs of the hybrid energy storage, including economic and low-carbon operational costs, while maintaining power quality. The constraints in the design take into account the linearized DistFlow power flow model for the underlying distribution network, the operational models of the cooling and heating equipment, the constraints on the output of the devices in the system, and the operational models of hybrid energy storage devices.

3.1. Optimization Objective Function for Distributed Multi-Energy Storage Configuration

The optimization objective for the hybrid energy storage configuration considers its cost-effective and low-carbon performance during system operation, reflecting the costs associated with purchasing external energy and the construction costs of the hybrid energy storage system over its entire lifecycle, including the initial costs of investing in the power of its design and its capacity and the cumulative costs of CAES over its lifespan. Finally, the power flow is calculated to tailor the optimization objective to voltage fluctuations, highlighting the support provided by the design of the energy storage in terms of the system voltage.

3.1.1. Economic Cost

The integrated energy system's operational cost is based on the external purchase of electricity and natural gas, accounting for time-of-use pricing to maximize the utilization of its storage capacity and exploit peak–valley energy pricing. The optimization timeframe is based on the lifespan of CAES, summing up all daily operational events and then aggregating them across its entire lifecycle considering the frequency of specific scenarios occurring.

$$f_{\text{Trans}}=n_{\text{life}}\sum_{n=1}^{N_{\text{S}}}\sum_{t=1}^{T}n_n^{\text{scen}}\Delta T\left[C_{\text{P}}^{\text{G}}V_n^{\text{G}}(t)+C_{\text{P}}^{\text{E}}(t)p_n^{\text{GP}}(t)\right] \tag{12}$$

where n_{life} represents the designed usage time for the energy storage system, N_{S} is the number of scenarios, T is the total number of time intervals, n_n^{scen} is the frequency of the nth scenario, ΔT is the time interval within each unit of time, C_{P}^{G} is the purchase cost per unit of natural gas, V_n^{G} is the consumption of natural gas per unit time, C_{P}^{E} is the time-of-use price of purchasing electricity from the grid, and p_n^{GP} is the quantity of electricity purchased from the grid per unit time.

3.1.2. Low-Carbon Cost

Carbon emission targets are part of operational cost optimization and are similar in form to economic cost. Carbon emissions from natural gas consumption and grid electricity generation are considered and then integrated with the carbon market trading price to unify the unit of measurement of the objective function.

$$f_{\text{CO2e}}=n_{\text{life}}\sum_{n=1}^{N_{\text{S}}}\sum_{t=1}^{T}n_n^{\text{scen}}\Delta TC_{\text{P}}^{\text{C}}\left[m_{\text{gas}}V_n^{\text{G}}(t)+m_{\text{grid}}p_n^{\text{GP}}(t)\right] \tag{13}$$

where m_{gas} is the carbon emission factor for natural gas and m_{grid} is the carbon emission factor for the grid. These values are based on the carbon market trading prices.

3.1.3. Lifecycle Cost of Hybrid Energy Storage System

The lifecycle cost of energy storage encompasses all costs related to the energy storage system throughout its entire lifecycle, including its design and construction, maintenance and operation, and replacement and recycling. This comprehensive metric permits a more thorough evaluation of the system's economic efficiency and sustainability. The objective function designed accounts for the initial construction costs and the annual operating and maintenance costs, as well as the costs of replacing and recycling equipment, based on the system's lifespan, adjusted for the standard discount rate [26].

$$\begin{aligned} f_{\text{ES}} &= C_{\text{E}}^{\text{ES}}E_{\text{set}}+C_{\text{P}}^{\text{ES}}P_{\text{set}} \\ &+\sum_{y=1}^{Y}\frac{C_{\text{E}}^{\text{OM}}E_{\text{set}}+C_{\text{P}}^{\text{OM}}P_{\text{set}}+C_{\text{labor}}n_{\text{labor}}+C_{\text{C}}^{\text{R}}E_{\text{set}}+C_{\text{Rec}}}{(1+r)^y} \end{aligned} \tag{14}$$

where C_{E}^{ES} is the unit cost of storage capacity, E_{set} is the installed capacity of the energy storage solution, C_{P}^{ES} is the unit cost of storage power, P_{set} is the installed power of the energy storage solution, Y is the service life in years of the system design, C_{E}^{OM} is the maintenance cost per unit of energy storage capacity, C_{P}^{OM} is the maintenance cost per unit of power, C_{labor} is the cost per person per year of maintenance personnel, n_{labor} is the number of energy storage system operation and maintenance personnel, C_{C}^{R} is the cost of replacements per unit of energy storage capacity, C_{Rec} is the cost of energy storage recovery, and r is the standard discount rate or the benchmark rate of return.

The cost of recovery is expressed as

$$C_{Rec} = (C_{E}^{ES} E_{set} + C_{P}^{ES} P_{set}) \times F_{EOL} \tag{15}$$

where F_{EOL} is the end-of-life cost ratio, which is the ratio of the recovery cost to the initial investment and can be positive or negative.

3.1.4. Minimizing Voltage Deviation

In this study, we incorporate modeling the power flow and making calculations for the underlying grid into the integrated energy system, allowing us to obtain the voltage values at various nodes. By considering the per-unit (pu) system, the absolute deviations in each node's voltage from the standard reference voltage are summarized and then aggregated over multiple typical scenarios throughout the system's entire operational period, resulting in an objective function that reflects the quality of the system's power supply.

$$f_{Volt} = n_{life} \sum_{n=1}^{N_S} \sum_{t=1}^{T} n_{n}^{scen} \Delta T \left[\sum_{i=1}^{\mathcal{N}} |V_i^*(t) - 1| \right] \tag{16}$$

where V_i^* is the normalized value of the voltage at node i.

3.1.5. Renewable Energy Utilization

We also consider the system's actual ability to utilize renewable energy in this study. By calculating the difference between the renewable energy actually used by the system during its operation and the maximum amount of renewable energy theoretically able to be generated, we derive an objective function that reflects the system's capacity to utilize renewable energy.

$$f_{Utilization} = n_{life} \sum_{n=1}^{N_S} \sum_{t=1}^{T} n_{n}^{scen} \Delta T \left[P_{energy}^{theory}(t) - P_{energy}^{real}(t) \right] \tag{17}$$

where P_{energy}^{real} is the total new energy output power of the actual operation of the system, and P_{energy}^{theory} is the total maximum new energy output power that can theoretically be generated by the system.

3.1.6. The Composite Objective

The first three indicators proposed in this paper are unified into a cost function measured by price, while the fourth and fifth indicators used are operational voltage quality and renewable energy utilization. Given that these indicators reflect different types of utility and that there are conflicts and complementarities between them, based on utility fusion theory, we designed a utility function to characterize to what extent these various objectives are satisfied. This ultimately represents a comprehensive evaluation objective function that can be used to guide the optimization planning process. First, the three objective functions that have already been unified are combined:

$$f_1 = f_{Trans} + f_{CO2e} + f_{ES} \tag{18}$$

And the fourth and fifth objective functions reflecting the system's operational effectiveness indicators are

$$\begin{aligned} f_2 &= f_{\text{Volt}} \\ f_3 &= f_{\text{Utilization}} \end{aligned} \tag{19}$$

Since the three types of objectives in this study are all minimization optimization problems with differing units and magnitudes, the difference ratio is used as the basic utility function for the optimization objectives. In transforming each optimization metric into a dimensionless basic utility value within the range of 0 to 1, it can thereby normalize these metrics and eliminate the incommensurability between them. The difference ratio effectively reflects the discrepancy between the current value for each optimization metric and its theoretical optimal value.

$$F_k = \frac{f_k^{\max} - f_k}{f_k^{\max} - f_k^{\min}} \tag{20}$$

where F_k is the underlying utility function for the kth objective, f_k is the kth objective function, $f_k^{\max}$ is the theoretical maximum of the kth objective function, and $f_k^{\min}$ is the theoretical minimum of the kth objective function.

Considering the negative correlation between the objective of the minimum voltage deviation, reflective of the level of the system's operational optimization, and the cost indicators for construction and operation, we use a distance function to combine it with these cost indicators. Additionally, the utility function designed is inversely proportional to the original objective function. Therefore, the final multi-dimensional utility integration objective function for the configuration of the hybrid energy storage system is formulated as a maximization problem:

$$\min W = \omega_1 F_1 + \omega_2 F_2 + \omega_3 F_3 \tag{21}$$

where ω_1, ω_2, and ω_3 are the weights corresponding to each type of utility function.

3.2. *Distributed Hybrid Energy Storage Configuration Optimization Constraints*

3.2.1. Power Flow Operation Constraints

The voltage constraints for the node design specify the upper and lower bounds of voltage fluctuations:

$$V_{i,n}^{\text{lower}} \leq V_{i,n} \leq V_{i,n}^{\text{upper}}, \forall i \in \mathcal{N} \tag{22}$$

where $V_{i,n}^{\text{lower}}$ is the lower bound of the voltage magnitude at node i and $V_{i,n}^{\text{upper}}$ is the upper bound of the voltage magnitude at node i.

3.2.2. Cold/Heat Load Demand Energy Balance Constraints

Ensuring a balance between the cooling and heating load requirements in buildings is crucial. Based on models of the electric chillers, electric heaters, micro gas turbines, gas boilers, and absorption chillers, a thermal load energy balance constraint is established for the system:

$$\begin{cases} r_{i,n}^{\text{EC}}(t) + r_{i,n}^{\text{ACH}}(t) = r_{i,n}^{\text{load}}(t) \\ q_{i,n}^{\text{MT}}(t) + q_{i,n}^{\text{BLR}}(t) + q_{i,n}^{\text{EH}}(t) = q_{i,n}^{\text{load}}(t) \end{cases} \tag{23}$$

where $r_{i,n}^{\text{load}}$ is the cooling load demand at node i in the nth scenario and $q_{i,n}^{\text{load}}$ is the heating load demand at node i in the nth scenario.

3.2.3. Equipment Unit Output Constraints

Among the constraints on the output of various forms of equipment in the integrated energy system are the non-negative constraints on natural gas consumption and the operational output constraints on different energy supply devices, such as micro gas turbines, natural gas boilers, electric chillers, electric heaters, and absorption chillers. In the hybrid energy storage system, constraints on the charging and discharging power limits are also considered. To avoid energy being lost due to the energy storage simultaneously charging and discharging, for each energy storage device in the system, a binary variable U is introduced to indicate its operating state (charging or discharging), constrained between 0 and 1. Therefore, the overall equipment operation constraint can be expressed as

$$
\begin{aligned}
& P_{\min} \leq A P_{\text{run}} \leq U P_{\max} \\
& P_{\text{run}} = \begin{bmatrix} P_{\text{run}}^{\text{EU}} & P_{\text{run}}^{\text{ES}} \end{bmatrix}^{\text{T}} \\
& P_{\text{run}}^{\text{EU}} = \begin{bmatrix} q_{i,n}^{\text{BLR}}(t) & p_{i,n}^{\text{GP}}(t) & p_{i,n}^{\text{EC}}(t) & p_{i,n}^{\text{EH}}(t) & p_{i,n}^{\text{MT}}(t) & r_{i,n}^{\text{ACH}}(t) \end{bmatrix} \\
& P_{\text{run}}^{\text{ES}} = \begin{bmatrix} p_{i,n}^{\text{cCAES}}(t) & p_{i,n}^{\text{dCAES}}(t) & p_{i,n}^{\text{cBATT}}(t) & p_{i,n}^{\text{dBATT}}(t) & p_{i,n}^{\text{cEVCS}}(t) & p_{i,n}^{\text{dEVCS}}(t) \end{bmatrix} \\
& P_{\min} = \begin{bmatrix} 0 & 0 & 0 & 0 & 0 & 0 & 0 & 0 \end{bmatrix}^{\text{T}} \\
& P_{\max} = \begin{bmatrix} B & G & C & H & M & R & P_i^{\text{CAES}} & P_i^{\text{CAES}} & P_i^{\text{BATT}} & P_i^{\text{BATT}} & P_i^{\text{EVCS}} & P_i^{\text{EVCS}} \end{bmatrix}^{\text{T}} \\
& U = \begin{bmatrix} \mathrm{I}_{6\times 6} & 0 & 0 & 0 & 0 \\ 0 & U_{i,n}^{\text{cCAES}}(t) & 0 & 0 & 0 \\ 0 & 0 & U_{i,n}^{\text{dCAES}}(t) & 0 & 0 \\ 0 & 0 & 0 & U_{i,n}^{\text{cBATT}}(t) & 0 \\ 0 & 0 & 0 & 0 & U_{i,n}^{\text{dBATT}}(t) \end{bmatrix} \\
& A = \begin{bmatrix} n_{\text{h}} L_{\text{ng}} & 0 \\ 0 & \mathrm{I}_{7\times 7} \end{bmatrix}
\end{aligned}
\tag{24}
$$

where P_{run} is the set of operational state variables for each device in the system; A is the matrix of the system equipment's operating parameters; U is the matrix of the 0–1 variables indicating the operating states of the hybrid energy storage devices; $P_{\min}$ is the set of the lower limits of the output of each piece of equipment; $P_{\max}$ is the set of the upper limits of the output of each piece of equipment; $P_{\text{run}}^{\text{EU}}$ is the set of the operational state variables for the system's energy supply-related equipment; $P_{\text{run}}^{\text{ES}}$ is the set of the operational state variables for the system's hybrid energy storage devices; P_i^{CAES} is the installed power of the CAES device designed; P_i^{BATT} is the installed power of the BESS device designed; B is the upper limit on the heat provided by the boiler; C is the upper limit on the electric chiller's power; H is the upper limit on the electric heater's power; G is the upper limit of power purchases per unit of time on the grid; M is the upper limit on the electrical power of the microturbine; R is the upper limit on the refrigeration power of the absorption chillers; $U_{i,n}^{\text{cCAES}}$ is a 0–1 variable, which is the state of charge flag for the CAES system; $U_{i,n}^{\text{dCAES}}$ is a 0–1 variable, which is the state of discharge flag for the CAES system; $U_{i,n}^{\text{cBATT}}$ is a 0–1 variable, which is the state of charge flag for the BESS; and $U_{i,n}^{\text{dBATT}}$ is a 0–1 variable, which is the discharge state flag for the BESS.

In order to avoid simultaneously discharging and charging an energy storage device and to impose constraints on the siting of the hybrid energy storage installation, constraints for charging and discharging the energy storage system that vary in sign are designed as follows:

$$
\begin{cases}
U_{i,n}^{\text{cCAES}}(t) + U_{i,n}^{\text{dCAES}}(t) \leq 1 \\
U_{i,n}^{\text{cBATT}}(t) + U_{i,n}^{\text{dBATT}}(t) \leq 1 \\
U_{i,n}^{\text{cCAES}}(t) \leq X_i^{\text{CAES}} \\
U_{i,n}^{\text{dCAES}}(t) \leq X_i^{\text{CAES}} \\
U_{i,n}^{\text{cBATT}}(t) \leq X_i^{\text{BATT}} \\
U_{i,n}^{\text{dBATT}}(t) \leq X_i^{\text{BATT}}
\end{cases}
\tag{25}
$$

where X_i^{CAES} is a 0–1 flag variable for the siting of the CAES device indicating whether or not a CAES system is installed at the *i*th node, with this value equal to 1 if it is, and X_i^{BATT} is a flag variable for whether the *i*th node has battery storage installed or not.

The constraints on the siting flag variables are as follows:

$$\begin{cases} X_i^{\mathrm{CAES}} \leq X_i^{\mathrm{canset}} \\ X_i^{\mathrm{BATT}} \leq X_i^{\mathrm{canset}} \\ \sum\limits_{i=1}^{n} X_i^{\mathrm{CAES}} = N_{\mathrm{caes}} \\ \sum\limits_{i=1}^{n} X_i^{\mathrm{BATT}} = N_{\mathrm{batt}} \end{cases} \quad (26)$$

where X_i^{canset} is a flag for whether energy storage devices can be installed at the *si*th node, which takes 1 if so and 0 otherwise; N_{CAES} is the total number of nodes selected for CAES installation; and N_{BATT} is the total number of nodes selected for battery installation.

3.2.4. Operational Constraints on the Hybrid Energy Storage System

The effect of the depth of discharge is taken into account in the constraints on the upper and lower limits of the energy level of the storage system:

$$\begin{cases} e_{i,n}^{\mathrm{CAES}}(T) \geq E_{i,n}^{\mathrm{CAES0}} \\ e_{i,n}^{\mathrm{BATT}}(T) \geq E_{i,n}^{\mathrm{BATT0}} \\ n_{\mathrm{depth}}^{\mathrm{CAES}} \mathrm{E}_i^{\mathrm{CAES}} \leq e_{i,n}^{\mathrm{CAES}}(t) \leq \mathrm{E}_i^{\mathrm{CAES}} \\ n_{\mathrm{depth}}^{\mathrm{BATT}} E_i^{\mathrm{BATT}} \leq e_{i,n}^{\mathrm{BATT}}(t) \leq E_i^{\mathrm{BATT}} \end{cases} \quad (27)$$

where $\mathrm{E}_i^{\mathrm{CAES}}$ is the installed capacity of the CAES unit, E_i^{BATT} is the installed capacity of the BESS unit, $d_{\mathrm{depth}}^{\mathrm{CAES}}$ is the lower limit of the discharge depth for CAES, and $d_{\mathrm{depth}}^{\mathrm{BATT}}$ is the lower limit of the discharge depth for the BESS.

We set the power, as well as the capacity constraints, installed for the hybrid energy storage system devices:

$$\begin{cases} P_{\mathrm{min}}^{\mathrm{CAES}} \leq P_i^{\mathrm{CAES}} \leq P_{\mathrm{max}}^{\mathrm{CAES}} \\ P_{\mathrm{min}}^{\mathrm{BATT}} \leq P_i^{\mathrm{BATT}} \leq P_{\mathrm{max}}^{\mathrm{BATT}} \\ E_{\mathrm{min}}^{\mathrm{CAES}} \leq E_i^{\mathrm{CAES}} \leq E_{\mathrm{max}}^{\mathrm{CAES}} \\ E_{\mathrm{min}}^{\mathrm{BATT}} \leq E_i^{\mathrm{BATT}} \leq E_{\mathrm{max}}^{\mathrm{BATT}} \\ P_{\mathrm{set}}^{\mathrm{CAES}} = \sum\limits_{i=1}^{N} P_i^{\mathrm{CAES}} \\ P_{\mathrm{set}}^{\mathrm{BATT}} = \sum\limits_{i=1}^{N} P_i^{\mathrm{BATT}} \\ E_{\mathrm{set}}^{\mathrm{CAES}} = \sum\limits_{i=1}^{N} E_i^{\mathrm{CAES}} \\ E_{\mathrm{set}}^{\mathrm{BATT}} = \sum\limits_{i=1}^{N} E_i^{\mathrm{BATT}} \end{cases} \quad (28)$$

where $P_{\mathrm{MAX}}^{\mathrm{CAES}}$ is the upper limit of the installed power of the CAES device; $P_{\mathrm{MIN}}^{\mathrm{MAX}}$ is the lower limit of the installed power of the CAES device; $P_{\mathrm{MAX}}^{\mathrm{BATT}}$ is the upper limit of the installed power of the BESS device; $P_{\mathrm{MIN}}^{\mathrm{BATT}}$ is the lower limit of the installed power of the BESS device; $E_{\mathrm{MAX}}^{\mathrm{CAES}}$ is the upper limit of the installed capacity of the CAES device; $E_{\mathrm{MIN}}^{\mathrm{CAES}}$ is the lower limit of the installed capacity of the CAES device; $E_{\mathrm{MAX}}^{\mathrm{BATT}}$ is the upper limit of the installed capacity of the BESS device; and $E_{\mathrm{MIN}}^{\mathrm{BATT}}$ is the lower limit of the installed capacity of the BESS device.

Overall, the planning model considers time-series variables across multiple typical scenarios, including the actual renewable energy output, grid electricity purchases, natural gas purchases, the operational status of the CCHP devices, and the charging and discharging power of the hybrid energy storage devices with different variation cycles. It also considers overall decision variables such as the location of the energy storage devices, their rated charging and discharging power, and their storage capacity. By optimizing for a composite objective that includes operational economic costs, the full lifecycle cost of the energy storage systems, carbon emission costs, power quality, and renewable energy utilization, the model ultimately determines the optimal location, rated power, and storage capacity configuration for each energy storage device, providing a valuable reference for the design and planning of IESs in large-building microgrids.

4. Results: A Case Study of a Hybrid Energy Storage Configuration in Integrated Energy Systems for Large-Building Microgrids

This paper is based on an improved IEEE 13-bus test case to which a hybrid energy storage system is added and into which renewable energy generation and a CCHP system are integrated. The renewable energy output and building load data cover four typical scenarios for spring, summer, autumn, and winter. In practically planning for large buildings, local, multi-year historical meteorological data can be used to calculate and generate actual photovoltaic and wind power generation data. By applying clustering analysis (such as K-means, DBSCAN, etc.) to the output curves for each season, representative output curves can be selected for the typical daily output, serving as foundational scenarios for planning. The system parameters are designed as shown in Table 1, and the time-of-day tariff data are shown in Table 2.

Table 1. Parameters of the integrated energy system.

Equipment Type	Equipment Parameter	Parameter Value
Microturbine	Electrical efficiency η_e^{MT}	0.3
	Heat loss coefficient η_l^{MT}	2 %
	Upper limit of the electrical power M	150 kW
Natural gas boiler	Boiler efficiency n_h	0.9
	Upper limit of heat B	100 kW
Electric heater	COP C_{op}^{EH}	3.5
	Upper limit of power H	400 kW
Electric chiller	COP C_{op}^{EC}	3.5
	Upper limit of power C	400 kW
Absorption chiller	COP C_{op}^{ACH}	1.38
	Upper limit of refrigeration power R	200 kW
Battery	Ratio of the power variation cycle N_T	4
	Charging efficiency n_c^{BATT}	0.9
	Discharging efficiency $n_d^{ES,BATT}$	0.9
	Unit cost of storage capacity $C_E^{ES,BATT}$	168 USD/kWh
	Unit cost of storage power $C_P^{ES,BATT}$	70 USD/kW
	Maintenance cost per unit of capacity $C_E^{OM,BATT}$	4.2 USD/kWh
	Maintenance cost per unit of power $C_P^{OM,BATT}$	3.5 USD/kW
	Replacement cost per unit of capacity $C_C^{R,BATT}$	168 USD/kWh
	End-of-life cost ratio F_{EOL}^{BATT}	0.05
	Standard discount rate r_{BATT}	0.08
	Lower limit of discharge depth d_{depth}^{BATT}	0.2
CAES	Charging efficiency n_c^{CAES}	0.6
	Discharging efficiency n_d^{CAES}	0.6
	Unit cost of storage capacity $C_E^{ES,CAES}$	14 USD/kWh
	Unit cost of storage power $C_P^{ES,CAES}$	1120 USD/kW
	Maintenance cost per unit of capacity $C_E^{OM,CAES}$	0.28 USD/kWh
	Maintenance cost per unit of power $C_P^{OM,CAES}$	2.8 USD/kW
	End-of-life cost ratio F_{EOL}^{CAES}	0.1
	Standard discount rate r_{CAES}	0.08
	Lower limit of discharge depth d_{depth}^{CAES}	0
EVCS	Charging efficiency n_c^{EVCS}	1
	Discharging efficiency n_d^{EVCS}	1
Natural gas	Purchase cost C_P^G	0.14 USD/m³
	Carbon emission factor m_{gas}	1 $kgCO_2/m^3$
	Lower heating value of natural gas L_{ng}	9.78 (kW·h)/m³

Table 2. Time-share tariff.

Time	Prices
0:00–7:00	0.0434 USD/kWh
7:00–10:00/15:00–18:00/21:00–24:00	0.084 USD/kWh
10:00–15:00/18:00–21:00	0.1386 USD/kWh

The load demands of the CCHP system and the output of renewable energy are illustrated in the following series of figures. Figure 2 illustrates the maximum photovoltaic and wind power output that can theoretically be generated for a building at time intervals of 30 min. These data were derived based on the typical renewable energy output generation scenarios designed in [37] considering the influence of seasonal variability and short-term intermittency. Based on the variations in these curves, we factored the typical scale of large buildings into generating the renewable energy output data, as shown in the figure. Figure 3 shows the fluctuations in the electric load, heating load, and cooling load over a 24 h period.

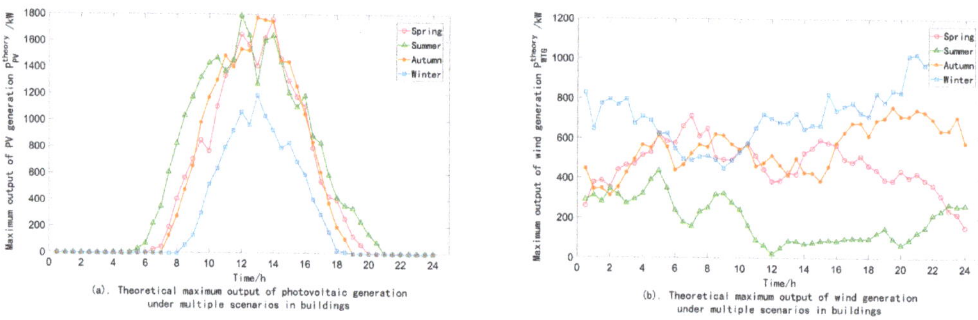

Figure 2. Theoretical total new energy output of integrated energy system .

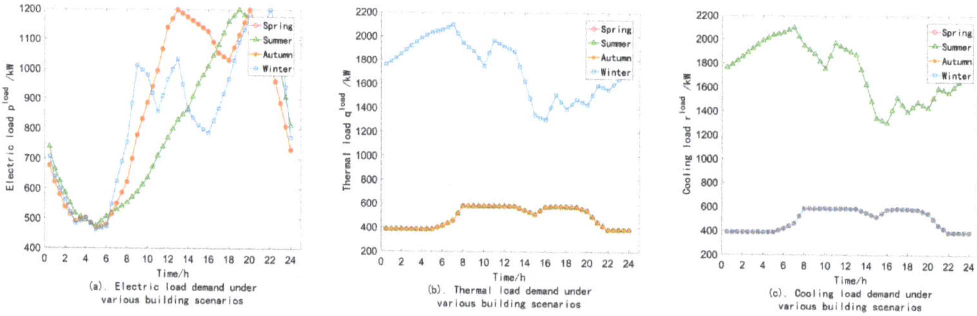

Figure 3. Total load demands of various types in integrated energy system.

For the hybrid energy storage optimization planning model proposed, the constraints and integrated utility objectives were modeled by using MATLAB 2021b and Yalmip on a Windows computer equipped with a 12th Gen Intel(R) Core(TM) i7-12700 processor and 16 GB of RAM. Optimization was performed by using the Gurobi solver, with an actual solving time of 362 s. The optimization results for the siting and sizing of the hybrid energy storage system are shown in Figure 4.

Figures 5 and 6 illustrate the 24 h optimized operation results for the BESS and the centralized CAES system, respectively, in different scenarios within a distributed unit of a building. It is worth mentioning that since the model in this paper focused on long-term planning and was optimized over a longer time scale, a relatively long time interval of 30 min was selected. Given that the response times of the battery energy storage and small-scale compressed-air energy storage systems designed are within the range of seconds and minutes, respectively, ample time is allocated for these actual energy storage systems to respond to the power variation demands shown in the response curve. From the comparison, we can see that the BESS experiences larger power fluctuations and operates at a higher power limit, reflecting its role in managing the variability in the renewable energy output. In contrast, the CAES system operates at a lower power but has a larger capacity, emphasizing its function in peak shaving and applying cost-effective electricity strategies based on time-of-use pricing.

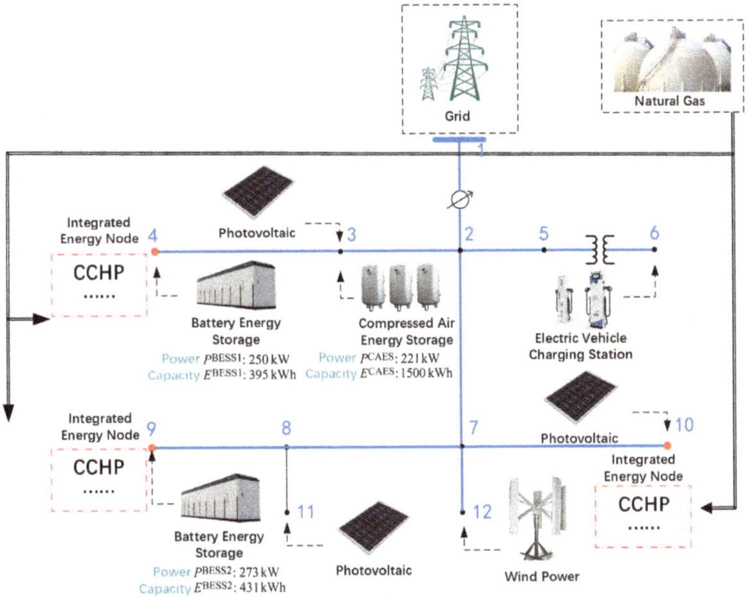

Figure 4. Results of optimizing siting and capacity planning for hybrid energy storage system equipment.

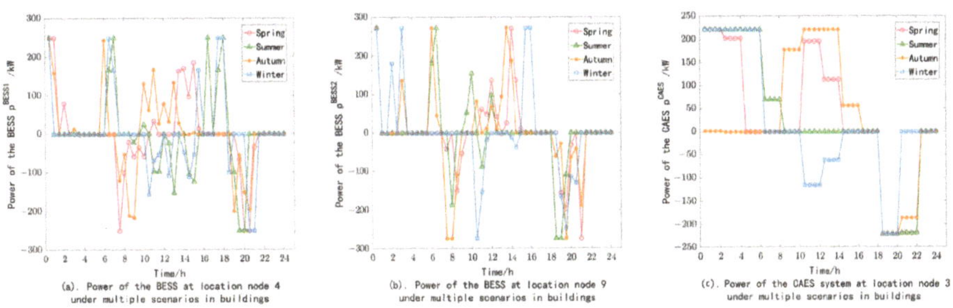

Figure 5. Charging and discharging power of hybrid energy storage system devices.

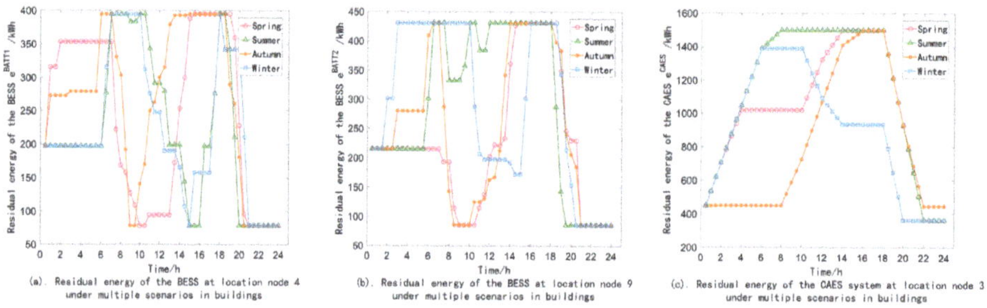

Figure 6. Remaining energy during charging and discharging processes of hybrid energy storage system devices.

In designing and comparing different combinations of objectives, we consider the following: Combination 1 represents the independent use of the basic utility function F1; Combination 2 represents an objective function combining F1 and F2; Combination 3 represents an objective function composed of F1 and F3; and Combination 4 represents an objective function composed of F1, F2, and F3, with the final optimization results compared as shown in Figure 7. Subfigure (a) compares the economic costs, including the system's operational costs, the carbon trading costs, and the full lifecycle cost of the hybrid energy storage system. Subfigure (b) compares the voltage quality, reflected in the total voltage deviation across nodes. Subfigure (c) compares the renewable energy utilization, showing the actual renewable energy output against the theoretical maximum output. It is evident that in integrating different utilities for multi-objective optimization, we can control the economic costs while enhancing the operational quality of the system.

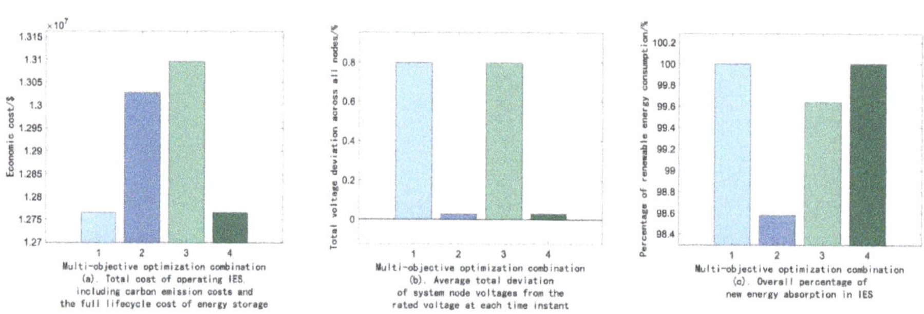

Figure 7. The distributed integrated energy system.

5. Conclusions

This paper established the operational constraints on the equipment and power flow in an IES in a microgrid for a large building. We proposed an optimization model that integrated multiple objectives, including its operational economic and low-carbon costs, the construction costs for the energy storage devices, power quality, and renewable energy utilization. Optimizing the siting and operational configuration of the hybrid energy storage system enhances the renewable energy consumption within the IES in this context. With the aims of constructing zero-energy buildings with an improved power quality and accelerating the transition to a higher-quality power supply system in mind, this study applied hybrid energy storage technology within the IES in a large-building microgrid. Its main conclusions are as follows:

- In designing the IES in this paper, a power flow model, a CCHP model, and a hybrid energy storage model were considered. The installation node locations and power capacities of the hybrid energy storage devices were planned out, providing a valuable

reference for integrating renewable energy systems into large buildings and ensuring a higher-quality power supply.
- Optimizing the configuration of the hybrid energy storage accounted for the comprehensive needs of its cost-effective and low-carbon operation, its lifecycle construction costs, the power quality in the grid's operation, and renewable energy utilization. By establishing a utility integration function, the siting and sizing of hybrid energy storage are optimized on the basis of multiple objective dimensions, providing a reference for planning more comprehensive IESs in research on large-building microgrids.
- In planning the optimization of the proposed model, we derived the location and sizing of hybrid energy storage devices, as well as the power and capacity variation curves for different storage systems. This further validated the effectiveness of the multi-period approach to designing hybrid energy storage presented in this study. Comparing various multi-objective combinations of our chosen indicators demonstrated that the planning results positively impact economic efficiency, power quality, and renewable energy utilization. These findings indicate that the multi-objective integrated planning method is effective in improving the overall performance and sustainability of IESs in microgrids for large buildings.

The model proposed in this paper focused on overall planning, meaning that some of the operational processes of various devices were simplified. For instance, the power transition processes in the energy storage devices and voltage changes due to battery cycle aging were omitted, and the models of the CCHP system's components were linearized. Although we made these simplifications in order to reduce the computational complexity of the overall planning process, this represents a limitation of our study. In the future, we will consider more precise modeling methods and advanced algorithms, such as hierarchical optimization and machine learning-based intelligent optimization algorithms, to enhance the accuracy of the planning results. Additionally, we will extend the simulation-based work carried out in this paper to more practical experimental validation, further investigating the potential of planning optimized energy storage for IESs in microgrids for large buildings.

Author Contributions: Conceptualization, C.H., X.T. and J.A.; methodology, C.H., X.T. and J.A.; software, Z.L.; validation, Z.L., R.Z. and X.T.; formal analysis, J.A.; investigation, Z.L.; resources, X.L.; data curation, X.L.; writing—original draft preparation, Z.L.; writing—review and editing, R.Z. and G.L.; visualization, Z.L.; supervision, C.H.; project administration, X.T.; funding acquisition, C.H. All authors have read and agreed to the published version of the manuscript.

Funding: This research was funded by scientific and technological projects from State Grid Hebei Electric power CO., LTD., grant number SGHEJY00GHJS2200052.

Data Availability Statement: The dataset is available upon request from the authors.

Conflicts of Interest: The authors He Chunguang, Tan Xiaolin, and An Jiakun are employed by the company State Grid Hebei Economic Research Institute. The author Li Xuejun is employed by State Grid Handan Electric Power Supply Company. The authors Liu Zixuan, Li Gengfeng, and Zhang Runfan are from the School of Electrical Engineering, Xi'an Jiaotong University. The authors declare that this study received funding for scientific and technological projects from the State Grid Corporation of China under grant number SGHEJY00GHJS2200052. The funder was involved in this study in providing research direction and ideas.

Appendix A

The following table lists all the variable and parameter symbols used in this article.

Table A1. Variable symbols and their descriptions.

Symbol	Description	Symbol	Description
$q_{i,n}^{\mathrm{MT}}$, $q_{i,n}^{\mathrm{BLR}}$, and $q_{i,n}^{\mathrm{EH}}$	Heat output of the microturbine, natural gas boiler, and electric heater	V_n^{G}	Consumption of natural gas per unit time
$p_{i,n}^{\mathrm{MT}}$	Electrical output power of microturbine	p_n^{GP}	Total electricity purchased from grid per unit time
$v_{i,n}^{\mathrm{MT}}$ and $v_{i,n}^{\mathrm{BLR}}$	Amount of natural gas consumed by microturbine and gas boiler	E_{set} and P_{set}	Installed capacity/power of energy storage
$p_{i,n}^{\mathrm{EH}}$ and $p_{i,n}^{\mathrm{EH}}$	Electrical power consumption of electric heater and chiller	C_{Rec}	Cost of energy storage recovery
$r_{i,n}^{\mathrm{EC}}$ and $r_{i,n}^{\mathrm{ACH}}$	Cooling output of electric chiller and absorption chiller	V_i^*	Normalized value of voltage at node i
$q_{i,n}^{\mathrm{ACH}}$	Heat input of absorption chiller	$P_{\mathrm{energy}}^{\mathrm{real}}$	Total new energy output power of actual operation of system
$p_{i,n}^{\mathrm{dCAES}}$, $p_{i,n}^{\mathrm{cCAES}}$, $p_{i,n}^{\mathrm{dBATT}}$, and $p_{i,n}^{\mathrm{cBATT}}$	Discharge/charge power of CAES/BESS	f_k and F_k	kth objective function and underlying utility function for kth objective
$e_{i,n}^{\mathrm{CAES}}$ and $e_{i,n}^{\mathrm{BATT}}$	Remaining energy of CAES system/BESS	P_{run}	Set of operational state variables for each device in system
$p_{i,n}^{\mathrm{cEVCS}}$ and $p_{i,n}^{\mathrm{dEVCS}}$	Charging/discharging power provided by charging station to electric vehicle	U	Matrix of 0–1 variables for operating states of hybrid energy storage devices
$e_{i,n}^{\mathrm{EV}}$	Remaining energy of electric vehicle	$P_{\mathrm{run}}^{\mathrm{EU}}$ and $P_{\mathrm{run}}^{\mathrm{ES}}$	Set of operational state variables for equipment related to system's energy supply and hybrid energy storage devices
$P_{i,n}$ and $Q_{i,n}$	Active and reactive power flows into node i in underlying distribution network	P_i^{CAES} and P_i^{BATT}	Designed installed power of CAES/BESS devices
$p_{i,n}$ and $q_{i,n}$	Injected power of all devices accessing node i	$U_{i,n}^{\mathrm{cCAES}}$, $U_{i,n}^{\mathrm{dCAES}}$, $U_{i,n}^{\mathrm{cBATT}}$, and $U_{i,n}^{\mathrm{dBATT}}$	State of charge and discharge flags of CAES/BESS devices
$p_{i,n}^{\mathrm{GP}}$	Power purchased from grid	X_i^{CAES} and X_i^{BATT}	Flag variable for siting of CAES/BESS device
$p_{i,n}^{\mathrm{PV}}$ and $p_{i,n}^{\mathrm{WTG}}$	Output power from photovoltaic and wind power systems	E_i^{CAES} and E_i^{BATT}	Installed capacity of CAES/BESS units
$V_{i,n}$	Voltage magnitude at node i		

Table A2. Parameter symbols and their descriptions.

Symbol	Description	Symbol	Description
$\eta_{\mathrm{e}}^{\mathrm{MT}}$ and $\eta_{\mathrm{l}}^{\mathrm{MT}}$	Electrical efficiency and heat loss coefficient of microturbine	m_{gas} and m_{grid}	Carbon emission factor for natural gas and grid
L_{ng}	Lower heating value of natural gas	$C_{\mathrm{E}}^{\mathrm{ES}}$ and $C_{\mathrm{P}}^{\mathrm{ES}}$	Unit cost of storage capacity and power
n_{h}	Boiler efficiency	Y	Service life in years of system design
$C_{\mathrm{op}}^{\mathrm{EH}}$, $C_{\mathrm{op}}^{\mathrm{EC}}$, and $C_{\mathrm{op}}^{\mathrm{ACH}}$	Coefficients of performance (COPs) of electric heater, electric chiller, and absorption chiller	$C_{\mathrm{E}}^{\mathrm{OM}}$ and $C_{\mathrm{P}}^{\mathrm{OM}}$	Maintenance cost per unit of energy storage capacity and power
T_{BATT} and ${}^{l}T_{\mathrm{CAES}}$	Power variation cycle of BESS and lth power variation cycle of CAES system	C_{labor}	Cost per person per year of maintenance personnel
N_{T}	Ratio of power variation cycle of CAES system to BESS	n_{labor}	Number of energy storage system operation and maintenance personnel
$E_{i,n}^{\mathrm{CAES0}}$ and $E_{i,n}^{\mathrm{BATT0}}$	Initial energy of CAES/BESS units	$C_{\mathrm{C}}^{\mathrm{R}}$	Replacement cost per unit of energy storage capacity
$n_{\mathrm{c}}^{\mathrm{CAES}}$, $n_{\mathrm{d}}^{\mathrm{CAES}}$, $n_{\mathrm{c}}^{\mathrm{BATT}}$, and $n_{\mathrm{d}}^{\mathrm{BATT}}$	Charging and discharging efficiency of CAES/BESS units	r	Standard discount rate or benchmark rate of return
$E_{i,n}^{\mathrm{EV0}}$	Initial energy (initial state of charge) of electric vehicle	F_{EOL}	End-of-life cost ratio
$n_{\mathrm{c}}^{\mathrm{EVCS}}$ and $n_{\mathrm{d}}^{\mathrm{EVCS}}$	Charging and discharging efficiency of EV charging station	$P_{\mathrm{energy}}^{\mathrm{theory}}$	Total theoretical maximum new energy output power
$\mathcal{S}_i$	Set of child nodes for grid node i	$f_k^{\max}$ and $f_k^{\min}$	Theoretical maximum and minimum of kth objective function
$\mathcal{N}$	Set of all nodes of microgrid	ω_1, ω_2, and ω_3	Weights corresponding to each type of utility function

Table A2. *Cont.*

Symbol	Description	Symbol	Description
$P_{i,n}^{\text{LOAD}}$	Electrical load	$V_{i,n}^{\text{lower}}$ and $V_{i,n}^{\text{upper}}$	Lower and upper bounds of voltage magnitude at node i
$\theta(i)$	Previous node connected to node i	$r_{i,n}^{\text{load}}$ and $q_{i,n}^{\text{load}}$	Cooling and heating load demands
r_i and x_i	Line resistance and reactance with node i as endpoint	A	Matrix of system equipment operating parameters
V_0	Reference voltage for entire microgrid	P_{min} and P_{max}	Set of lower and upper limits of output of each piece of equipment
n_{life}	Designed usage time of energy storage system	B, C, H, M, R, and G	Upper limit of boiler heat power, electric chiller's power, electric heater's power, microturbine's electrical power, absorption chiller's cooling power, and power purchases per unit of time on grid
N_S	Number of scenarios	X_i^{canset}	Flag for node being installable
T	Total number of time intervals	N_{CAES} and N_{BATT}	Total number of nodes selected for installation of CAES/BESS
n_n^{scen}	Frequency of nth scenario	$d_{\text{depth}}^{\text{CAES}}$ and $d_{\text{depth}}^{\text{BATT}}$	Lower limit of CAES/BESS discharge depth
ΔT	Time interval within each unit of time	$P_{\text{MAX}}^{\text{CAES}}$, $P_{\text{MIN}}^{\text{MAX}}$, $P_{\text{MAX}}^{\text{BATT}}$, and $P_{\text{MIN}}^{\text{BATT}}$	Upper and lower limits of installed power of CAES/BESS device
C_P^G and C_P^E	Purchase cost per unit of natural gas and time-of-use price from grid	$E_{\text{MAX}}^{\text{CAES}}$, $E_{\text{MIN}}^{\text{CAES}}$, $E_{\text{MAX}}^{\text{BATT}}$, and $E_{\text{MIN}}^{\text{BATT}}$	Upper and lower limits of installed capacity of CAES/BESS device

References

1. Zhang, J.; Chen, J.; Ji, X.N.; Sun, H.Z.; Liu, J. Low-carbon economic dispatch of integrated energy system based on liquid carbon dioxide energy storage. *Front. Energy Res.* **2023**, *10*, 1051630. [CrossRef]
2. Wang, L.Y.; Lin, J.L.; Dong, H.Q.; Wang, Y.Q.; Zeng, M. Demand response comprehensive incentive mechanism-based multi-time scale optimization scheduling for park integrated energy system. *Energy* **2023**, *270*, 126893. [CrossRef]
3. Li, Y.; Wang, B.; Yang, Z.; Li, J.Z.; Chen, C. Hierarchical stochastic scheduling of multi-community integrated energy systems in uncertain environments via Stackelberg game. *Appl. Energy* **2022**, *308*, 118392. [CrossRef]
4. Moghadam, E.A.; Gord, F.M.; Abadi, K.M. Sensitivity analysis and working fluid selection for a biogas-fueled hybrid energy system based on nearly zero energy building concept: A case study. *J. Clean. Prod.* **2024**, *435*, 140499. [CrossRef]
5. Yao, G.; Chen, Y.; Lin, Y.X.; Wang, Y.G. Energy-Saving Design Strategies of Zero-Energy Solar Buildings-A Case Study of the Third Solar Decathlon China. *Buildings* **2023**, *13*, 405. [CrossRef]
6. Guo, J.C.; Wu, D.; Wang, Y.Y.; Wang, L.M.; Guo, H.Y. Co-optimization method research and comprehensive benefits analysis of regional integrated energy system. *Appl. Energy* **2023**, *340*, 121034. [CrossRef]
7. Kingston, T.; Guada, B.A. Simulated Use Performance of an Integrated Energy System for Thermal and Power Management with Micro-combined Heat and Power in Nano-grid Environments. *ASHRAE Trans.* **2023**, *129*, 536–544.
8. Mianaei, P.K.; Aliahmadi, M.; Faghri, S.; Ensaf, M.; Ghasemi, A.; Abdoos, A.A. Chance-constrained programming for optimal scheduling of combined cooling, heating, and power-based microgrid coupled with flexible technologies. *Sustain. Cities Soc.* **2022**, *77*, 103502. [CrossRef]
9. Hu, J.F.; Shan, Y.H.; Yang, Y.; Parisio, A.; Li, Y.; Amjady, N.; Islam, S.; Cheng, K.W.; Guerrero, J.M.; Rodriguez, J. Economic Model Predictive Control for Microgrid Optimization: A Review. *IEEE Trans. Smart Grid* **2024**, *15*, 472–484. [CrossRef]
10. Liu, J.; Chen, X.; Yang, H.X.; Shan, K. Hybrid renewable energy applications in zero-energy buildings and communities integrating battery and hydrogen vehicle storage. *Appl. Energy* **2021**, *290*, 116733. [CrossRef]
11. Hu, J.J.; Wang, Y.D.; Dong, L. Low carbon-oriented planning of shared energy storage station for multiple integrated energy systems considering energy-carbon flow and carbon emission reduction. *Energy* **2024**, *290*, 130139. [CrossRef]
12. Sunny, M.R.; Ali, T.; Aghaloo, K.; Wang, K. Techno-economic feasibility of stand-alone hybrid energy system with battery storage in educational buildings: A case study of Uttara University. *Energy Build.* **2024**, *304*, 113852. [CrossRef]
13. Belkhier, Y.; Oubelaid, A. Novel design and adaptive coordinated energy management of hybrid fuel-cells/tidal/wind/PV array energy systems with battery storage for microgrids. *Energy Storage* **2024**, *6*, e556 . [CrossRef]
14. Mussetta, M.; Le, X.C.; Trinh, T.H.; Doan, A.T.; Duong, M.Q.; Tanasiev, G.N. An Overview of the Multilevel Control Scheme Utilized by Microgrids. *Energies* **2024**, *17*, 3947. [CrossRef]
15. Zhu, M.S.; Fang, J.K.; Ai, X.M.; Cui, S.C.; Feng, Y.; Li, P.; Zhang, Y.H.; Zheng, Y.L.; Chen, Z.; Wen, J.Y. A comprehensive methodology for optimal planning of remote integrated energy systems. *Energy* **2023**, *285*, 129443. [CrossRef]
16. Li, Z.C.; Xia, Y.H.; Bo, Y.L.; Wei, W. Optimal planning for electricity-hydrogen integrated energy system considering multiple timescale operations and representative time-period selection. *Appl. Energy* **2024**, *362*, 122965. [CrossRef]
17. Lei, Z.J.; Yu, H.; Li, P.; Ji, H.R.; Yan, J.Y.; Song, G.Y.; Wang, C.S. A compact time horizon compression method for planning community integrated energy systems with long-term energy storage. *Appl. Energy* **2024**, *361*, 122912. [CrossRef]

18. Wang, T.L.; Huo, T.Y.; Li, H.H. Bi-Layer Planning of Integrated Energy System by Incorporating Power-to-Gas and Ground Source Heat Pump for Curtailed Wind Power and Economic Cost Reduction. *Energies* **2024**, *17*, 1447. [CrossRef]
19. Chen, L.; Yang, D.D.; Cai, J.; Yan, Y. Robust optimization based coordinated network and source planning of integrated energy systems. *Int. J. Electr. Power Energy Syst.* **2024**, *157*, 109864. [CrossRef]
20. Dong, Y.C.; Wang, C.; Zhang, H.L.; Zhou, X.J. A novel multi-objective optimization framework for optimal integrated energy system planning with demand response under multiple uncertainties. *Inf. Sci.* **2024**, *663*, 120252. [CrossRef]
21. Dong, Y.C.; Zhang, H.L.; Ma, P.; Wang, C.; Zhou, X.J. A hybrid robust-interval optimization approach for integrated energy systems planning under uncertainties. *Energy* **2023**, *274*, 127267. [CrossRef]
22. Dong, W.K.; Lu, Z.G.; He, L.C.; Geng, L.J.; Guo, X.Q.; Zhang, J.F. Low-carbon optimal planning of an integrated energy station considering combined power-to-gas and gas-fired units equipped with carbon capture systems. *Int. J. Electr. Power Energy Syst.* **2022**, *138*, 107966. [CrossRef]
23. Li, P.; Jiang, L.; Zhang, C.Y.; Wang, J.H.; Wang, N.; Liu, H.T.; Dou, Z.L.; Pan, Y.P.; Zhou, X.C.; Zeng, P.L. Coordinated planning of integrated electricity-heat-gas energy system considering renewable energy consumption. *IET Gener. Transm. Distrib.* **2022**, *17*, 3232–3244. [CrossRef]
24. Li, F.; Lu, S.R.; Cao, C.W.; Feng, J. Operation Optimization of Regional Integrated Energy System Considering the Responsibility of Renewable Energy Consumption and Carbon Emission Trading. *Electronics* **2021**, *10*, 2677. [CrossRef]
25. Gao Y.; Zhang, X.; Xu, X.; Liu, L.; Zhao, Y.; Zhang, S. Application and research progress of phase change energy storage in new energy utilization. *J. Mol. Liq.* **2021**, *343*, 117554. [CrossRef]
26. He, Y.; Guo, S.; Zhou, J.X.; Ye, J.L.; Huang, J.; Zheng, K.; Du, X.R. Multi-objective planning-operation co-optimization of renewable energy system with hybrid energy storages. *Renew. Energy* **2022**, *184*, 776–790. [CrossRef]
27. Dong, H.L.; Fang, Z.J.; Ibrahim, A.; Cai, J. Optimized Operation of Integrated Energy Microgrid with Energy Storage Based on Short-Term Load Forecasting. *Electronics* **2022**, *11*, 22. [CrossRef]
28. Wu, G.; Li, T.; Xu, W.T.; Xiang, Y.; Su, Y.C.; Liu, J.W.; Liu, F. Chance-constrained energy-reserve co-optimization scheduling of wind-photovoltaic-hydrogen integrated energy systems. *Int. J. Hydrogen Energy* **2023**, *48*, 6892–6905. [CrossRef]
29. Kiryanova, N.G.; Matrenin, P.V.; Mitrofanov, S.V.; Kokin, S.E.; Safaraliev, M.K. Hydrogen energy storage systems to improve wind power plant efficiency considering electricity tariff dynamics. *Int. J. Hydrogen Energy* **2022**, *47*, 10156–10165. [CrossRef]
30. Zeng, X.Q.; Xiao, P.; Zhou, Y.; Li, H.J. Hybrid energy storage for the optimized configuration of integrated energy system considering battery-life attenuation. *J. Eng.-JOE* **2023**, *2023*, e12331. [CrossRef]
31. Gao, M.F.; Han, Z.H.; Zhao, B.; Li, P.; Wu, D.; Li, P. Optimal planning method of multi-energy storage systems based on the power response analysis in the integrated energy system. *J. Energy Storage* **2023**, *73*, 109015. [CrossRef]
32. Wang, Y.L.; Zhang, Y.L.; Xue, L.; Liu, C.; Song, F.H.; Sun, Y.L.; Liu, Y.; Che, B. Research on planning optimization of integrated energy system based on the differential features of hybrid energy storage system. *J. Energy Storage* **2022**, *55*, 105368. [CrossRef]
33. Chen, C.M.; Liu, C.; Ma, L.Y.; Chen, T.W.; Wei, Y.Q.; Qiu, W.Q.; Lin, Z.Z.; Li, Z.Y. Cooperative-game-based joint planning and cost allocation for multiple park-level integrated energy systems with shared energy storage. *J. Energy Storage* **2023**, *73*, 108861. [CrossRef]
34. Zhang, L.Y.; Guo, J.X.; Yu, X.R.; Hui, G.; Liu, N.; Ren, D.D.; Wang, J.J. Optimization of Integrated Energy Systems Based on Two-Step Decoupling Method. *Electronics* **2023**, *13*, 2045. [CrossRef]
35. Hou, E.G.; Xu, Y.L.; Tang, J.R.; Wang, Z. Strategy of Flywheel-Battery Hybrid Energy Storage Based on Optimized Variational Mode Decomposition for Wind Power Suppression. *Electronics* **2024**, *13*, 1362. [CrossRef]
36. Xing, C.; Xiao, J.J.; Li, P.Q.; Xi, X.Z.; Chen, Y.H.; Guo, Q. Adaptive Virtual Inertial Control and Virtual Droop Control Coordinated Control Strategy for Hybrid Energy Storage Taking into Account State of Charge Optimization. *Electronics* **2024**, *13*, 1228. [CrossRef]
37. Wang, J.X.; Wei, J.D.; Zhu, Y.C.; Wang, X.L. The Reliability and Operational Test System of a Power Grid with Large-scale Renewable Integration. *CSEE J. Power Energy Syst.* **2020**, *6*, 704–711.

MDPI AG
Grosspeteranlage 5
4052 Basel
Switzerland
Tel.: +41 61 683 77 34

Electronics Editorial Office
E-mail: electronics@mdpi.com
www.mdpi.com/journal/electronics

www.ingramcontent.com/pod-product-compliance
Lightning Source LLC
Chambersburg PA
CBHW040316270926
51850CB00007BC/1496
* 9 7 8 3 7 2 5 8 6 8 0 0 1 *